Sonar Bangla?

Sonar Bangla?

Agricultural Growth and Agrarian Change in West Bengal and Bangladesh

edited by
**Ben Rogaly
Barbara Harriss-White
Sugata Bose**

Sage Publications
New Delhi   Thousand Oaks   London

Copyright © Ben Rogaly, Barbara Harriss-White and Sugata Bose, 1999

All rights reserved. No part of this book may be reproduced or utilised in any form or by any means, electronic or mechanical including photocopying, recording or by any information storage or retrieval system, without permission in writing from the publisher.

First published in 1999 by

Sage Publications India Pvt Ltd
M-32 Market, Greater Kailash-I
New Delhi-110 048

Sage Publications Inc
2455 Teller Road
Thousand Oaks, California 91320

Sage Publications Ltd
6 Bonhill Street
London EC2A 4PU

Published by Tejeshwar Singh for Sage Publications India Pvt Ltd, lasertypeset by Excel Pages, Pondicherry, and printed at Chaman Enterprises, Delhi.

Library of Congress Cataloging-in-Publication Data

Sonar Bangla? : agricultural growth and agrarian change in West Bengal and Bangladesh / edited by Ben Rogaly, Barbara Harriss-White, Sugata Bose.
     p. cm.   (cloth)
  Includes bibliographical references and index.
  1. Agriculture—Economic aspects—India—West Bengal.
2. Agriculture—Economic aspects—Bangladesh. 3. Agriculture and state—India—West Bengal. 4. Agriculture and state—Bangladesh. I. Rogaly, Ben. II. Harriss-White, Barbara. 1946–III. Bose, Sugata.
HD2075. W4S66   338. 1'0954'14—dc21   1999   98-50749

ISBN: 0–7619–9307–X (US–hb)
       81–7036–771–9 (India–hb)

Sage Production Team: Aruna Ramachandran, M.S.V. Namboodiri and Santosh Rawat

Contents

List of Tables 7
List of Figures 9

1. Introduction: Agricultural Growth and Agrarian Change in West Bengal and Bangladesh 11
   BEN ROGALY, BARBARA HARRISS-WHITE AND SUGATA BOSE

PART 1
Agricultural Growth, Poverty and Well-Being

2. Agricultural Growth and Agrarian Structure in Bengal: A Historical Overview 41
   SUGATA BOSE
3. Agricultural Growth and Recent Trends in Well-Being in Rural West Bengal 60
   HARIS GAZDAR AND SUNIL SENGUPTA
4. Slowdown in Agricultural Growth in Bangladesh: Neither a Good Description Nor a Description Good to Give 92
   RICHARD PALMER-JONES
5. Agricultural Growth Performance in Bangladesh: A Note on the Recent Slowdown 137
   QUAZI SHAHABUDDIN
6. Why is Agricultural Growth Uneven? Class and the Agrarian Surplus in Bangladesh 147
   BEN CROW

PART 2
Policies and Practices

7. Agrarian Structure and Agricultural Growth Trends in Bangladesh: The Political Economy of Technological Change and Policy Interventions ... 177
   SHAPAN ADNAN
8. Panchayati Raj and the Changing Micro-Politics of West Bengal ... 229
   GLYN WILLIAMS
9. From Untouchable to Communist: Wealth, Power and Status among Supporters of the Communist Party (Marxist) in Rural West Bengal ... 253
   ARILD ENGELSEN RUUD
10. Politics of Middleness: The Changing Character of the Communist Party of India (Marxist) in Rural West Bengal (1977–90). ... 279
    DWAIPAYAN BHATTACHARYYA

PART 3
Changing Agrarian Structures

11. From Farms to Services: Agricultural Reformation in Bangladesh ... 303
    GEOFFREY D. WOOD
12. Institutions, Actors and Strategies in West Bengal's Rural Development—A Study on Irrigation ... 329
    NEIL WEBSTER
13. Dangerous Liaisons? Seasonal Migration and Agrarian Change in West Bengal ... 357
    BEN ROGALY
14. Agricultural Growth and the Structure and Relations of Agricultural Markets in West Bengal ... 381
    BARBARA HARRISS-WHITE

Notes on Contributors ... 415
Index ... 418
About the Editors ... 425

Tables

2.1	Rate of Growth of Per Capita Agricultural Output, 1949–80	46
3.1	Head-Count Ratio of Rural Poverty in India and the Eastern States	73
3.2	Estimated Trend Growth Rates in West Bengal	73
3.3	Rural Infant Mortality Rate—Three-Year Moving Averages	76
3.4	Literacy Rates by Gender and Caste	80
3.5	Literacy Rates by Caste and Age Cohort	80
3.6	Predicted Probability of a Child Being Well-Nourished	82
3.7	Predicted Probability of a Child Being Well-Nourished for Various Household Characteristics	83
4.1	Estimates of Population in Poverty in Bangladesh from Household Economic Surveys, 1973–74 to 1991–92	115
4.2	Rice Wages of Agricultural Labourers and Food and Non-Food Rural Cost of Living Indices, Bangladesh, 1981–95	118
5.1	Trend Growth Rate of Fertiliser Use by Type in Bangladesh	138
5.2	Trend in Irrigation Coverage and Growth Rates	139
5.3	Ranking of Regions According to Trend Rate of Growth of Rice Production	141
6.1	Numbers of Households by Class and Income Group	154
6.2	Gross Paddy Output, Agricultural Households by Class	155
6.3	Proportion of Output Sold by Class and Region	156
6.4	Kind Repayment of Loans: Frequency by Class and Region	166
6.5	Appropriation and Redistribution of Surplus through Non-Tied Trade	168
8.1	Party Affiliation of Village Gram Panchayat Members	233
8.2	Statistical Comparison of Agriculture in the Survey Villages	234

8.3	Landownership Structure in the Survey Villages	241
8.4	Impact of Land Redistribution on the Survey Villages	241
9.1	Recipients of Redistributed *Khas* by Jati Group, Udaynala, 1993	262
9.2	Recipients of IRDP Loans by Jati Group, Udaynala, 1993	262
9.3	Landownership by Person and Jati, Gopinathpur, 1960 and 1993	263
9.4	Landownership by Person and Jati, Udaynala, 1957 and 1993	264
9.5	Representation by Jati in the Gram Panchayat to which Udaynala and Gopinathpur Belong	266
9.6	Gram Panchayat Representation by Jati, Udaynala and Gopinathpur, 1978–93+	267
9.7	Education and Age Groups (in Percentages) of Men and Women of Bagdi Jati Above 16 Years of Age, Udaynala and Gopinathpur, 1993	271
9.8	Education and Age Groups (in Percentages) of Men and Women of Muchi Jati Above 16 Years of Age, Udaynala and Gopinathpur, 1993	272
12.1	Rice Production in Bardhaman District, 1969–92	337
12.2	Irrigation Intensity and Irrigation by Source, West Bengal and Districts, 1971–72	338
12.3	Crop-Wise and Gross Area Irrigated in Bardhaman District from the Damodar Valley Canal System	339
12.4	Jayanapur Village: Land Ownership	341
12.5	Jayanapur Village: Land Ownership by Ownership of MSTW	343
12.6	Caste by Ownership of MSTW	343
12.7	Jayanapur Village: Caste by Land Ownership	344
13.1	Uses of Remittances by Seasonal Migrants from a Purulia Locality in the Aman Harvest, 1991–92	363
13.2	Summary of the Caste and Class Composition of Para in the Purulia Locality	365
13.3	Number of Seasonal Migrant Workers Leaving the Purulia Locality for the Aman Paddy Harvest, 1991	365
13.4	Number of Migrant Labourers Working for Sampled Employers in a Bardhaman Locality by Place of Origin for the Aman Paddy Harvest, 1991	367
13.5	How Migrants from the Purulia Locality were Controlled through Late and/or Withheld Payment in the Aman Paddy Harvest, 1991	373
14.1	Market System Structure	391
14.2	Approximate/Indicative Net Profits	392
14.3	Traders' Credit	397
14.4	Co-operative Crop Loans in Study Areas	398
14.5	Regulation	399

Figures

3.1	Trends in Foodgrain Output in West Bengal and India, 1970–71 to 1991–92	62
3.2	Trends in Foodgrain Output in West Bengal, Bihar and Orissa, 1970–71 to 1991–92	64
3.3	Trends in Foodgrain Output and Real Wages of Agricultural Labourers in West Bengal, 1970–71 to 1991–92	72
3.4	Rural Mean Per Capita Consumption in West Bengal and India, 1958–92	74
3.5	Head-Count Ratio of Rural Poverty in West Bengal and India, 1958–92	74
3.6	Trends in Rural Infant Mortality Rates, 1982–91	76
3.7	Trends in Rural Literacy Rates, 1961–91	77
4.1a	Production of Major Foodgrains in Bangladesh, 1990–97	96
4.1b	Indexes of Major Foodgrains Production in Bangladesh, 1990–97	97
4.2	Numbers of STW and Areas Irrigated by STW in Bangladesh, 1972–96	98
4.3	Distribution of Minor Irrigation in Bangladesh, 1962–88	100
4.4	*Aman* production in Bangladesh, 1960–96	101
4.5	*Boro* production in Bangladesh, 1960–96	103
4.6	Actual and Predicted Wheat Production in Bangladesh, 1961–97	104
4.7	Actual and Predicted *Aus* Production in Bangladesh, 1961–97	107
4.8	Actual and Predicted Total Grains Production in Bangladesh, 1960–97	109

4.9	Bangladesh Rice Price Index, 1979–96	111
4.10	Rural and Urban Poverty in Bangladesh, 1973–74 to 1991–92	116
4.11	Rice Wages of Agricultural Labourers in Bangladesh, 1981–92	119
4.12	Real Wage Rate of Agricultural Labourers in Bangladesh, 1981–91	120
6.1	Proportion of Grain Output Paid for Land	157
6.2	Loan Repayment as Proportion of Output	159
6.3	Proportion of Output Paid in Grain for Labour, Commodities and Services	160
6.4	Output, Grain Consumption and Surplus Product	161
6.5	Appropriation of Surplus Product in Bogra Region	163
6.6	Appropriation of Surplus Product in Noakhali *Chars*	164
12.1	Jayanapur Village: Land Ownership	342

1

Introduction: Agricultural Growth and Agrarian Change in West Bengal and Bangladesh

BEN ROGALY, BARBARA HARRISS-WHITE and SUGATA BOSE

INTRODUCTION

When in the mid-1980s, after decades of stagnation, the output of Bengali agriculture at last began to grow faster than population, the change was heralded as a breakthrough. In both communist West Bengal and in rapidly liberalising Bangladesh there was no shortage of claims to credit for the achievement. It seemed as though the implications that had been drawn from the then recent classic explanation of the obstacles to agrarian change in the two Bengals for the period 1949–80 (Boyce, 1987) had been contradicted. Inequality in the countryside and rivalry between better-off individuals had not in the end stood in the way of the unprecedented expansion of groundwater irrigation. Water control had been identified by Boyce as the central constraint impeding expansion of production in both Bengals in earlier years. Agrarian structures, in particular the distribution of owned land and other assets, were central to Boyce's explanation of why agriculture in Bengal had continued to stagnate after 1947. The enormous groundwater potential of the region had not been harnessed precisely because collective ways of doing so, such as investment in deep tubewells, involved cooperation between rivals among the relatively rich, and were likely to be sabotaged by those excluded from the process (ibid.).

The present collection of research is the first attempt to bring together analyses of agrarian change in West Bengal and Bangladesh in one volume since the apparent turnaround. Some of the papers were presented in draft form at the Workshop on Agricultural Growth and Agrarian Structure in Contemporary West Bengal and Bangladesh in Calcutta in January 1995. The others were written especially for the book.* There are many awkward questions around the cryptic *Sonar Bangla?* of the title which in this context should be taken to mean: are the recent changes in rural West Bengal and Bangladesh as good as they appear to be? Questions are raised about the reliability of the agricultural growth data itself, about the roles of the contrasting agrarian policy regimes in West Bengal and Bangladesh in bringing about that growth, about the interrelation between agricultural growth and agrarian structures, and about the outcomes in terms of well-being.[1]

Foodgrains production[2] in West Bengal had grown by between 4.3 and 6.5 per cent per annum between the early 1980s and the early 1990s.[3] This exceeded the growth rates of neighbouring states in eastern India, which also appeared to be reversing earlier stagnation. Agricultural growth in Bangladesh over the same period exceeded population growth. Much of this book is concerned with debates over the causes and consequences of the recent relatively rapid agricultural growth in the two Bengals. However, the temperature of the debates has been raised by an apparent slowdown in the production of rice in both West Bengal and Bangladesh

---

*The Workshop was hosted by the Centre for Studies in Social Sciences, Calcutta, in collaboration with Queen Elizabeth House, University of Oxford, UK, and the Center for South Asian and Indian Ocean Studies, Tufts University, USA. The editors of this book are deeply grateful to Nripen Bandyopadhyay who co-directed the Workshop, to Aruna Ramachandran at Sage who went over the manuscript in great detail and, through her queries on each chapter, improved it significantly, and to the Department for International Development, UK, for providing funding for the Workshop and dissemination of the papers and, through its Livelihood Trajectories research project at the University of East Anglia, UK, for subsequent research by Rogaly which has fed into this introduction. Views expressed in each paper are those of the authors alone and do not necessarily represent those of the funders.

in the first half of the 1990s. With regard to Bangladesh, Shapan Adnan argues that the downturn in the early 1990s provides evidence that impoverishment resulting from *unequal growth* renders that growth unsustainable by bringing about failure in domestic demand for rice. Richard Palmer-Jones, on the other hand, brings in data for 1995–96 indicating an overall increase in production of rice, led by renewed growth in *boro* rice output, to argue that *long term* growth trends are still high and unhindered by agrarian structures.

West Bengal too experienced a slowdown in agricultural growth in the three crop years following 1991–92 (GoWB, 1995–96: Statistical Appendix, Table 5.0a. See also Rawal and Swaminathan, 1998). Total production of cereals, which in 1991–92 was reported as 12.7 million tonnes (GoWB, 1994–95: Statistical Appendix, Table 5.6), 84 per cent higher than in the triennium ending 1981–82, declined in the following year to 12.2 million tonnes, recovering in 1993–94 to 12.9 million tonnes and showing only very incremental upward movement in 1994–95 (13.1 million tonnes). The estimates for 1995–96 suggest that total cereals production in that year was almost identical to production in 1994–95 (GoWB, 1995–96: Statistical Appendix, Table 5.0).

Clearly then, long term growth trends in West Bengal should remain a matter for debate, just as they do with respect to Bangladesh. Perhaps not surprisingly—given his conviction about the link between agrarian reforms 'betting on the weak' and rapid growth in agriculture—Lieten, although making use of production data up to 1994–95, ignores the plateauing of rice production after 1991–92 (1996b: 33). Both Palmer-Jones' and Quazi Shahabuddin's contributions in this volume concede that their interpretations of the importance of the slowdown in Bangladesh rice production in the early 1990s, which are very different indeed, are speculative. Lieten, although obviously aware of more up-to-date trends, does not admit to the possibility of a different reading from his own of West Bengal's agricultural production trends in the 1990s.

Whatever the disputes over the figures and in spite of the recent slowdown, agricultural growth in West Bengal was highly impressive; indeed, together with the reports of increased rates of growth in Bangladesh (Palmer-Jones, 1992), it provided the inspiration for this book. The agrarian reforms of the state government, led by the Communist Party of India (Marxist) (CPI[M]), made the task of bringing together the most recent research seem even more important. In particular, West Bengal's early and wholehearted implementation of panchayati raj in 1978 (see Webster, 1992) set it apart from other states in India and made it an international leader in the decentralisation of political representation. Since 1978 elections have continued to be held regularly every five years to the gram panchayat (approximately ten villages) at block and district levels on party political tickets. Other important reforms included legislation for secure tenancy rights for sharecroppers (Operation Barga) and the active redistribution of land held over the legal ceiling (Mukarji and Bandyopadhyay, 1993; Lieten, 1996a; Sengupta and Gazdar, 1997).

In reporting the apparent simultaneity of its agrarian reforms and rapid agricultural growth in the state, the government of West Bengal has asserted a connection between the two (GoWB, 1995–96: 19). They are indeed likely to be connected, and the nature of the connection is an important area of research. Sengupta and Gazdar (1997) point out that rapid agricultural growth began around 1983–84, six years after the reforms began. Mukarji and Bandyopadhyay (1993) carefully sift the evidence on Operation Barga and the land redistribution. Security of tenure, while clearly important for tenants, is unlikely in itself to have made a big difference to overall agricultural performance in spite of claims to the contrary (see, for example, Banerjee and Ghatak, 1995). Even in September 1995 the area cultivated by registered sharecroppers was only 8 per cent of net sown area (GoWB, 1995–96: Statistical Appendix, Tables 5.3a and 5.18a). Much of the land redistribution took place under the Congress governments of the 1970s (Mukarji and Bandyopadhyay, 1993), although it is important also to emphasise that the process began in earnest during the communist-led United Front governments of the late 1960s, and that land redistributed in the 1970s included the titling of some of the land seized in the preceding years. While wanting to draw attention to the innovative aspects of the West Bengal model of agrarian change, this book sets out neither to deride it (as Mallick has, 1993), nor to oversell it (as Lieten has, 1996a and b), but rather to understand better (and in a comparative context) its relation to agricultural growth.

In Bangladesh a very different set of policies emerged during the 1980s. Agricultural input markets were liberalised. Levies were removed from imports of irrigation pumpsets and their siting ceased to be regulated. By the early 1990s fertiliser subsidies had been removed. The degree to which these changes away from government involvement in price-setting and regulation influenced investment in groundwater and other inputs required for the production of high yielding varieties of rice, the resulting growth or stagnation, and the outcome (intended, stated or otherwise) for different people in rural Bangladesh are the subjects of major dispute—a dispute carried on, in the contributions to this volume, between Adnan and Palmer-Jones.

In their attempts to find the links between policies, growth and changes in well-being, the contributors to this book have kept agrarian structures central to the problematic. On close inspection agrarian structure is about much more than the distribution of land and other assets among rural people and the organisation of agricultural production. Even these factors varied widely across Bengal (for a typology of landholding structures and tenurial arrangements in pre-independence Bengal, see Bose, 1986). Moreover, structures of rural industry, trade and commerce may be so closely interwoven with agricultural activity that the questions requiring study concern the causal links between rural social and economic structures and growth. Did structures act as a constraint on growth; did they shape the pace and social distribution of the growth rather than simply restricting it? Causality is likely to be complex, interactive and lagged rather than unidirectional, with growth impacting on structures as much as vice versa. The papers in this book include investigations focusing on both directions of causality and varying concepts of structure, including land and other productive asset distribution, ownership and mode of allocation of irrigation water, agricultural product markets, contractual arrangements for hired workers (especially migrants) and ideologies of social rank.

The rhetoric and practice of state 'policy' may also be seen as structural. Policies may, for example, reflect the interests of particular classes or they may change the relative power of one class or another (intentionally or otherwise). Policies as understood here include local action by political parties and their elected representatives as well as the more rhetorical statements and legislative processes of those in power at the state (West Bengal) and national (Bangladesh and India) levels. Policies are subject to external as well as internal influences and reflect international as well as local and national power relations. This book seeks to provide some answers to the questions of how structures, 'policies' and growth/stagnation influence and are influenced by each other in contemporary West Bengal and Bangladesh.

The convictions which have polarised recent debates on agrarian change in the two Bengals indicated the need for a volume which would seek to explore the evidence and include contested interpretations. Divergent views will thus be found on (*a*) the future growth in Bangladeshi agriculture, (*b*) the role of particular policies and of agrarian structures in bringing about the current situations in rural West Bengal and Bangladesh, and (*c*) whether the incidence of poverty has declined. The readers are left to make up their own minds. Whichever view is taken, however, the differences in interpretation are of potentially great significance to the livelihoods of rural Bengalis because of their political and policy implications.

Attempts to link micro level change with national (and state government) 'policies' do not have a particularly strong record. One of the reasons for this is an excessive reliance on crudely specified and aggregated data from large scale surveys—or, more recently, quick 'participatory assessments' (UNDP, 1996). This book leavens official data with extended primary research in rural Bengal by the great majority of contributors as well as by detailed archival research. It represents a unique project of interdisciplinary synthesis.

The remainder of this introduction acts as a guide for the reader of the book, but also makes reference to papers and discussants' comments from the Workshop. As indicated above, the editorial line has been to encourage evidence-based argument, including from ideologically opposed perspectives, rather than pushing any single interpretation of change. It is for the latter reason that papers presented at the Workshop but not included here are cited in this introduction. The book has three parts corresponding to the major themes of the Workshop: growth, policy and structure. This lends the organisation of papers a certain symmetry although some of the contributions range over more than one theme and could have been placed differently. Moreover, while Part 1, on 'Agricultural Growth, Poverty and Well-Being', is weighted more towards Bangladesh than West Bengal, the opposite is the case in Parts 2 and 3, which focus respectively on 'Policies and Practices' and 'Changing Agrarian Structures'.

The section which follows first summarises the historical background to agricultural development in Bengal, which is expanded on in Chapter 2 of the book. It then reports the character and extent of agricultural growth in recent years together with a stringent critique of the method for collecting agricultural production data in West Bengal by James Boyce. The third section highlights the contrast between the Bangladesh and West Bengal policy regimes, summarising the several contributions in the book on how policy has been played out in practice at the micro level. In the final section, reference is made to papers which focus on the main theme of the book: the relation between agricultural growth/stagnation and changing agrarian structures.

THE DEVELOPMENT OF AGRICULTURE IN BENGAL AND THE EVIDENCE FOR A TURNAROUND

Until the middle of the 19th century crop output in Bengal grew, mostly at the extensive margin and largely driven by population increase. However, despite some continued expansion of the net cropped area in east Bengal from 1860 to 1890 and despite demographically induced intensive[4] techniques thereafter, Bengal as a whole experienced stagnation in agricultural production between 1860 and 1920. In west Bengal over that period, malaria and the decay of rivers led to a decline in both area and output (Bose, 1993).

Between 1920 and 1980 the regional trends converged: stagnation in agricultural production coexisted with rapidly increasing population. In particular there was slow growth in the yields of *aman* paddy, which accounted for about half the gross cropped area in both east and west Bengal (Bose, 1993).

Sugata Bose argues here that research into the structural obstacles that may have held back growth in production should involve analysis of the role of the state—including, in the case of the colonial state in Bengal, its acts of omission in irrigation and drainage development and of commission of the modes of appropriation of the agrarian surplus over which it could exert influence. Saugata Mukherji and Manoj Sanyal (1995) re-examine cropping patterns and output trends for the period 1900–88, locating Bengal's agricultural stagnation until the late 1970s in conditions created by the colonial state: ill-defined property rights, differential access to inputs and resources and the failure of the state to revive canal irrigation and promote private investment in land and technological inputs. Bose also argues that agricultural commercialisation in the colonial era was as much the outcome of state imperatives as of expanding market forces. And it was made possible by the continuing non-commodification of labour, i.e., the reliance on family labour in peasant production.

In resolving the contradiction between expansion of commodity production for capitalist markets and low levels of productive investment, Bose shows that the surplus from production had been appropriated by a number of class fractions including the colonial state, local landlords, metropolitan capitalists, indigenous merchants and moneylenders, and rich peasants. However, the principal modes of exploitation and relations of appropriation have shifted over the last two centuries from being rent-based (in the context of the colonial revenue system), through credit and debt (from the late 19th century until the 1930s) to the land-lease basis of later decades. For Rajat Ray, commenting on Bose's paper at the Workshop, the key issue provoked by this argument was why the domestic market for rice had not been expanding in the period prior to the Great Depression of the 1930s. The answer lies in structures of surplus appropriation which prevented surplus from accruing to peasants and which thus dampened local demand. Bengal had had to achieve any increases in output through sales on the highly competitive international market.

Bose's contribution also shows how, in the decades following independence, the impacts of agrarian structures on growth were indirect and context-specific. Those who appropriated surplus aimed to do just that. The holding back of growth was a by-product. In many eastern districts, by this time in East Pakistan/Bangladesh, where smallholdings were predominant, richer smallholders leased out land in small parcels to ensure increased labour input. Production continued to be based on family labour. In western districts (by now in the state of West Bengal), also characterised by smallholdings but with a longer established agricultural labourer class and a more substantial rich peasantry, the latter maintained a wider range of economic activities and had access to higher prices for their produce. Increasingly, richer smallholders in Bangladesh have used leverage to gain the major share of resources originating in foreign aid, while the same group in West Bengal have better access than others to capital intensive agricultural technology. In north Bengal (characterised by a rich farmer–sharecropping system), despite rapid growth between the 1940s and 1960s, class disparities widened, and agricultural labourers and sharecroppers became further impoverished. The question of the relationship between growth in agricultural production on the one hand, and absolute as well as relative poverty on the other, thus remains an empirical one for the present period as well. First, however, it is necessary to ask whether the extraordinary rates of growth reported, especially in West Bengal over the decade from the early 1980s, are based on sound data.

Haris Gazdar and Sunil Sengupta summarise the aggregated growth data reported by Abhijit Sen and Ranja Sengupta (1995). Sen and Sengupta report an annual growth rate in foodgrains output of 6.9 per cent per annum between 1981–82 and 1991–92 at a time when population growth in the state was 2.2 per cent per annum (Census of India, 1991). As Gazdar and Sengupta imply in their paper and conclude explicitly elsewhere (Sengupta and Gazdar, 1997), even if the production figures produced by the Government of West Bengal's Department of Agriculture are assumed to be reliable, there is a serious problem involved in choosing the base year. In 1981–82 and 1982–83 harvests were unusually low. Choosing the base year as 1983–84 with the same data yields an annual growth rate of 4.3 per cent per year (ibid.: 165). Moreover, at the Workshop, James

Boyce suggested that a major source of bias was the changing methodology for the collection and preparation of crop output data. Until the 1980s the state Bureau of Applied Economics and Statistics (BAES) conducted independent sample surveys for acreage estimation and crop cuts for yield estimation, based on the statistical methodology developed under the direction of P.C. Mahalanobis at the Indian Statistical Institute in the 1940s. The subjective estimates of the Directorate of Agriculture (DoA)—the state agency charged with implementing policies to boost agricultural production—could be checked against the BAES figures, and the BAES data were generally used as the basis of official state government estimates. In the early 1980s the DoA unilaterally began to 'adjust' some BAES estimates. In the mid-1980s the integrity of the data was further compromised as BAES sample surveys for acreage estimation were abandoned altogether, and official yield figures were converted to a simple average of the often quite divergent BAES and DoA estimates.

Boyce speculated that the true rate of agricultural growth in West Bengal in the 1980s may have been around 4 per cent per year. While not as spectacular as the rates estimated from the official data, this would still represent an impressive performance, far surpassing prior growth rates in the state and comparing quite favourably with agricultural growth elsewhere in India. The fact that this growth occurred during a period of modest but significant agrarian reforms is noteworthy, suggesting that greater equity is compatible with efficiency and growth. Boyce regretted that questionable changes in West Bengal's agricultural output estimation procedures have made it difficult to assess objectively the state's performance and to analyse these accomplishments. He strongly recommended that the modus operandi of BAES acreage and yield estimation be restored to its pre-1980s status, that separate publication of the resulting BAES data be resumed, and that official figures be restored to the earlier BAES basis.

Replying to Boyce at the Workshop, Biplab Dasgupta argued that if the bias in yield reports had been constant over the 1980s, the rate of growth, which was the main focus of dispute, would have remained unchanged (an argument supported by Lieten, 1996b: 32; see also Dasgupta, 1995). Whatever the actual rate of growth in agricultural production in West Bengal, the Left Front government has gained credit from the very high levels implied by official data (see *Business India*, 1994).[5]

Data indicating the plateauing of foodgrains production cited at the start of this introduction were not available to Gazdar and Sengupta at the time of writing. However, using Sen and Sengupta's quantitative analysis (1995), the outstanding growth rates of West Bengal agriculture up to 1991–92 are described in the context of a general turnaround in agriculture in the neighbouring states of Bihar and Orissa, both of which experienced stagnation or slow growth in the 1970s and higher growth in the 1980s. As West Bengal's growth was the highest of the three it merits special attention.[6]

The main proximate causes of growth were the adoption of higher yielding varieties of monsoonal *aman* paddy, still the most important single crop in both West Bengal and Bangladesh, and the cultivation of summer *boro* paddy in rotation with *aman*. Both of these forms of intensification were enabled by the rapid spread of groundwater irrigation, mainly in the form of privately owned shallow tubewells (STWs), but later, as water tables dropped, through mini-submersible tubewells (MSTWs).[7] Other trends included a rapid increase in the cultivation of potato, a potentially high value winter *rabi* season crop that could be cultivated in rotation with *aman* paddy.

Shahabuddin (1995) shows that Bangladesh also experienced impressive growth in the output of rice in the 1980s. His data suggest in fact that average annual growth in output of rice at 2.5 per cent (consistently the most important crop accounting for three-quarters of total acreage) exceeded the annual population growth of 2.1 per cent for the entire period 1972–73 to 1993–94. However, disaggregating the data into the 1970s and 1980s, average growth rates showed a slight decline from 2.8 to 2.3 per cent between the two decades. More strikingly, the production of rice between 1989–90 and 1993–94 appeared to have stagnated at just 0.48 per cent per annum.

According to Shahabuddin, just as in West Bengal, the proximate cause of the growth in output in Bangladesh for the period upto the end of the 1980s was an increase in area under high yielding varieties of rice which, in the early 1990s, accounted for 75 per cent of the gross cropped area and 60 per cent of the total value of crop output. Although in his contribution here Shahabuddin attributes the recent growth in production to intensification of input use, this raises rather than answers questions about the direction of growth. Both the total consumption of chemical fertiliser and total irrigated area increased into the early 1990s. The annual rate of growth of total irrigated area was 2.85 per cent in the 1970s, 6.21 per cent in the 1980s and 8.26 per cent in the early 1990s. Why then were output growth rates in the 1990s so low? Indeed, why was there a decline in the acreage of *boro* paddy in the early 1990s? Has there been compensating growth in the production of other crops?

Palmer-Jones completes the growth trend picture for Bangladesh up to 1995–96. He argues that the most plausible reading of available data suggests that the apparent slowdown in the early 1990s did not affect long term growth trends in foodgrains production; in fact the levels of foodgrains production attained in 1989–90 and 1990–91 were higher than longer term trends due to high *aman* paddy production following two years of floods, and high *boro* paddy production due to the effects of liberalisation and favourable groundwater and economic conditions. Both Adnan and Palmer-Jones pay much attention to changes in the number of tubewells purchased each year, which are linked directly to the production of *boro* paddy. However, while Adnan attempts to show a decline in the profitability of *boro* paddy related to falling rice prices (with causal links back to a failure of demand brought about by uneven distribution of the returns to growth), Palmer-Jones argues that, although incentives to production of *boro* paddy were undermined by drastically lower rice prices between mid-1992 and the end of 1994, and in any case the scope for increasing area cultivated in *boro* paddy was limited by the quantity of suitable land, the subsequent revival in the sales of shallow tubewells in 1994–95 and in the growth of *boro* paddy production in 1995–96 suggests a return to trend.

Inter-district differences in growth performance in Bangladesh are illustrated by Ben Crow in his paper on surplus appropriation. Working in parts of Noakhali experiencing sluggish growth and comparing the findings with field research in the more dynamic area of Bogra, Crow relates the growth prospects of agriculture back to differences in *the ways surplus is appropriated* at the levels of both production and distribution. In the Bogra study area where environmental conditions favoured double cropping of rice and where own-account cultivation was predominant, investment in agriculture (especially in tubewells) was high. By contrast, in Noakhali, a less favourable environment and much greater incidence of sharecropping resulted in a smaller marketed surplus, as much more of the harvest was required for rental payments and repayment of production loans. Crow concedes that there are organic connections between the 'backward' and 'advanced' areas of agricultural production which require much more research. However, he asserts that historical explanations for differences in the forms and uses of surplus are required to understand the regionally diverse character of agricultural growth in Bangladesh.

Despite their differences, all those contributing papers on the subject at the Calcutta Workshop agreed that there had been a significant turnaround in production in agriculture, especially rice production, both in West Bengal from the mid-1980s and, albeit less strikingly, in Bangladesh from the mid-1970s. How could this be reconciled with the very different policy regimes in the two Bengals?

POLICIES AND PRACTICES

Given the similarities in the directions of change in West Bengal and Bangladesh it is surprising that they followed divergent agricultural policy regimes both in rhetoric and in practice; policies of course have unintended as well as intended consequences and actual intentions may be different from stated ones.

Policies include action as well as inaction. The main contrast between Bangladesh and West Bengal in legislation directly affecting rural sectors over the last twenty years is that West Bengal attempted to intervene directly in the agrarian structure (as conventionally understood, as well as through local government reform; see Lieten, 1996a), while the policy focus in Bangladesh was on the liberalisation of prices and on deregulation (for example of the siting of tubewells).

The distinction between the policies adopted in Bangladesh and in West Bengal is blurred by the favourable input–output price ratios for owner-cultivators of rice and the credit for tubewells available in West Bengal. Harriss (1993) attempts to assess separately the influences of institutional policies (agrarian reforms) on the one hand and price movements and technical changes on the other in bringing about growth in the state. This separation is rejected by Sengupta and Gazdar (1997) on the grounds that the agrarian reforms, in particular the reform of local government, created the conditions for more effective distribution of inputs, including electricity for powering groundwater irrigation, credit for the purchase of shallow tubewells, and fertilisers. They speculate that improved electricity supply may have enabled greater use of tubewells. However, in at least some areas most shallow tubewells are diesel powered (e.g., for Birbhum district: S.N. Chatterjee, 1993).

Nripen Bandyopadhyay (1995) emphasises the importance of initial conditions in West Bengal, which helped bring about the agrarian reforms. Key among these were:

1. The relatively large proportion of landholdings of less than 5 acres in the state; and
2. the relatively long history of organised peasant struggle.

The reforms have been documented meticulously and commented on in numerous publications so there was no need for contributions to repeat that exercise in the present volume. However, several of the contributions which follow put a magnifying glass to policy practice at the local level. They serve to illustrate relations between the state and agrarian society but make no claim to be representative.

In particular, Glyn Williams examines the changes in micropolitics associated with panchayati raj in three villages in Birbhum district. Using Davis' distinction between formal politics (*sorkari kaj*) and informal politics (*gramer kaj*), Williams investigates the combined effect of the introduction of much greater local level enfranchisement and the growth in agricultural production on the political power of poor people. He finds that in the research area, the patronage politics characterising *gramer kaj* remains, but that the dominance of local 'big men' has been dramatically challenged by their need to win votes. Combined with increased demand for agricultural labour, the result has been a degree of economic and political empowerment of previously marginalised people.

The theme of patronage and its changing character is also important to Arild Ruud's paper on cultural change. Ruud explores the culture represented by the ruling CPI(M) party officials in the village he studied in Bardhaman district. The practices of the local cadres of the ruling party are found to be embedded in the socio-cultural stratum of society to which they belong: lesser landlords and middle-to-rich peasant households, mainly of the peasant castes. The association of the CPI(M) with anti-casteism, teetotalism and literacy derives from specific ideas about morality and respectability, which are partly the product of class position. In terms of the practice of local CPI(M) politicians what is important here is the way in which they pay disproportionate degrees of attention to particular groups of rural poor people delineated by *jati*. Both Ruud and Williams find that while resources distributed by the panchayats (whether originating from central or state coffers) did find their way to poor rural people as intended, those with loyalty to the ruling party benefitted much more.

Dwaipayan Bhattacharyya also studied the practice of the CPI(M) at the micro level. The focus of his contribution is on the policy rhetoric of the Krishak Sabha (the CPI[M]'s peasant union) and its efforts to keep a lid on class conflict through the adoption of the 'peasant unity' line. Bhattacharyya illustrates the electoral pragmatism in the practices of the CPI(M) which overrode most other considerations (see also P. Chatterjee, 1997: Chapter 10). The Krishak Sabha, whose documents Bhattacharyya analyses in detail, is dominated by middle and rich peasants. But at the same time the Sabha is the union of all 'peasants', including agricultural workers. Indeed, the latter were not permitted to form a separate union. Despite this the CPI(M) has succeeded, election after election, in keeping the votes of agricultural workers, and this is where the pragmatism comes in. Ritual strikes are used to demonstrate radicalism and the selection of candidates is on the basis of the individuals' local vote-pulling power rather than their history. Local issues were above all manoeuvred to electoral advantage. It is reasonable to hypothesise that the management of rural social relations by the CPI(M) and the longevity of the regime have been influential in bringing about favourable conditions for private investment in groundwater irrigation, and were thus partly responsible for agricultural growth (see the paper by Gazdar and Sengupta, this volume).

In relation to other states in India, more has been done by the government of West Bengal to change the structure of property relations in favour of less well-off people (though not necessarily leading to exceptional achievements in terms of well-being—see below and Gazdar and Sengupta, this volume). However, Bhattacharyya's work suggests that much of the radicalism of the Left Front government in West Bengal is rhetorical. Their unprecedented political success may reflect their consummate skill in understanding the electorate. A more unkind view might be that like the Partido Revolucionario Institucional (PRI), which came to power in the Mexican revolution in 1910, the party has become more of a club than an organisation motivated by radical ideologies.[8] The class make-up of West Bengal undoubtedly changed between the 1950s and 1970s with the rapid decline of ex-zamindar *bhadralok* and the rise of the *chashis*.[9] What happened at the end of the 1970s was the consolidation of the power of middle peasants, a key part of which was the incorporation of agricultural workers and poor peasants (including sharecroppers) through a role in conflict resolution and prevention which sought to reflect rather than challenge local balances of power.

While West Bengal's innovative reform policies justifiably attracted the attention of scholars, they have come in for criticism on the grounds of gender bias.[10] In particular the titling of land redistributed under the land reform programme was almost always in the name of a male household head. This may have strengthened patriarchal relations in the state (Basu, 1992; Sengupta and Gazdar, 1997). Jayoti Gupta (1997) draws out the critical direct links between unequal property rights in land for men and women on the one hand and unequal gender relations in the household (and at wider levels) on the other (see also Agarwal, 1994).

Drawing on evidence from Medinipur district of West Bengal in the late 1980s, Gupta also highlights the importance of dowry in the reproduction of the land-owning structure. The non-land resources of dowry enable land to be retained by males in the natal household. Equally, dowry in movable goods and money may be used to capitalise land controlled by males in the bride-receiving household.[11] Greater coordination was required between the women's movement and the 'peasant movement' in pushing the gender distribution of land-owning rights onto the policy agenda. At the Workshop, Naila Kabeer also emphasised how dowry became a key means of economic accumulation for those with sons, showing the importance of studies of family size and composition to studies of household mobility.

Basu's comparison of women's activism in West Bengal under the CPI(M) and the much more radical (but less sustained) grassroots activism in Maharashtra explores the links between democratic centralism in the organisation of the CPI(M) and democratic centralism in the Bengali home. She provides an archetype of Bengali caste Hindu ideas about women and their roles as 'repositories of tradition', which she argues are, 'central to the construction and maintenance of lineages, kinship networks and caste boundaries' (Basu, 1992: 13), which may partly explain the exceptionally low percentage of women in West Bengal's rural workforce. The attempt to live up to archetypal roles is likely to entail abstinence from paid manual work and unwaged work outside the homestead; ideas about such roles may also be used by women to avoid drudgery (see Rogaly, 1997).

It is not surprising that policy practices at the local level appear to be far removed from policy rhetoric at the apex of the party in the state, or that their radicalism has declined in the name of electoral realities. This may be seen as responding to popular demand.[12] The Left Front government has managed to use its intuition about the equations between social forces in the countryside to remain in power (in the 1996 assembly elections, the decline in the share of votes achieved by the Left Front parties was largely an urban phenomenon) (see Rogaly, 1998).

The policy climate in Bangladesh can be contrasted with that in West Bengal. Relatively recently liberated from Pakistani rule, Bangladesh is a new country. In the 1970s Bangladesh had more pervasive state control over modern agricultural technology, yet it was at the forefront of experimental grassroots attempts to alter the distribution of power in the countryside (see, for example, Wood, 1994). These were fragmented rather than tied to a single political movement and, together with foreign aid money, gave rise to the mushrooming of non-governmental development organisations. Adnan's structural analysis of agricultural policy clearly links together the interests of some of the dominant rural classes with those of aid donors. Palmer-Jones notes the extent to which donor consortia based on different ideological approaches to rural change vied for dominance in the competition for influence over the direction of Bangladesh's agricultural policies.

Policies associated with the World Bank and other large donors, notably the United States Agency for International Development (USAID), including liberalisation of agricultural input markets, became the hallmark of agricultural policy in Bangladesh much earlier than in India (see Hariss-White and Janakarajan, 1997). In his paper here, Shahabuddin narrates the process by which subsidies on fertilisers were lifted, speculating that rises in fertiliser prices (following the withdrawal of subsidies in 1991) were connected—along with declining rice prices—with the slowdown in agricultural growth in the following three years. It is not just the aggregate dosage of fertiliser but also the mix which affects productivity. Since the subsidy withdrawal has affected the prices of Triple Super Phosphate (TSP) and Murate of Potash (MP) more than the price of urea, there is a greater risk of imbalance emerging along with underutilisation.[13]

Apart from the removal of fertiliser price subsidies, the government of Bangladesh also intervened to increase private sector involvement in the market for irrigation equipment. Initially, private sector importers were permitted to buy foreign-made shallow tubewells which satisfied certain standards. As in India, the government also attempted to control the siting of shallow tubewells. Minimum standards and siting regulations were removed along with subsidies on minor irrigation equipment in 1988. Shahabuddin attributes the rapid expansion of investment in shallow tubewells in the following three years to this policy. However, he also notes a subsequent decline in the purchase of tubewells, which according to Palmer-Jones was temporary and could be explained by Gulf War–related diesel price increases and physical shortages of diesel and fertiliser.

Mahabub Hossain (1995) traces the history of agricultural policy reform in Bangladesh. He shows liberalisation to be a drawn-out procedure, in some cases beginning in the late 1970s and spreading over fourteen years. Each sector was liberalised in unique ways, there evidently being no set template. While in the fertiliser sector changes in permitted forms of ownership were initially confined to the retail end and festooned with quantitative price restrictions so that private manufacture, storage, imports and domestic trade took over a decade to be fully deregulated, the deregulation and privatisation of irrigation equipment and its management took far less time. Some sectors have been reformed only at the margins, for example the public distribution system of food. It has proved possible to reduce targeting of the public distribution system to the rural sector and increase private trade only by open market sales by the state trading corporation. In some sectors the unintended outcome of one phase of deregulation affected subsequent policy implementation. In credit, for example, the proliferation of loan repayment delinquency during the first phase of reform provoked a new phase of regulation. Further unforeseen political interference in the commercialisation of credit resulted in a revived wave of overdues and regulative backwash.

Adnan offers a powerful deconstruction of agricultural policies. His examination of unstated objectives of agricultural policy and the divergence between actual outcomes and stated intentions bears some similarity to Bhattacharyya's more specific analysis of the discourse around CPI(M) policy in West Bengal. However, Adnan discusses not just one political regime but the nexus of interests between rural elites, the Bangladeshi state and foreign donors. He identifies the mediating structures through which policies are structured and played out. Nevertheless, the paper is not a diatribe against input market liberalisation. Adnan notes that both the growth of the late 1980s and the slowdown of agricultural production in the early 1990s occurred in the wake of the deregulation policy regime.

Crucially, the policy interventions for all their stated intentions interacted with what Adnan calls the mediating structures to bring about shifts in income distribution in the direction of larger land- and water-owners. The result of this was the continuing impoverishment of other classes and a subsequent failure of demand for rice causing the fall in the rice price, reductions in the profitability of *boro* paddy and a slowdown in the growth of *boro* paddy production.

Adnan uses case study evidence from ten villages in Daripalla to show that social classes with greater wealth and influence renegotiated contracts in the markets for both inputs and for agricultural products so as to increase the share of the surplus accruing to them. Irrigation cooperatives were manipulated in this way, as were transactions in the land lease, credit and product markets. The forms of contractual arrangements which emerged were those which suited the dominant classes. The implications for the poverty of other people necessitate a more detailed examination of the central problematic: the role of agrarian structures.

CHANGING STRUCTURES

Agrarian structures may shape growth, while growth itself influences structure. Structures influence policies at the national level, while the latter themselves play a role in bringing about structural change.

Geoffrey Wood reflects on the structural trends towards smaller holdings in rural Bangladesh and the strong intrusion of agrarian capital and new technology. Increasing landlessness has only partly been absorbed by non-agricultural employment, and the intensification of land use has brought about a proliferation of roles, including the provision of water and 'other' services such as mechanised ploughing. Wood advocates a conceptual shift in thinking, away from the 'farm' as unit towards 'command area' of a shallow tubewell as unit. Wood terms this shift 'agricultural reformation'. The central importance of groundwater irrigation in the organisation of agricultural production has led to a consolidation of management around tubewells. The more land a household owns, the greater the number of tubewell command areas the land will fall under, and thus the greater the number of transactions entered into for water and other services. Water-holding is as important an element in agrarian structure as land.

In West Bengal agriculture too, according to Neil Webster, there has been consolidation of management around tubewell command areas. Earlier tenancy patterns have been reversed for the irrigation-based *boro* paddy cultivation season. Water-owners who, especially in the case of mini-submersible tubewells (MSTWs), tend to be the richest and most powerful in the village, rent in land for *boro* paddy cultivation. According to Webster the siting of MSTWs is decided according to social rank and caste identity. If there are competing claims for sinking an MSTW in a particular area, there is an implicit legitimacy to be gained from high local status ranking. Sinking an MSTW requires the acceptance of others with land in the command area.

Webster's warning that the increasing power of tubewell owners could be leading to greater inequality in the West Bengal countryside is backed up by Beck (1995) and is consonant with Adnan's analysis of unequal gains from the introduction of shallow tubewells in an area of Bangladesh. Beck analyses the differential returns to the rapid adoption of green revolution technology, including irrigation by shallow tubewells, in a village in Medinipur district. He shows that although there was an improvement in employment and wages, this improvement was marginal compared to the gains in earnings accruing to the tubewell owners, whom he portrays as a small elite.[14]

It would seem that changes in the landholding structure under the Left Front government's agrarian reforms were not as significant as might have been implied by the attention they received. A shift in power away from absentee landlords towards rich and middle peasants took place in the late 1960s in West Bengal. The first United Front regime was elected in 1967, and, followed by the second United Front government which was elected in 1969, determined to rid those with very large landholdings of their ceiling-surplus land. Although initially frustrated by the bureaucracy and the police, they presided over a significant shift in the structure of land control (Ruud, 1994: Table 3). These were the defining moments of change in the distribution of land in post-independence West Bengal (Mukarji and Bandyopadhyay, 1993: 2.3). The much-discussed redistribution of ceiling surplus land under the Left Front government after 1977 was quantitatively less significant than that which preceded it (Sengupta and Gazdar, 1997).

The slowdown in production reported in West Bengal and Bangladesh in the early-to-mid-1990s (and which Palmer-Jones disputes as not representative of longer run trends), may have been caused by a rapid drop in the water table. The intensive use of groundwater has been blamed for shortages in the state's three main canal-feeding reservoirs (*The Statesman*, Calcutta, 11 December, 1995). The 1997 *boro* paddy season was in crisis in at least four districts: Haora, Hugli, Murshidabad and Birbhum. The state was being urged to purchase water from Bihar (*Asian Age*, 6 December, 1996). A further serious environmental threat which has been linked to intensive use of groundwater is the widespread arsenic poisoning in drinking water in West Bengal and Bangladesh. Research has found that over 200,000 people in seven districts of West Bengal suffer from arsenic-related diseases attributable to drinking water from contaminated wells, and that the likely cause of this contamination was heavy groundwater withdrawal (Mandal et al., 1996).

At least as far as the villages were concerned the new 'big men' in rural West Bengal—as elsewhere in rural India—were the middle and rich peasantry.[15] This trend is quite in line with the rise of the rural 'backwards' in Bihar, UP, Tamil Nadu and elsewhere in India. Indeed, the communist party cadres so active in rural West Bengal in the late 1960s were mainly drawn from the ranks of the middle and rich peasant classes, and these classes were not targeted in the land seizures. Moreover, in order to retain their support, there was relatively little agitation on the issue of wages for agricultural workers (Ruud, 1994: 364).

Despite the Left Front government's having been in power for twenty years, West Bengal lacks effective regulation regarding the terms and conditions of agricultural employment. The only states with separate Agricultural Workers Acts were the two others which have been led at least periodically by the CPI(M): Kerala and Tripura (Egger, 1996: 57). Government statistics in this regard are particularly weak. For example, the state government's annual *Economic Review* has no section on agricultural wage workers. The employment section concerns only formal sector workers; the agriculture section has nothing on labour employment, wages or types of contract. A 272-page report produced on labour in West Bengal by the government in 1994 hardly mentions agricultural workers. That little attention appears to have been paid in legislation or in official statistics to agricultural wage workers' terms and conditions puts West Bengal on a par with most of the rest of India. The government of West Bengal did not take direct action to ensure minimum wages—certainly not with anything like the vigour put into land redistribution and registration of tenants.

The increase in seasonal and other temporary migration for employment, which has consistently been underestimated in large scale surveys in India (Breman, 1996: 18; see also Srivastava, 1997; Singh 1995; Pai et al., 1997), is of great importance in West Bengal. Ben Rogaly's paper draws on reports from bus drivers and from village level Integrated Child Development Scheme workers, suggesting that migration for agricultural work involves hundreds of thousands of men and women (as well as some children) moving south and east into the intensive rice cultivation areas of the central southern districts. Reports of migration for agricultural work in these areas stretch back at least 100 years; the scanty data currently available indicate that it increased rapidly with the recent fast growth in production of rice (the transplanting and harvesting of which remain highly labour intensive). Rogaly's paper recognises seasonal migration (and the contractual arrangements involved) as part of the agrarian structure. Its central problematic is how seasonal migration, boosted by the growth in demand for hired workers associated with increased production, in turn brings about wider changes in agrarian structures. Drawing on evidence from a pair of simultaneous microstudies in Purulia and Bardhaman districts, Rogaly argues that labour migration changes relations between employers and workers in both source and destination localities. Workers (migrants and others) in source areas benefit in local employment negotiations from their own and local employers' knowledge of the potential employment and relatively high wages available if they migrate. Employers in destination areas, boosted by the potential as well as actual supply of migrant workers, are able to continue social control of 'their' local workforce. Other forms of structural change brought about by increasing migration include conflict within the class of rural employers when efforts to recruit migrant workers clash with the labour requirements of employers in source areas. Moreover, the accommodation provided by workers in source areas to recruiting employers involves sharing cooked food and sleeping arrangements across boundaries of social rank. It may thus be part of the process of shifting those boundaries.

Changing contractual arrangements, including increased seasonal labour-tying as well as hiring of migrants, can be interpreted as part of the efforts of accumulating employers to contain a workforce which, partly due to agricultural growth, has gained in power. Williams reports on changes in the micropolitics following panchayati raj in three villages in Birbhum district. It appears that in Williams' study area panchayati raj has led to increased assertiveness (in the wake of political representation) on the part of workers. In contrast, Ruud shows that in Raina *thana* of Bardhaman district employers have used the CPM as a way of maintaining control over formerly 'untouchable' workers through differential patronage along *jati* lines.

Real wages have risen in West Bengal during the period of rapid agricultural growth (Sengupta and Gazdar, 1997). According to Palmer-Jones they also increased significantly during the 1990s in Bangladesh, coinciding with falling rural poverty according to head-count measures. Zillur Rahman (1995) has suggested that while 'moderate' poverty did indeed decline between 1990 and 1992, bringing about an overall reduction in poverty in rural Bangladesh, 'extreme' poverty continued at the same level. Poor people gained from falls in rice price in 1992–93, as there was no parallel decline in wages. Rahman adds that the trend away from casual towards piece-rate contracts has also increased the earnings of poor people.

Barbara Harriss-White's paper on agricultural markets in West Bengal also focuses on the relation between growth and structure, particularly the impact of the former on the latter. Harriss-White's framework for analysing market structure is built on a systems approach to agricultural markets which also attends to class differentiation within the system, to the functional differentiation of component firms, to variation in the competitive environment for contract formation and to the social and political embeddedness of exchange. She examines transactions in market systems for three staples (mustard oil, potato and rice) in three market centres in West Bengal (Memari, Gulsi and Katwa). Trading firms revealed complex internal structures taking the form of a high degree of uniqueness of combinations of activities. These activities included buying, selling, brokerage, storage, processing, transport, and the finance of trade and production. Asset ownership revealed a highly skewed market structure mirroring to a large extent the distribution of land, and including urban land and storage and processing sites. However, non-local, non-land based capital was as much involved as local commercial capital, and qualified any simple relations of determination of market structure by the structure of landholdings.

Harriss-White finds that growth in the marketed surplus in West Bengal was associated with an explosion of petty trade by a political constituency generally favourable to the Left Front government. However, the state has effectively blocked accumulation by petty traders (and therefore competition from them) by rationing licences to them and continuing to try to protect mercantile magnates through legal arrangements guaranteeing local monopoly control over wholesale trade, by restricting processing technology to large scale and through targeting subsidised credit away from small traders. With the state's failure to implement regulated market law, much of such regulation has been left in the hands of collective institutions of civil society with oppressive, arbitrary and extra-legal penalties on (usually weaker) outsiders.

Harriss-White asks whether this economic empowerment of commercial magnates was intentional, or had research problematising the development of small scale petty trade simply not reached the regime? The Left Front government's focus on reforms in the structure of land ownership and in tenurial relations represented a concentration on structures of production to the exclusion of property relations in exchange and circulation. In the meanwhile a compromising accommodation had been developed between a powerful commercial elite and the state.

Suman Sarkar and Kali Shankar Chattopadhyay (1995) tackle the issue of whether growth and technical change in agriculture could dissolve the constraining institutions in which market exchange is embedded or whether market institutions survive notwithstanding growth. Village level field research in the mid-1970s and in the 1990s in localities undergoing a variety of growth trajectories was compared. In the 1970s markets were physically fragmented and socially segmented such that a spectrum of prices could obtain in a given location and point in time for a given crop, and such that inter-village price differences were greater than justified by transport costs alone. Agricultural growth has brought about considerable change in marketing systems: a reduction in the use of periodic marketplaces, a great increase in itinerant trade and the creation of chains of agents bound by credit, the conversion of village traders into rice mill agents and increased price collusion by commercial rice millers. Sarkar and Chattopadhyay conclude that commercialisation was relatively economically differentiating, that growth in the marketed surplus accentuated the economic power of the mercantile capitalist class and that liberalisation would increase their capacity for bias towards traders and away from suppliers in the gains from trade.

Drawing on a comparative study of villages in Noakhali and Bogra districts of Bangladesh, Crow shows how different environmental conditions working in tandem with the structure of landholdings and tenancy relations are associated with contrasting outcomes in terms of growth. Crow contrasts the proportion of output sold by each economic class in each region. The main contrast is between the *chars* of Noakhali, where even rich peasants generate only a small surplus and small and middle peasants are deeply in deficit, and the positive surplus accruing to rich peasants and landlord households in Bogra district. In Noakhali only a small proportion of agricultural output is sold because of extensive kind payments necessary in the land-lease and credit markets. Collusive arrangements among controllers of transport physically constrain the freedom of producers of small surpluses to transact. The contrasting modalities of surplus appropriation (with a small minority of transactions on money and grain markets being tied or interlocked in Bogra) affects investible surplus and thus growth. Crow concludes that historical explanations are needed to explain differences in the forms and uses of surplus.

Using the same study as his base, K. A. S. Murshid (1995) examines the effects of the structure of finance of agricultural production on the pace of agricultural output growth. He presents a typology of links between trade, credit and production and argues that there has been significant continuity in the internal organisation of institutions associated with grain trading and processing, despite a big increase in their number. Economically weaker producers experience forced commerce and price-fixing loans. Moreover, Murshid argues controversially that personalised finance, like that which prevailed in the Noakhali area, has higher transaction costs and therefore higher overall costs than formal financial intermediation would have had.

The emphasis on agrarian structures, albeit reconceptualised to include structures of water-holding, commerce, bureaucracies, contractual arrangements, and shifting ideological structures, could be read as denying the importance of people's agency in processes of change. This is not the intention of the editors. However, we do assert that a better understanding of agrarian structures, how they are shaped and themselves impact other processes of change including changes in agricultural output, is necessary for a useful and relevant policy analysis.

NOTES

1. The volume is a thematically connected collection of different pieces of research; it includes elements of comparison. However, it does not claim in itself to be a comparative study. This is a matter of regret for the editors. The different scholarly traditions in and historical trajectories of West Bengal and Bangladesh made the Workshop output less symmetrical than would perhaps be ideal. For example, in West Bengal there is a stronger tradition of examining the workings of the state on the ground, although in one of the papers included here (by Shapan Adnan) a similar type of detailed fieldwork is reported from Bangladesh.
2. Although, in analysing the causes and consequences of agricultural growth, it is necessary to disaggregate by crop for the purpose of illustrating the magnitude of growth/stagnation in West Bengal, not much is lost in switching between rice, cereals and foodgrains production. In 1994–95, for example, rice production was 93 per cent of all cereals, which in turn made up 99 per cent of all foodgrains.
3. Depending on which base year was used and whether the government of West Bengal's own figures were to be taken at face value. See Gazdar and Sengupta (this volume) and the queries raised by Boyce about government statistics summarised later in this introduction.
4. There was a significant switch to jute—a labour intensive cash crop; its demographic determinants were evidenced by, among other things, large scale migration to Assam.
5. Conversely in Bangladesh, it could be argued that in conditions of aid-dependency it works to the advantage of the regime to demonstrate low levels of growth in food production.
6. Sawant and Achuthan similarly report a *general* increase in foodgrains production in eastern India, with West Bengal in the lead (1995).
7. Mini-submersible tubewells are shallow tubewells with submersible pumpsets which are able to access water from more than 15 metres below ground compared to 8–10 metres for shallow tubewells (Webster, this volume) and have greater command areas (than shallow tubewells).
8. Chatterjee reports a 'considerable churning inside the (CPI[M])' following the coming to power of the Left Front in 1977, the subsequent quintupling of the membership and the new members' 'almost exclusive concern with the politics of winning elections' (1997: 154).
9. *Bhadralok* here refers to rural gentry living off rental income, while *chashis* are mainly own-account producers. These can also be used as cultural categories (see Ruud, this volume)

10. As indeed most of the contributions to this book and the collection as a whole may do for neglecting gender analysis. This has frustrated the editors who, as Workshop organisers together with Nripen Bandyopadhyay, invited a number of gender specialists with well-established work on gender and agrarian change in Bangladesh and West Bengal. None of them produced papers for publication.
11. However, Gardner suggests that, at least among Muslim households in rural Sylhet, dowry can be broken down into three types of goods: gold given directly to the bride by her parents; goods sent with the bride to her marital household; and goods given directly to the groom and his blood relatives. Gardner argues that increasing dowry has more to do with competition between households and marriage as a means of upward mobility than with the worsening subordination of women (1995: 178–9).
12. The Left Front government's self-presentation as the best available of all those with any chance of power in the state (see Lieten, 1996b) is akin to the line adopted by New Labour in the UK in relation to the more radical end of its constituency.
13. Palmer-Jones argues that the wage–rice price ratio is more significant as an incentive/disincentive to production than the fertiliser–rice price ratio because wages are the single largest cost item in paddy cultivation in Bangladesh. However, his contention that real wages have not declined would appear to contradict his view that relatively high growth rates are unlikely to continue unless there is a greater and sustained increase in rice prices. But the money wage–rice price ratio also has implications for the consumption of agricultural workers. Rising rice prices are not in themselves good news for poor rural Bangladeshis (see Rahman, 1994).
14. In Beck's interpretation, the small wage increases were much less important than the decline in self-respect and quality of life; Beck emphasised the importance of *relative* poverty in poor people's experience.
15. It has been argued that the middle peasants of West Bengal, who now dominate agriculture, are themselves 'immiser[ised]' (Basu, 1992: 11; see also Harriss-White, this volume). Nevertheless, the point made here remains that middle peasants are politically powerful and economically dominant at the village level.

REFERENCES

Agarwal, Bina. 1994. *A Field of One's Own: Gender and Land Rights in South Asia.* Cambridge: Cambridge University Press.

Asian Age. 1996. 6 December. New Delhi.

Bandyopadhyay, Nripen. 1995. 'Agrarian Reforms in West Bengal—An Enquiry into Their Impact and Some Problems.' Paper presented at the Workshop on Agricultural Growth and Agrarian Structure in Contemporary West Bengal and Bangladesh, Centre for Studies in Social Sciences, Calcutta, 9–12 January 1995.

Banerjee, Abhijit V., and Maitreesh Ghatak. 1995. 'Empowerment and Efficiency: The Economics of Tenancy Reform.' Mimeograph, Department of Economics, Harvard University.

Basu, Amrita. 1992. *Two Faces of Protest: Contrasting Modes of Women's Activism in India.* Berkeley: University of California Press.

Beck, Tony. 1995. 'The Green Revolution and Poverty in India – A Case Study of West Bengal', *Applied Geography*, 15(2): 161–81.

Bose, Sugata. 1986. *Agrarian Bengal: Economy, Social Structure and Politics, 1919–1947.* Cambridge: Cambridge University Press.

Bose, Sugata. 1993. *Peasant Labour and Colonial Capital: Rural Bengal Since 1770*. The New Cambridge History of Inida, Volume 3(2), Cambridge: Cambridge University Press.

Boyce, James K. 1987. *Agrarian Impasse in Bengal: Institutional Constraints to Technological Change*. Oxford: Oxford University Press.

Breman, Jan. 1996. *Footloose Labour: Working in India's Informal Economy*. Cambridge: Cambridge University Press.

Business India. 1994. Interview with Jyoti Basu. June–July 1994.

Census of India. 1991. New Delhi: Government of India.

Chatterjee, Partha. 1997. *The Present History of West Bengal: Essays in Political Criticism*. Delhi: Oxford University Press.

Chatterjee, S.N. 1993. 'Spatial Survey for Block Development (Bolpur-Sriniketan),' *Visva-Bharati Annals* (New Series V), Santiniketan: Visva-Bharati.

Dasgupta, Biplab. 1995. 'West Bengal's Agriculture since 1977.' Paper presented at the Workshop on Agricultural Growth and Agrarian Structure in Contemporary West Bengal and Bangladesh, Centre for Studies in Social Sciences, Calcutta, 9–12 January 1995.

Egger, Philippe. 1996. *Wage Workers in Agriculture: Conditions of Employment and Work*. Geneva: International Labour Office (ILO).

Gardner, Katy. 1995. *Global Migrants, Local Lives: Travel and Transformation in Rural Bangladesh*. Oxford: Clarendon Press.

Government of West Bengal (GoWB). 1994–95. *Economic Review 1994–95*.

———. 1995–96. *Economic Review 1995–96*.

Gupta, Jayoti. 1997. 'Voices Break the Silence,' Nitya Rao and Luise Rürup (eds) *A Just Right: Women's Ownership of Natural Resources and Livelihood Security*. New Delhi: Friedrich Ebert Stiftung.

Harriss, John. 1993. 'What is Happening in Rural West Bengal? Agrarian Reform, Growth and Distribution,' *Economic and Political Weekly*, 28(24): 1237–47.

Harriss-White, Barbara, and S. Janakarajan. 1997. 'From Green Revolution to Rural Industrial Revolution in South India,' *Economic and Political Weekly*, 32(25): 1469–77.

Hossain, Mahabub. 1995. 'Agricultural Policy Reforms in Bangladesh: An Overview.' Paper presented at the Workshop on Agricultural Growth and Agrarian Structure in Contemporary West Bengal and Bangladesh, Centre for Studies in Social Sciences, Calcutta, 9–12 January 1995.

Lieten, G.K. 1996a. 'Land Reforms at Centre Stage: The Evidence on West Bengal,' *Development and Change*, 27(1): 111–30.

———. 1996b. *Development, Devolution and Democracy; Village Discourse in West Bengal*. New Delhi: Sage.

Mallick, Ross. 1993. *Development Policy of a Communist Government: West Bengal since 1977*. Cambridge: Cambridge University Press.

Mandal, B., T. Chowdhury, G. Samanta, G. Basu, P. Chowdhury, C. Chanda, D. Lodh, N. Karan, R. Dhar, D. Tamili, D. Das, K. Saha and D. Chakraborti. 1996. 'Arsenic in Groundwater in Seven Districts of West Bengal, India—The Biggest Arsenic Calamity in the World,' *Current Science*, 70(11): 976–86.

Mukarji, Nirmal, and Debabrata Bandyopadhyay. 1993. *New Horizons for West Bengal's Panchayats*. Government of West Bengal.

Mukherji, Saugata, and Manoj Kumar Sanyal. 1995. 'Growth and Institutional Change in West Bengal Agriculture 1901–1988.' Paper presented at the Workshop on

Agricultural Growth and Agrarian Structure in Contemporary West Bengal and Bangladesh, Centre for Studies in Social Sciences, Calcutta, 9–12 January 1995.

Murshid, K.A.S. 1995. 'Market Institutions and Market Exchange: Insights from the Bangladesh Paddy Market.' Paper presented at the Workshop on Agricultural Growth and Agrarian Structure in Contemporary West Bengal and Bangladesh, Centre for Studies in Social Sciences, Calcutta, 9–12 January 1995.

Pai, Madhukar, Anand Zachariah, Winsley Rose, Samuel Satyajit, Santosh Verghese and **Abraham Joseph.** 1997. 'Malaria and Migrant Labourers: Socio-Epidemiological Inquiry,' *Economic and Political Weekly*, 32(16): 839–42.

Palmer-Jones, Richard. 1992. 'Sustaining Serendipity? Groundwater Irrigation, Growth of Agricultural Production, and Poverty in Bangladesh,' *Economic and Political Weekly*, 27(39): A–128 to A–140.

Rahman, Hossain Zillur. 1994. *Low Price of Rice: Who Loses, Who Gains? Findings from a Recent Survey of Rural Bangladesh*. Report for Analysis of Poverty Trends Project, Working Paper, New Series No. 5. Dhaka: Bangladesh Institute of Development Studies.

———. 1995. 'Rural Poverty in Bangladesh: Evidence and Explanations.' Paper presented at the Workshop on Agricultural Growth and Agrarian Structure in Contemporary West Bengal and Bangladesh, Centre for Studies in Social Sciences, Calcutta, 9–12 January 1995.

Rawal, Vikas, and **Madhura Swaminathan.** 1998. 'Changing Trajectories: Agricultural Growth in West Bengal, 1950–1996,' *Economic and Political Weekly*, 33(40): 2595–602.

Rogaly, Ben. 1997. 'Linking Home and Market: Towards a Gendered Analysis of Changing Labour Relations in Rural West Bengal,' *IDS Bulletin*, 28(3): 63–72.

———. 1998. 'Containing Conflict and Reaping Votes: Management of Rural Labour Relations in West Bengal,' *Economic and Political Weekly*, 33(42–43): 2729–39.

Ruud, Arild Engelsen. 1994. 'Land and Power: The Marxist Conquest of Rural Bengal,' *Modern Asian Studies* 28(3): 357–80.

Sarkar, Suman, and **Kali Shankar Chattopadhyay.** 1995. 'Agrarian Market Characteristics and Agricultural Development: Lessons from West Bengal's Case.' Paper presented at the Workshop on Agricultural Growth and Agrarian Structure in Contemporary West Bengal and Bangladesh, Centre for Studies in Social Sciences, Calcutta, 9–12 January 1995.

Sawant, S.D., and **C.V. Achuthan.** 1995. 'Agricultural Growth Across Crops and Regions: Emerging Trends and Patterns,' *Economic and Political Weekly*, 30(12): A–2 to A–13.

Sen, Abhijit, and **Ranja Sengupta.** 1995. 'The Recent Growth in Agricultural Output in Eastern India, with Special Reference to the Case of West Bengal.' Paper presented at the Workshop on Agricultural Growth and Agrarian Structure in Contemporary West Bengal and Bangladesh, Centre for Studies in Social Sciences, Calcutta, 9–12 January 1995.

Sengupta, Sunil, and **Haris Gazdar.** 1997. 'Agrarian Politics and Rural Development in West Bengal,' Jean Drèze and Amartya Sen (eds) *Indian Development: Selected Regional Perspectives*. Oxford: Oxford University Press.

Shahabuddin, Quazi. 1995. 'Agricultural Growth Performance in Bangladesh: Some Recent Evidence.' Paper presented at the Workshop on Agricultural Growth and Agrarian Structure in Contemporary West Bengal and Bangladesh, Centre for Studies in Social Sciences, Calcutta, 9–12 January 1995.

Singh, Manjit. 1995. *Uneven Development in Agriculture and Labour Migration (A Case of Bihar and Punjab)*. Shimla: Indian Institute of Advanced Study.

Srivastava, Ravi. 1997. 'Rural Labour in Uttar Pradesh: Emerging Features of Subsistence, Contradiction and Resistance.' Paper presented at a Workshop on Rural Labour Relations in India Today, London School of Economics, 19–20 June 1997.

The Statesman. 1995. Calcutta, 11 December, 1995.

United Nations Development Programme (UNDP). 1996. *Report on Human Development in Bangladesh.* Dhaka.

Webster, Neil. 1992. *Panchayati Raj and the Decentralisation of Development Planning in West Bengal.* Calcutta: K.P. Bagchi.

Wood, Geoffrey D. 1994. *Bangladesh: Whose Ideas, Whose Interests?* Dhaka: University Press.

PART I

Agricultural Growth, Poverty and Well-Being

2

Agricultural Growth and Agrarian Structure in Bengal: A Historical Overview

SUGATA BOSE

The aim of this paper is to provide a historical and comparative context for the exploration of the relationship between agricultural growth and agrarian structure in contemporary West Bengal and Bangladesh. First, I will delineate the broad temporal trends and regional variations in the history of agricultural growth in Bengal. Second, I will revisit and re-examine the more important arguments and counter-arguments concerning the relationship between agrarian structure and agricultural growth. Third, I will try to make a case against a recent tendency to concentrate narrowly on the question of the growth performance of Bengal agriculture, and for the need to refocus attention on the implications of the nature of this growth for agrarian relations, especially the entitlements of subordinate social groups.

THE HISTORY OF GROWTH, DECLINE AND REVIVAL OF BENGAL AGRICULTURE

The buoyancy of agricultural production since the early 1980s, especially

---

An earlier version of this paper was presented at the Workshop on Agricultural Growth and Agrarian Change in West Bengal and Bangladesh, Calcutta, 9–12 January 1995. I am grateful to the discussants Rajat Kanta Ray and Willem van Schendel as well as the other participants in the Workshop for their comments.

in West Bengal, has been understandably hailed as a remarkable departure from the trend of decline and stagnation that had characterised the late colonial and the early post-colonial eras. On a more extended historical view, however, the 100 or so years preceding 1980 would appear to be a rather brief and unhappy interlude in the success story of the growth performance of Bengal agriculture. A long agrarian cycle of growth in Mughal and Nawabi Bengal came to a catastrophic end with the great famine of 1770. Agricultural production recuperated rapidly, however, in subsequent decades. Whatever the other negatives associated with the Company raj (and there were many), the ninety years from 1770 to 1860 were a period of growth in almost all parts of rural Bengal.

A high concordance between population and production can be noticed in this phase of mostly, though not exclusively, extensive growth.[1] In the immediate aftermath of 1770, however, a number of constraints including the high revenue demand of the early colonial state as well as the malaise in the grain market had ensured that output increased at a slower rate than population. These negative impulses were countered by a rising demographic trend in a context of absolute deficit labour which enabled the negotiation of advantageous rental rates for reclamation of fallow land. In west Bengal districts, which had reached a more advanced stage of cultivation in 1770 than eastern ones, short distance migration by *paikasht raiyats* played an important part in the process of recovery until about 1820.

In the mature parts of west Bengal peasants were already nibbling away at the extensive margins by the 1820s. An estimate of the population of Burdwan in 1816 gave a count of more than one-and-a-half million people with a very high density of more than 600 persons per square mile.[2] By the mid-1830s it was being reported that west and central Bengal were 'far too populous to admit of tracts of land remaining uncultivated'.[3] The first census report of 1872 was quite categorical about the steady increase of the population in the western parts of Burdwan division during the course of the century. The levelling noticed in the eastern parts was probably of very recent origin. The tightening of the land–person ratio in much of west Bengal is quite evident from the 1828 report of collector Halhead.[4] So it would be incorrect to emphasise only the extensive nature of growth in west Bengal between 1830 and 1860. Districts such as Burdwan and Hooghly had reached densities of more than 700 per square mile by the mid-19th century—nearly twice that of urbanised, industrial Belgium, the world's most heavily populated country at the time (Klein, 1972: 156). These densities could only be sustained through some resort to intensive techniques including limited switching to cash-crops like sugarcane, cotton and mulberry. But it would be a mistake to assume that the move to cash-cropping in this period was mainly demographically driven. Indigo, the most important cash-crop of this period, was largely a forced cultivation even if the compulsion was facilitated by an emerging framework of demographic pressures.

In the more expansive delta of east Bengal, laterally proliferating small peasant cultivation provided the main impetus to increased agricultural production. Early evidence of the spread of cultivation came in the replies of judges and district collectors to the inquiry initiated by Wellesley in 1801. In the southern *parganas* of Dhaka, for instance, cultivation was said to have increased by some 12.5 per cent since the Permanent Settlement. The extension of the arable, even accounting for the district collector's obvious over-optimism, was striking in Chittagong.[5] In Tippera, where in the 1760s hills and jungles predominated over cultivated land, Buchanan-Hamilton saw in 1798 'one continuous field yielding the richest crops'; in 1801 seven-ninths of the district was reported to be under the plough. The proportion might well have been an overestimate but the district settlement officer in the 1910s, after weighing the available evidence, suggested almost a doubling of the cultivated area between 1793 and the revenue survey of 1860–64.[6] In Faridpur and Mymensingh the scale of the new cultivation was far greater than the pockets of decline. Rennell's survey of 1770–78 had revealed 1,128 square miles of unoccupied waste in Bakarganj; that was down to 925 square miles in 1793 and to 526 square miles by the time of the revenue survey of 1859–65.[7] The ecology of this littoral district cut across by numerous streams gave rise to an extraordinary pattern of subinfeudation under substantial farmers known as *haoladars* to facilitate the work of reclamation.

In the northern tracts of jungle in Rangpur and Dinajpur, population and production increased moderately in this period. It would appear that the cultivated area in Rangpur in the early 1870s was about 15 per cent greater than in 1809. Much of the reclamation organised and financed by large *jotedars* was recent.[8]

Overall, a high birth rate outdistanced a high death rate between 1770 and 1860. Demographic growth was a strong, but not the sole, factor in the spread of cultivation and rising gross output. The agricultural surplus was mainly appropriated through the mechanisms of revenue and rent, both depending for their enhancement primarily on the expansion of acreage. By the middle of the 19th century peasants in west Bengal had exhausted the extensive margins at the current low level of technology and had begun to exploit new intensive strategies. Open spaces were still being brought under smallholding peasant cultivation in much of east Bengal. The jungles of the frontier regions of north and south were being reclaimed under big farmer supervision by sharecroppers and undertenants.

Trends of population and production in west and east Bengal diverged sharply from the middle of the 19th century to about 1920. The decay of the rivers and a high mortality rate owing to malaria epidemics resulted in a demographic arrest and a reduction in output and in the area under cultivation in west Bengal. During the first half of this period, c. 1860–90, east Bengal witnessed a secular rise in population and rapid expansion of cultivation through the extensive proliferation of peasant smallholdings. Between 1890 and 1920, although the extensive margins were reached in many parts, population growth induced new intensive techniques, which more than offset the diminishing returns from new lands in terms of productivity. The overall picture, however, during these three decades shows stagnation in agricultural production (Blyn, 1966: 102–7).

The performance of agriculture was adversely affected in west Bengal both by the exhaustion of the land of the moribund delta and by the mortality as well as morbidity of labour as a consequence of malarial infection. In the Magura and Jhenida subdivisions of Jessore, for instance, a survey in 1876 had reported 75 per cent of the gross area to be under cultivation; by 1920 the proportion had shrunk to under 40 per cent. In addition to the 'dearth of labour', 'the agricultural population stricken by malaria [had] lost their physical vigour and energy' and become 'incapable of hard work in the field.'[9] With depopulation reducing acreage and debilitation affecting yields, it is likely that gross output in west Bengal declined between 1860 and 1920 in a context of stagnant population. This bleak picture was broken by small pockets of growth. One was provided by the slow but steady retreat of the Sunderbans in the 24-Parganas, Khulna and parts of Midnapur, effected by the money and supervision of enterprising farmers and the labour of small peasants. Another was opened up by the withdrawal of the government's salt monopoly in the 1860s in the Contai and Tamluk subdivisions of Midnapur which became a safe haven for those fleeing the scourge of malaria.

In east Bengal, by contrast, as Binay Chaudhuri (1969: 166) shows clearly, growth was the norm and decline the exception. The secular trend of expanding population and cultivation of the pre-1860 period was maintained in the sixty years that followed with only a slight slackening of pace towards the end of the period and brief interruptions caused by a major cyclone in 1876 and an earthquake in 1897. In addition to the atomistic extension of cultivated acreage by smallholding peasants, there was some limited resort to intensive techniques in the first two decades of the 20th century. In Bakarganj, where peasants had settled 'wholly without reference to any future village community',[10] the 'unoccupied waste' shrank from 526 square miles in the early 1860s to 184 square miles in 1905, despite the accretion of 180 square miles of new alluvial land. Cultivation in the 'occupied area' posted a 23 per cent increase.[11] In Mymensingh the cultivated area increased from 3,562 square miles to 4,292 square miles between 1872 and 1910, which included 470 square miles of stiff clay in the Madhupur jungle on the Dhaka border. Extensive cultivation had 'almost reached its full limits'. By the second decade of the 20th century nearly all land cultivable at the current level of technology was being cultivated in most east Bengal districts. In Noakhali 'every inch of land . . . fit for cultivation' was reportedly growing crops or fruit-bearing trees. The settlement officer of Dacca wondered 'to what extent the land [could] be induced to provide the rapidly increasing numbers with employment'.[12]

The crunch might have been felt sooner had it not been for two developments: first, the exploitation of the intensive margins partly through a switch to a high-value and labour-intensive cash-crop, and second, the utilisation of the escape hatch of migration particularly to the floodplains of the Brahmaputra in neighbouring Assam. Around 1920 double-cropping accounted for 34 per cent of the net cropped area in east Bengal; the corresponding figure for west Bengal was only 18 per cent. Moreover, the value of the second crop was generally much greater in the east. Jute had assumed some importance as a cash-crop from the early 1870s onwards, but experienced a major spurt in its acreage in 1906–07. But the migration of nearly a million peasants, often in search of jute lands, from overcrowded east Bengal districts to Assam during the first three decades of the 20th century was an indication that even the intensive margins were being exhausted. The northern frontier regions continued to witness immigration of tribal labour and some expansion of cultivation in this phase.

If the sixty-year period 1860–1920 saw a sharp divergence in population and production trends in west and east Bengal, the next sixty-year period from 1920 to 1980 was an era of rough convergence. From the 1920s, better control of disease brought about a sharp fall in the death rate in west Bengal, so that the population graph once again swung upwards but failed on this occasion to have any commensurate impact on production.

In east Bengal, too, demographic growth appeared to acquire an autonomous, self-sustaining character while output stagnated. Between 1920 and 1980 population rose almost inexorably, barring the setback during the great famine of 1943, spurred along by a high fertility regime.[13] On the production side, cultivated area displayed 'near-constancy' and yield per acre 'a near-zero trend' between 1920 and 1946 (lslam, 1979: 203). In the period 1949–80 the performance of the total output improved, but figures of per capita growth (see Table 2.1) indicated that production was trailing way behind population in both West and East Bengal (Boyce, 1987: Chapters 3, 4 and 5).

Table 2.1
Rate of Growth (in Percentages) of Per Capita Agricultural Output, 1949–80[1]

	1949–64	1965–80	1949–80
West Bengal	−1.42	0.27	−0.57
	(−1.49)	(0.10)	(−0.69)
Bangladesh	−0.59	0.38	−0.11
	(−0.76)	(−0.27)	(−0.51)

Source: Boyce, 1987: 141–2.
[1]The main figures denote rates of growth of output per capita of rural population; the rates of growth per capita of total population are given in parentheses.

Between 1920 and 1946, according to M.M. Islam's estimate, total agricultural output grew at a mere 0.3 per cent per annum and food crops output stagnated, while population increased at an annual rate of 0.8 per cent. Burdwan division, the heart of rural west Bengal, actually showed a negative output growth rate of −1.08 per cent. The relatively high growth rate in agricultural output of 1.1 per cent in the Presidency division can probably be explained by reclamation in the Sunderbans tracts in Khulna and the 24-Parganas. In east Bengal output rose at 0.4 per cent per annum in Dacca division and declined at the rate of −0.7 per cent in Chittagong division. Output in the Rajshahi division, which included the north Bengal districts, grew at an annual rate of 0.5 per cent (Islam, 1979: 50–6). Even in the areas of aggregate agricultural growth, per capita output declined.

In the post-independence period, the growth performance of agriculture improved but still lagged behind population. Between 1949 and 1980, James Boyce estimated that total agricultural output grew at the rate of 2.03 per cent in Bangladesh and 2.74 per cent in West Bengal. A positive break in the trend was noticeable in the mid-1960s. Output grew in Bangladesh at the rates of 1.57 per cent and 2.49 per cent during the sub-periods 1949–64 and 1965–80 respectively; a similar acceleration occurred in West Bengal from 1.2 per cent to 2.27 per cent. The northern districts of both Bangladesh and West Bengal put up a relatively strong showing based mainly on area growth in the first sub-period but fell back, particularly in West Bengal, in the second, performing relatively poorly on the productivity front. Five core districts of West Bengal and the south-eastern and north-western districts in Bangladesh displayed the fastest growth rates, based on yields and cropping pattern in the west and yields alone in the east. The painfully slow growth in the yield of *aman* rice, which accounted for about 50 per cent of gross acreage in both the east and the west, acted as a principal constraint on any better performance of output. It was the higher growth rates of secondary crops, namely *aus* and *boro* rice, wheat and potato, which enabled production to stay in the race with population at all. As for jute, since partition output had increased in the west but stagnated in its natural habitat in the east.

Boyce's careful sifting of the data relating to the 1949–80 period revealed that the process of induced innovation stemming from rural population growth was by no means absent even though it failed to make itself felt forcefully on agricultural performance. A static analysis of the linkage between population density and agricultural productivity showed positive results. A dynamic analysis also suggested that population growth had positive associations with subsequent performance of agricultural output. This could not have occurred simply as a function of labour intensity without innovative labour-utilising and land-saving techniques. A close examination of the population–production linkage in all its ramifications led Boyce to conclude that population growth in Bengal, contrary to the alarms raised by Malthusian doomsayers, was 'not an unmitigated evil' (1987: 159). The blockage to induced innovation in post-partition Bengal, especially in the field of water-control development which was the leading input, was found by Boyce to be in the character of social and political institutions (ibid.: Chapters 6 and 7).

As Gazdar and Sengupta report in their contribution to this volume, there was a significant acceleration of the agricultural growth rate in West Bengal during the decade 1981–82 to 1990–91. In Bangladesh over the same period any improvement in agricultural performance was quite modest by comparison (see Palmer-Jones' and Shahabuddin's contributions to this volume). Whatever the doubts about the quality of the data, some of the points that emerge from the analysis of Anamitra Saha and Madhura Swaminathan (1994a) are plausible and noteworthy.[14] First, there seems to have been a significant positive break in the trend of growth in 1983.

Second, although expansion of area under *boro* rice has made a major contribution to new growth, the historical sluggishness in *aman* yields also appears to have been overcome. Third, the fast rates of growth have been widespread with the two northern districts of Darjeeling and Jalpaiguri being the only exceptions. Overall, agricultural production in West Bengal seems to have made a momentous departure from its own past record of the preceding 120 years and also performed better than Bangladesh and other major Indian states during the 1980s. The change remains worthy of attention despite the slowdown across eastern India and Bangladesh in the early–mid-1990s (see Rogaly, Harriss-White and Bose, this volume).

AGRARIAN STRUCTURE AND AGRICULTURAL GROWTH

In an interview with a business magazine the Chief Minister of West Bengal naturally boasted about this impressive achievement and had no hesitation in identifying its cause. 'Our land reforms,' he asserted, 'are responsible for it' (interview with Jyoti Basu, *Business World*, July 1994).

Since the acceleration of growth had coincided with 'the implementation of a program of limited land reform and the establishment of new democratic institutions in the form of the three-tier panchayat system', Saha and Swaminathan (1994a), among others, have correctly pointed out the need for research on 'the socio-economic and political determinants' of the recent agricultural growth (see also Gazdar and Sengupta, this volume). There is, however, another argument put forward by John Harriss that the nature of the new growth 'casts doubt on Boyce's thesis'. Since the 'dramatic spurt in agricultural production' has been based on private sector development of tubewell irrigation and the establishment of a water market, earlier views about institutional constraints are deemed to be misconceived. Harriss asks why irrigation-powered growth has taken place 'in the absence of any [sic.] reform of the agrarian structure.' The question is based on a curious premise since all but the most cynical of the Left Front government's critics would accept that there has been some, limited, structural reform. Anyhow, the answer is that it is the result of 'the availability of suitable technology—appropriate MVs [modern varieties] for *boro* cultivation, and shallow tubewells' which in turn has come about, Harriss concedes, 'as a result of certain interventions by the state' in the form of government credit and incentive prices. State intervention would presumably not have been necessary in the absence of structural obstacles. So, having acknowledged the critical role of the state, it is rather odd to hold believers in the need for structural change rather than appropriators of the agrarian surplus responsible for having 'obstructed the progress of Bengal agriculture' (Harriss, 1992: 219–20). What Harriss advances as a dialectical view of change ends up being no more than a circular argument.

It has been pointed out by Haris Gazdar and Sunil Sengupta that '[e]xplanations for West Bengal's agricultural growth in the 1980s do not need to be viewed within "institutional versus technological", or "state action versus private incentives" dichotomies'. First, private investments and incentives have to be seen in the context of political empowerment and government intervention at the local level. Second, and more important, the 'institutional versus technological' debate is derivative of efficiency arguments in favour of agrarian reforms—arguments that have been 'largely incidental' to political mobilisation generally framed around redistributive agrarian issues in West Bengal (Sengupta and Gazdar, 1997; see also Gazdar and Sengupta, this volume).

The attempt to frame a debate on the relationship between agrarian structure and agricultural growth in the 1990s around Thorner's 1950s shorthand called a 'built-in depressor' largely ignores the historical research of recent decades on agrarian India in general and agrarian Bengal in particular. A number of insights drawn from interpretations of agrarian history might deepen our understanding of the contemporary dialectic between structure and growth.

First, the colonial and post-colonial states have generally occupied key locations within the complex of institutions that have been seen to have presented structural obstacles in the path of agricultural development, if not the narrower goal of growth. So it makes little sense to posit a state-centred argument against explanations emphasising institutional constraints. In the historical case of Bengal even a cursory glance at the history of irrigation reveals the glaring acts of omission in the field of public investment where decisions were dictated by the colonial state's order of strategic and economic priorities. More important, the acts of commission in the sphere of appropriation of the agrarian surplus took place squarely within the colonial matrix of power relations. In an important article published in 1973 Rajat Ray had underlined the colonial state's role in fashioning the dynamics of immobility that underlay the crisis of Bengal agriculture in the late 19th and early 20th centuries (Ray, 1973). At a more general level, I have tried to analyse the state–market nexus in the project of colonial capitalism. The story of agricultural commercialisation in the colonial era was not simply an economic one of expanding market forces which surely and steadily engulfed rural Bengal and redirected the thrust of its people's productive activities. It was equally a political story of state imperatives which, notwithstanding the rhetoric of free trade, imposed rules and restrictions in the marketplace (see Bose, 1993: Chapter 2).

In explaining the apparent mismatch between the rapid expansion of commodity production for capitalist markets on the one hand and low levels of productive investment in agriculture and the resilience of non-capitalist forms of production on the other, the one institution that the more perceptive agrarian historians have shifted the spotlight away from is the famous zamindari system. The dissonance between large-scale property rights in revenue collection and small-scale production and possessory rights based largely on peasant smallholding had something to do with the zamindars' incapacity to play the role of improving landlords in the early phase of the Permanent Settlement. However, the framework of the land-revenue structure was quite inadequate for fully addressing the question about the ways in which the relations of production might have fettered the forces of production. The theme of agrarian labour relations, having long suffered scholarly confinement within the shackles of the colonial land-revenue administration, has by now been placed within the broader and more relevant context of the colonial political economy and capitalist markets. In my own work, I sought to unravel the relationship between capitalist development and agrarian continuity or change at the point of primary production and appropriation by integrating the histories of land and capital. The counterpoint of land and capital, I felt, was crucial to the writing of labour history in the agrarian context. On the basis of an examination of the property–production dialectic I developed a typology of the main forms of material production and social reproduction of labour. The typology I proposed was structural rather than regional in character, even though I did demonstrate that each of the types predominated in certain regions. This typology, which included peasant smallholding, a demesne labour–peasant smallholding complex, a rich farmer–sharecropping system, the plantation, and a tribal, communitarian arrangement, does not require further elaboration here.[15]

The historical experience of colonial Bengal reveals the logical relationship between capitalist development and non-capitalist relations of production. The important strand of continuity in the social organisation of production rested on a labour process utilising the unpaid and underpaid work of family labour. Expanded commodity production for the capitalist world market was achieved efficiently and cheaply without resort to the formal commodification of labour. The refusal to be reduced to a commodity was itself a success of resistance by peasant labour determined to retain access to a combination of production-based and trade-based entitlement to consumption and subsistence. During the second century of colonial rule, the surplus produced by peasant labour was extracted largely through the economic circuits of debt at the end of the production cycle. But this had to be tempered by the sharing of the responsibility for assuring subsistence and minimal needs of peasant labour through adequate provision of credit by the owners of land and capital at the beginning of the production cycle. It was the lines of credit that tied together the domains of land, labour and capital, and geared agrarian society to undertake production that sustained colonial commerce. Over the long term, agrarian society 'peasantised' during the early 19th century may be seen to have fought a drawn-out rearguard action contesting and warding off the tentative forces of 'depeasantisation' since the late 19th century. The human and social costs of the contradictions between colonial and peasant labour in Bengal's agrarian context can be gauged by the direction of change in the different, otherwise resilient, types of social structure—the downward spiral of pauperisation, the slower but significant shift in favour of demesne lords and a richer peasantry, and the subtle change in gender and generational roles to the disadvantage of women and children. All of these altered entitlement relations, the consequences of which would be dramatically and cruelly evident in the differential impact of famines, especially the catastrophe of 1943, along lines of region, class, gender and generation (see Sen, 1981: Chapter 6; Greenough, 1982: Chapters 5 and 6; Bose, 1986: 87–97). Colonial capitalist development was successful in the sense that it attached the labour capacity of Bengal's working peasantry, robbed it of its creative power and reduced its share in the total social product.

The labour process in agrarian production marked by its predominantly familial character was encumbered by various forms of appropriation imposed upon it. The colonial state, local landlords, metropolitan capitalists, indigenous merchants and moneylenders, and richer peasants were among the many claimants of the surplus produced by the working peasantry. The aim of those who lorded over landed rights and controlled the circuits of credit was to hold down, with the assistance of those who wielded state power, the share of labour in the total social product. Several mechanisms of extracting the surplus were available and generally deployed simultaneously. Yet, while the social organisation of production displayed a strong strand of continuity despite important elements of qualitative change, the principal modes of exploitation and relations of appropriation underwent more decisive transitions over the two centuries following the imposition of colonial rule.

The simultaneous resort to multiple instruments of appropriation cannot obscure the role of rent, located within the context of the colonial revenue system, as the principal mode of exploitation in Bengal from the late 18th to the late 19th century. Within a framework of externally financed, expanded commodity production for the world market, the credit mechanism assumed pre-eminence in the extracting of surplus between the late 19th century and the onset of the great depression in 1930. During the tumultuous decades of the depression and the Second World War, the lease and land markets became the chief channels of expropriation.

The aim of surplus appropriators, it must be emphasised, was exactly that—to appropriate the surplus, not necessarily to hold down growth. Any impact of the structures of production and appropriation on the growth performance of agriculture was therefore indirect, mediated by other factors and dependent on particular historical contexts. For example, the revenue and rent offensive of the early 19th century in a framework of what Shapan Adnan (1983) called 'absolute deficit labour' accorded rather well with the trajectory of growth. By contrast, the principal mode of appropriation through debt from about 1880 to 1930 probably cancelled out any incentives provided by tenancy legislation to enhance productivity on the part of smallholding peasantry. The growth picture continued to be dismal in the 1930s and 1940s, even though it must be stressed that poverty and famines in the late colonial era were much more matters of relative deprivation than absolute dearth, and relational rather than parametric features had a more profound influence on the region's agrarian developments.

What, then, were the structural legacies in terms of production and appropriation for post-colonial East and West Bengal? And is it possible to ferret out the critical, new departures? Where peasant smallholding had prevailed, aggregate downward mobility within the peasantry and the proliferation of a class of landless labouers did not subvert the conditions of reproduction of the smallholding system as such. Willem van Schendel's (1982: 287–8) study of peasant mobility in Bangladesh reported that 'an increasingly crushed peasantry' had been 'producing a growing proportion of landless labourers', but 'well-to-do peasants' were not rural capitalists. Peasant smallholding society undoubtedly presented formidable structural obstacles, as Abu Abdullah and others showed (Abdullah, 1976: 217), as well as dogged political resistance to a capitalist transformation that might threaten the petty property base of small-scale agriculture. But equally, capital, conscious of its own imperatives and the minimum demands of peasant labour, persevered with a strategy of capitalist development resting squarely on peasant family labour which had paid such rich dividends under colonialism. Zillur Rahman noted in Noakhali in the 1970s, for instance, that richer peasants leased out their lands in small parcels to several tenants to 'ensure a greater labour input' as each tenant would 'have to devote his surplus family labour to the petty holding he gets'.[16] Even in a Comilla locality, which was the showpiece of state-aided capitalist agricultural development both before and after 1971, it was found that 'rich peasants... were interested in production increases only, not in replacing the existing relations of production with capitalist ones'. Their surpluses were reinvested not in agriculture but in usury, urban employment and trade, and education, as they aspired to join, according to van Schendel, 'the ranks of a shiftless bourgeoisie largely subsisting on foreign aid and concentrated in administrative, poorly industrialized urban centres.' The prognosis offered in the early 1980s was gloomy: there seemed 'very little room for breaking the involutionary downward spiral in which agriculture, and with it the entire society, [was] caught up' (van Schendel, 1982: 290–1).

In an erstwhile peasant smallholding–demesne labour complex, which had predominated in West Bengal, a richer peasantry initially had a more substantial presence in the 1950s and 1960s than in East Bengal. The ranks of this class were swelled by zamindars and *patnidars* of old, who lost their rentier rights in 1953 but held on to possessory rights to their *khas khamar* (personal demesne). The proliferating class of rural proletarians in West Bengal was not simply derived from a process of pauperisation of peasant smallholders but had a structurally distinct prior existence. At the same time, smallholding cultivation has been bolstered through powerful oppositional political campaigns since the mid-1960s and state initiatives since 1977.

A West Bengal district, of course, provided the empirical reference for a lively debate on the transition to capitalism in the late 1960s and 1970s. Although Amit Bhaduri's (1973: 120–37; see also Adnan, 1984) formulation on semi-feudalism was open to question from a variety of theoretical and empirical standpoints, there was plenty of evidence and argument to support his contention about the resilience of social relations, at the point of production, and even of primary appropriation to attempts at capital-intensive technological change. Since the 1970s a variety of state initiatives have contributed to a certain stabilisation of peasant smallholding. Smallholdings continued, however, to be at a disadvantage in relation to the richer peasants who were in command of a wider portfolio of investments in the agrarian economy and had a clear edge in manipulating developments in the product market. Their surplus lands were cultivated by sharecroppers and wage-labourers. Just prior to the launching of 'Operation Barga', Ashok Rudra's field research had revealed an array of cost-sharing practices for seeds, fertilisers and manure but, significantly, 'human labour cost' was 'always the responsibility of the tenant' and with a few exceptions so were the costs of bullocks and ploughs (Rudra, 1975: A–60). Far more important by now in terms of production relations was a class of agricultural labourers, accounting for over a third of the rural population (see Rogaly, this volume). In another significant development the proportion of *mahindars* or farm servants had dwindled all over West Bengal while that of *munishes* or day labourers had soared. According to an estimate by Pranab Bardhan and Ashok Rudra (1980: 1483) day labourers represented 83.6 per cent of the total number of agricultural laboures by 1979; only 10.9 per cent were attached and 5.5 per cent were semi-attached. The labour and credit markets, however, were closely interlinked and a great majority of labourers were dependent on subsistence loans.

While landlessness explained to a large extent the predicament of the exploited, landedness alone did not explain the dominance of the exploiters. In order to translate relatively small advantages in landholding into effective social dominance, involvement in new forms of capital and commodity markets had become indispensable. Although Bengal had not experienced a 'green revolution' in the later 1960s and 1970s on the scale of north-western India and parts of Pakistan, certain pockets had witnessed technological innovation in the form of new fertilisers and higher-yielding seeds. Richer peasants generally consolidated their edge in landholding by their ability to have better access to this capital-intensive agricultural technology. The same groups were better able to siphon off flows of aid in rural Bangladesh. Already by the late 1970s or early 1980s, control of new capital-intensive inputs, institutional credit and foreign aid provided important leverage in the relations of appropriation (see Adnan, this volume). These developments need to be borne in mind while investigating the interaction between structure and growth in the 1980s and early 1990s, a period which has witnessed considerable private investment in tubewell irrigation and the rapid development of a water market.

GROWTH, ENTITLEMENT AND DEPRIVATION

I had concluded my discussion of appropriation and exploitation in postcolonial Bengal in my book *Peasant Labour and Colonial Capital* with the following statement: 'Land and capital have always been inextricably intertwined in the agrarian history of Bengal. Yet in the context of narrowing differences in patterns of landholding, manipulation of a range of new sorts of capital investment in the agrarian economy has been critical in ensuring that the odds of life in rural Bengal remain titled against labour' (1993: 138).

I had written these lines before the data on rapid agricultural growth rates over the 1983–92 decade in West Bengal became available. The achievement in the field of agricultural production deserves analysis and even a round of applause, but it is more important to concentrate research efforts towards examining whether and to what extent the mass of the rural poor have benefited from this growth (see Gazdar and Sengupta, this volume). The chief failing of Indian development thinking and practice, as Amartya Sen (1989: 371) pointed out a few years ago, is an obsession with issues of means enhancement rather than means use. Higher agricultural output represents a welcome improvement in an instrumental variable which may or may not have a positive impact on what is intrinsically valuable, say, decline in rates of chronic malnutrition, depending on a range of intervening factors. The historical record makes clear that this is not just a philosophical statement, but a matter of practical concern. The period from the late 1940s to the mid-1960s, for instance, saw faster-than-average agricultural growth in the northern districts of Bengal. It was also a period of widening class disparities and deepening impoverishment of the class of sharecroppers and agricultural labourers.

Unlike that earlier period, post-1977 West Bengal has witnessed moderately progressive land reforms and the formalisation of democratically elected panchayats (see Williams, this volume). The general tenor of the Left Front government's policies including the early drive to register *bargadars* was the defence and extension of small-scale property-like rights in land. It can be plausibly hypothesised that such measures provided a set of incentives that had a positive impact on small-scale production. The structures of articulation of agrarian production with the output markets and with the market in the key input—water—will likely have an influence not only on the sustainability of growth, but on the patterns of distribution of the production gains (see Webster, this volume).

The question 'whether and to what extent the poor have been able to participate in the growth that has occurred' has already been addressed in insightful ways by Gazdar and Sengupta (this volume). Elsewhere (Sengupta and Gazdar, 1997) the same authors reach the conclusion, which needs to be underscored here, that 'the observed decline in poverty measures predates the recent rapid growth in agricultural productivity and output, and owes much to changes in the distribution of income'. They have also noted the 'general indifference' to health and education stemming from 'the narrow concern with establishing and redistributing rights to private property over and above establishing individual rights to a minimum level of education and health care.'

If we are at all concerned about the possibility of subordinate social groups being denied the benefits from new growth, three areas of future research seem to be especially relevant. The first, intra-family relations within Bengali agrarian society, is not tackled centrally by any of the contributions to this volume despite the best efforts of the editors (though see Gazdar and Sengupta's chapter). The greatest success of agrarian resistance over the past two centuries—the defence of the peasant small-holding—has until the closing decade of the 20th century involved a measure of implicit complicity in domination along lines of gender and generation. Being ardent believers in the labour-intensive economic efficiency and political resource base represented by the smallholding, middling peasantry, the CPI(M)-led Left Front in its policies has implicitly if not explicitly reinforced the patriarchal values of the landed strata and done little to address gender-based inequalities.[17] Whatever the negative effects on distribution at the level of the family and agrarian society, intensive and largely unremunerated family-based cultivation has historically had clear production advantages from the capitalist angle of vision. Capital and capitalists could concentrate their control over the product and credit markets and establish strategic footholds on the processing stage of production without incurring costs in the labour market. The inverse farm size–productivity ratio that came to light and was seized upon an as engine of growth in the 1950s had much to do with the kind of agrarian development based on labour intensity within family units that was fostered by colonial capitalism, even though it was not immediately recognised as such. As far as rural West Bengal is concerned, the often stark inequities along lines of gender and generation have been only recently brought into the spotlight of analysis (see Agarwal, 1994; Gupta, 1997; Rogaly, 1997).

Second, the relationship between agrarian production and the product market. While recent policies have focused on redistributing property rights in land, my impressionistic view is that gross inequities in the product market have been left untouched and may in fact have worsened. I had undertaken an analysis of how primary producers stood in relation to the marketing pattern and hierarchy in rice and jute during the period from 1919 to 1947 (Bose, 1986: Chapter 3). Exercises of this sort in unravelling the structure and trends in the markets for rice, potato and oilseeds over the past two decades may be extremely worthwhile (see Harriss-White, this volume).

Third, the absolutely crucial subject of water rights in agrarian Bengal. While collective action in water-control management of the kind advocated by Boyce may not have proved to be a necessary condition for the achievement of significant production gains during the 1983–92 decade, there may have been weighty equity arguments in favour of attempting such a course despite formidable political obstacles (see chapters by Webster and Wood, this volume). Given the limits of the project of land redistribution, some form of collective water rights in minor irrigation projects might have accorded landless and land-poor labour a critical measure of bargaining power vis-à-vis the landed. That such a path was not taken was ultimately a matter of political choice in the context of the class structures in rural Bengal. The development of a water market has been hailed in some circles on account of its apparent consequential achievements in powering rapid agricultural growth. But entitlement to water has now emerged as such a key distributional issue both at the macro- and micro-levels in Bangladesh and West Bengal that it calls for research on water-controlling structures on the lines of work on land-holding structures and their relationship to other markets conducted by agrarian historians. Class contentions over water rights have probably come to occupy the central place in the dialectic of domination and resistance in this deltaic land of great rivers.

NOTES

1. For a more detailed and critical analysis of the role of ecology and demography in influencing agrarian change from 1770 to 1990 see Bose (1993: Chapter 1).
2. W.B. Bayley, 1816, 'Statistical View of the Population of Burdwan,' *Asiatick Researches*, 12: 551.
3. N. Alexander, 1836, 'On the Cultivation of Indigo,' *Transactions of the Agricultural and Horticultural Society of India*, 2: 35.
4. Report No. 46 by collector Halhead, March 1828, Bengal Revenue Proceedings, Range 50, Vol. 54 (*India Office Records [IOR]*).
5. Bengal Judicial Civil Proceedings, 8 July 1802, and Parliamentary Papers (1812–13), 9, cited in Chaudhuri (1976: 305).

6. Notes by G. Ironside, Orme Manuscripts, 'India,' 17: 4950 (*IOR*); F. Buchanan-Hamilton, 'Account of a Journey through the Provinces of Chittagong and Tipperah,' Add. Ms. 19286, folio 3 (*IOR*); Parliamentary Papers (1812–13), 9: 416; W.H. Thompson, 1919, Report on the Survey and Settlement of Tippera District, 1915–19, Calcutta.
7. J.C. Jack, 1915, Report on the Survey and Settlement of Bakarganj District, 1900–08, Calcutta.
8. E.G. Glazier, 1873, A Report on the District of Rangpur, Calcutta; A.C. Hartley, 1941, Report on the Survey and Settlement of Rangpur District, 1931–38, Calcutta.
9. M.A. Momen, 1925, Jessore Settlement Report, 1920–24, p. 20, Calcutta.
10. J.E. Gastrell, 1868, Geographical and Statistical Report on the Districts of Jessore, Fureedpore and Backergunje, Calcutta, cited in Nicholas (1962: 67).
11. J.C. Jack, 1915, Bakarganj Settlement Report, 1900–08, pp. 10–11, Calcutta; on expansion of cultivation in Faridpur, see J.C. Jack, 1916, Faridpur Settlement Report, 1904–14, p. 5, Calcutta.
12. Mymensingh District Gazetteer (1919: 48); F.A. Sachse, 1919, Mymensingh Settlement Report, 1908–19, p. 29, Calcutta; W.H. Thompson, 1919, Noakhali Settlement Report, 1915–19, cited in Ganguli (1971: 50).
13. I have suggested explanations for the continuation of a high fertility regime through various agrarian cycles in Bose (1993: Chapter 1).
14. On problems with the methods of data collection, see S. Datta Ray (1994), and Saha and Swaminathan's reply (1994b) as well as our introduction to this volume.
15. See Bose (1993; Chapter 3); also Bose (1986: Chapters 1, 4 and 5). Crow, in this volume, examines the impact of regional differences in agrarian structure on the pace of agricultural growth in contemporary Bangladesh.
16. H. Zillur Rahman, Report from Raipur Thana (Copenhagen, 1982), cited in Adnan (1984).
17. See Gazdar and Sengupta (this volume). See also Amrita Basu's (1993) argument about democratic centralism in the home and the world.

REFERENCES

Abdullah, Abu Ahmed. 1976. 'Agrarian Structure and the IRDP: Preliminary Considerations,' *Bangladesh Development Studies*, 4(2).
Adnan, Shapan. 1980. 'Conceptualising Fertility Trends in Peripheral Formations,' C. Hohn and R. Mackensen (eds) *Determinants of Fertility Trends: Theories Re-examined*. Liege: IUSSP.
———. 1984. 'Peasant Production and Capitalist Development.' Ph.D. Dissertation, University of Cambridge.
Agarwal, Bina. 1994. *A Field of One's Own: Gender and Land Rights in South Asia*. Cambridge: Cambridge University Press.
Bardhan, Pranab, and **Ashok Rudra.** 1980. 'Types of Labour Attachment in Agriculture: Results of a Survey in West Bengal, 1979,' *Economic and Political Weekly*, 15(35): 1477–84.
Basu, Amrita. 1992. *Two Faces of Protest: Contrasting Modes of Women's Activism in India*. Berkeley: University of California Press.
Bhaduri, Amit. 1973. 'A Study in Agricultural Backwardness under Semi-Feudalism,' *Economic Journal*, 83: 120–37, March 1973.
Blyn, George. 1966. *Agricultural Trends in India, 1891–1947: Output, Availability and Productivity*. Philadelphia: Pennsylvania University Press.

Bose, Sugata. 1986. *Agrarian Bengal: Economy, Social Structure and Politics, 1919–1947.* Cambridge: Cambridge University Press.

———. 1993. *Peasant Labour and Colonial Capital: Rural Bengal Since 1770.* The New Cambridge History of India, Volume 3(2), Cambridge: Cambridge University Press.

Boyce, James K. 1987. *Agrarian Impasse in Bengal: Institutional Constraints to Technological Change.* Oxford: Oxford University Press.

Business World. 1994. Interview with Jyoti Basu. July 1994.

Chaudhuri, Binay Bhushan. 1969. 'Agricultural Production in Bengal: Co-existence of Decline and Growth,' *Bengal Past and Present,* 88(2): 152–206.

———. 1976. 'Agricultural Growth in Bengal and Bihar: Growth of Cultivation since the Famine of 1770,' *Bengal Past and Present,* 95: 305.

Ganguli, Birendra Nath. 1971. *Trends of Population and Agriculture in the Ganges Delta.* London. *1910–17.* Calcutta.

Greenough, Paul. 1982. *Prosperity and Misery in Modern Bengal.* New York: Oxford University Press.

Gupta, Jayoti. 1997. 'Voices Break the Silence,' Nitya Rao and Louise Rürup (eds) *A Just Right: Women's Ownership of Natural Resources and Livelihood Security.* Delhi: Friedrich Ebert Stiftung.

Harriss, John. 1992. 'Does the "Depressor" Still Work? Agrarian Structure and Development in India: A Review of Evidence and Argument,' *Journal of Peasant Studies,* 19(2): 189–227.

Islam, M.M. 1979. *Bengal Agriculture 1920–1946: A Quantitative Study.* Cambridge: Cambridge University Press.

Klein, Ira. 1972. 'Malaria and Mortality in Bengal, 1840–1921,' *Indian Economic and Social History Review,* 9(2): 132–60.

Nicholas, Ralph. 1962. 'Villages of the Bengal Delta.' D.Phil. dissertation, University of Chicago.

Ray, Rajat. 1973. 'The Crisis of Bengal Agriculture, 1820–1927—Dynamics of Immobility,' *Indian Economic and Social History Review,* 10(3): 244–79.

Ray, S. Datta. 1994. 'Agricultural Growth in West Bengal,' *Economic and Political Weekly,* 29(29): 1883–4.

Rogaly, Ben. 1997. 'Linking Home and Market: Towards a Gendered Analysis of Changing Labour Relations in Rural West Bengal,' *IDS Bulletin,* 28(3): 63–72.

Rudra, Ashok. 1975. 'Share-cropping Arrangements in West Bengal,' *Economic and Political Weekly,* 10(39): A–58 to A–64.

Saha, Anamitra, and Madhura Swaminathan. 1994a. 'Agricultural Growth in West Bengal in the 1980s: A Disaggregation by Districts and Crops,' *Economic and Political Weekly,* 29(13): A–2 to A–11.

———. 1994b. 'Agricultural Growth in West Bengal,' *Economic and Political Weekly,* 29(31): 2039–40.

Sen, Amartya. 1981. *Poverty and Famines: An Essay on Entitlements and Deprivation.* Oxford: Clarendon Press.

———. 1989. 'Indian Development: Lessons and Non-Lessons,' *Daedalus,* 118(4): 369–89.

Sengupta, Sunil, and Haris Gazdar. 1997. 'Agrarian Politics and Rural Development in West Bengal,' Jean Drèze and Amartya Sen (eds) *Indian Development: Selected Regional Perspectives.* Oxford and Delhi: Oxford University Press.

van Schendel, Willem. 1982. *Peasant Mobility: The Odds of Life in Rural Bangladesh.* Assen: van Gorcum.

3

Agricultural Growth and Recent Trends in Well-Being in Rural West Bengal

HARIS GAZDAR AND SUNIL SENGUPTA

INTRODUCTION

The recent rise in agricultural productivity in West Bengal following decades of stagnation and decline has sparked off a lively debate about the impediments to growth in the earlier period, and the causes of its acceleration since around the mid-1980s. The fact that West Bengal has experienced one of the most radical agrarian reform programmes in India in this period has attracted a great deal of attention, quite justifiably, to the possible connection between agrarian structure and productivity. Much of the debate has revolved around the various effects of institutional reform, market responses and technological change as factors explaining the agricultural take-off.

While this debate has been extremely useful in clarifying a number of important issues in the political economy of agricultural development, there is a danger that, taken out of context, it might obscure other equally

---

We are grateful to the Leverhulme Trust for supporting Haris Gazdar's contribution to this study.

important lessons that ought to be learnt from West Bengal's recent experience. We believe, for instance, that it is important to discuss not only the possibility of promoting growth through institutional reform, but also the impact of growth and institutional reform on the well-being of the population. Improvement of well-being is, after all, the main reason to be concerned about growth or institutional reform in the first place. We hope, therefore, to nudge the important and informative debate about West Bengal's agricultural performance—and the respective roles of markets, institutional reforms, public policy, and technological changes—towards a wider perspective.

The objectives of this paper are three-fold. First, we review the recent debate about the causes of agricultural growth in West Bengal and the possible relationship between agrarian reforms and agricultural productivity. We do this not as a prelude to entering this debate ourselves, but in order to summarise the key findings and to suggest some further areas of research on the subject. Second, an attempt is made here to place the recent experience of agricultural growth and institutional reforms within the perspective of changes in rural poverty and well-being. This we do largely with reference to state-level secondary data on wages, consumption, poverty indices, and indicators of education and health in rural West Bengal. Third, the impact of economic growth, public policy and social inequalities on various aspects of poverty is explored further with reference to micro-level village surveys. These data are from studies conducted in six West Bengal villages under a project of the World Institute for Development Economics Research (WIDER) at Visva-Bharati in Santiniketan.[1]

GROWTH IN AGRICULTURAL OUTPUT

From Stagnation to Rapid Growth

Following decades of stagnation and relative decline, agricultural output began to grow at a rapid rate in the 1980s (see Figure 3.1).[2] West Bengal achieved the fastest rates of growth in the output of foodgrains among all Indian states in this period. According to one estimate, while the output of foodgrains remained stagnant between 1971–72 and 1981–82, the annual growth rate was 6.9 per cent in the period 1981–82 to 1991–92.[3] This turnaround was noticed by Harriss (1993) primarily on the basis of micro-level observations, and by Saha and Swaminathan (1994) using state- and district-level secondary data.[4]

Ironically, the turnaround in agricultural output coincided with renewed interest in academic circles in the causes of agricultural stagnation in the region prior to the 1980s. In particular, the influential study by Boyce (1987) of agrarian stagnation in West Bengal and Bangladesh deserves mention. Boyce found that the agronomic potential for growth was extremely favourable in West Bengal and Bangladesh. According to him, the binding technical condition or, borrowing a term from Ishikawa (1974), the 'leading input', was the availability of irrigation. The main constraints in the development of the existing groundwater endowment in these regions were not technological, but distributional and institutional.

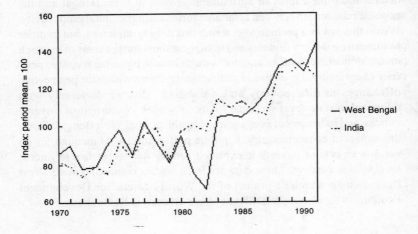

Figure 3.1: Trends in the output of foodgrains in West Bengal and India, 1970–71 to 1991–92 (three-year moving averages)
Source: CMIE, Performance of Agriculture in Major States, July 1993.

The high degree of fragmentation of ownership holdings combined with the small average farm area implied that individual holdings were too small to generate profitable returns to private investment in shallow tubewells. Furthermore, an unequal agrarian structure made it unlikely, according to Boyce, that scale economies could be achieved through cooperative collective action among farmers. Boyce was optimistic about the region's physical or technical potential for growth, but pessimistic, given his understanding of the institutional constraints, about the ability of the agrarian economy to realise such growth.

The upswing in West Bengal in the 1980s largely confirmed Boyce's optimism about growth potential in the region, and also the primacy he ascribed to irrigation in achieving this growth. A decomposition of West Bengal's growth in rice production between 1981 and 1993 shows that 40 per cent of the growth could be attributed to the expansion of the area under the *boro* crop, and another 35 per cent to higher yields in the main *aman* season (Banerjee and Ghatak, 1995). The expansion of the *boro* crop was directly related to the expansion of groundwater irrigation. The rise in the yield of *aman* was largely due to the introduction of newer higher yielding varieties (HYVs) which were particularly suited to the agronomic conditions of West Bengal. The adoption of these new HYVs was, in turn, facilitated by the improved availability of complementary inputs, particularly irrigation.

Although events have proven that Boyce's optimism in the technical potential for growth was justified, there is no longer much support for his pessimism about this potential being realised. The question that Boyce raised, however, does remain valid, albeit in a modified form: given the potential for growth, what had been the constraints up to the early 1980s, and how were these constraints finally overcome?

Regional Perspective

Since the neighbouring states of Bihar and Orissa share some of the agronomic conditions that exist in West Bengal, it is worthwhile to examine the latter's recent experience in a regional perspective. Bihar and Orissa displayed a similar pattern of stagnation (or, in the case of Orissa, slow growth) in the 1970s followed by higher growth rates in the 1980s. Output trends for the three eastern Indian states can be seen in Figure 3.2.[5] According to Sen and Sengupta's (1995) estimates, foodgrains output grew at the rate of 3.3 per cent annually in Bihar and 4.7 per cent per year in Orissa between 1981–82 and 1991–92, compared with 6.9 per cent in West Bengal.[6] In both these states, as in West Bengal, there was a statistically significant break in the trend growth rates between the two decades.

These comparisons reveal that although there were significant differences between these states in their respective rates of growth, there was a 'take-off' in agricultural output in all three. Given that there are some similarities in agronomic conditions in the region, some of the causes of stagnation as well as of subsequent growth are likely to be common to the three states. Having said that, however, there are also likely to be factors specific to each state that have influenced the extent and pattern of growth.

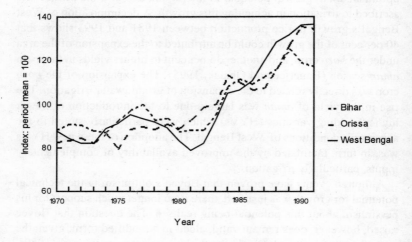

Figure 3.2: Trends in foodgrain output in West Bengal, Bihar and Orissa, 1970–71 to 1991–92 (three-year moving averages)
Source: CMIE, Performance of Agriculture in Major States, July 1993.

Both West Bengal and Orissa experienced growth in output as a result of the expansion of gross cropped area, as well as due to rising yields. In Bihar the cropped area remained stagnant. Between West Bengal and Orissa there are also notable differences (besides the obvious difference in their respective growth rates). First, much of the difference in the growth of output in the two states is due to West Bengal's higher growth in yields. Second, although the growth in gross cropped area between the two states was not very different—0.86 per cent per year in Orissa, compared with 1.08 per cent in West Bengal—the underlying processes were quite different. In particular, in West Bengal the growth in gross cropped area was largely the result of the rise in *boro* rice acreage which more than doubled in the 1980s. In fact, higher *average* yields were partly due to the greater contribution of *boro* rice to the total annual harvest.

In sum, while some of the factors that led to the agricultural take-off in West Bengal were probably common to its neighbouring states, others are likely to have been state-specific. It is appropriate to ask not one, but two sets of questions. First, what led to the take-off in agricultural output in the eastern region of India in the early 1980s? Second, what accounts for West Bengal's particularly good performance and the specific features of its agricultural growth in this period?

Causes of Growth

Much of the debate on the causes of growth in West Bengal agriculture has been dominated by arguments about the effects of political change on agricultural output. Since 1977 the state has been ruled by the Left Front coalition, led by the Communist Party of India (Marxist). This period has seen wide-ranging agrarian and political reforms in the countryside.

The main components of the agrarian reforms programme have been the redistribution of land to the poor and legislation for higher crop shares and security of tenure of sharecropper tenants. For the effective implementation of tenancy legislation, a campaign (Operation Barga) was launched for the registration of existing tenant leases. Political reforms have been based upon the establishment of democratic local bodies in the form of regularly elected Panchayati Raj Institutions (PRIs). Other institutional changes included the emergence of collective wage bargaining by agricultural labourers with intermediation by local panchayats and CPI(M) and other leftist party activists.[7] These reforms came at the end of a long period of political turmoil in West Bengal, and are undeniably associated with significant changes in the state's rural politics.[8]

The fact that the possible relationship between political change and agricultural output looms large in the discussion about West Bengal's recent agricultural growth is understandable. There is a long-standing view in agricultural economics that under a variety of conditions redistributive agrarian reforms are good for growth.[9] It is thought that land redistribution could lead to growth because small family-run farms are able to achieve higher land productivity than larger labour-hiring ones. The efficiency-and-growth case for tenancy reform (raising crop share and conferring secure occupancy rights) often rests upon the idea that a higher share would induce greater effort, and that security of tenure would improve incentives for investment. Furthermore, in the specific case of West Bengal, Boyce's thesis that there have been institutional (and distributional) constraints to the development of groundwater irrigation provides another possible link between agrarian reform and agricultural growth.

However, the idea of improved growth through smaller farm size is not, on its own, a strong candidate for explaining West Bengal's particularly good growth performance. Although West Bengal's land redistribution was very impressive in terms of the numbers of landless families who received land, the total area redistributed amounted to less than 6.5 per cent of cultivated land area in the state.[10] Harriss (1993: 1244) who had labelled the farm size–productivity argument as the 'classical view of the way agrarian reform may unleash productive forces' found little support for this view in his village studies in Birbhum and Bardhaman districts. The significant changes that he observed in his study villages were rather due to the development of groundwater irrigation. His observations corroborate the picture emerging from secondary data that there has been significant expansion in *boro* cultivation as well as rising yields all round.

The finding that irrigation has been a key engine of growth also confirms, as we have noted, the first part of Boyce's thesis that water is the 'leading input'. However, Harriss found that the factors leading up to increases in groundwater irrigation were quite different from those envisaged by Boyce. Rather than public intervention in the development of irrigation, or cooperative collective action by farmers, the investment appears to have been made by entrepreneurial individuals. Most of the institutional innovations that facilitated these investments were market-driven rather than cooperative. The enterprising farmers in Harriss' study were able to overcome the scale problems arising from small and fragmented holdings by selling water to neighbouring farmers, and by leasing in land seasonally from their neighbours.

The reasons for the take-off in the 1980s according to Harriss (1993) were most likely to have been the introduction of a new and more robust variety of higher yielding rice and the fall in the price of fertilisers relative to rice. Since an acceleration in growth also occurred in neighbouring Bihar and Orissa at the same time, and since factors such as the availability of new varieties of rice and fertiliser prices were present in those states as well, Harriss' account offers a possible explanation for growth acceleration in the entire region. This explanation, however, would still leave unanswered the question about the particularly strong performance of West Bengal compared even to its fast growing neighbours.

Sen and Sengupta (1995) pose this question in their analysis of agricultural growth rates. Controlling for inputs such as HYVs, fertilisers, irrigated land, and the consumption of electricity, Sen and Sengupta find that there was still a significant unexplained change in trend growth rates in West Bengal in the 1980s. The same was not found to have been true for Bihar and Orissa, where, once changes in inputs were taken into account, the break in trend in the 1980s was no longer statistically significant. The authors argue on the basis of this result that in addition to the growth in inputs there might have been unmeasured efficiency improvements in

West Bengal. They attempt to test the hypothesis that these efficiency improvements were due to various measures related to agrarian and political reform.

Their method was to conduct a statistical exercise using variations in inter-district growth rates in West Bengal in the 1980s.[11] The explanatory variables included a measure of tenancy reform, an indicator of land redistribution, as well as variables measuring local government investment and private investment in irrigation. Sen and Sengupta's results indicate that both agrarian reform variables (i.e., one measuring the extent of tenancy reform, and the other measuring the area of land redistributed) had statistically significant positive effects on growth.[12] Shallow tubewell acreage also had a positive and significant impact. There are, of course, limitations to what can be learnt about the differences between the growth experience of West Bengal and other states using inter-district variations within West Bengal alone. Despite this and its other shortcomings, the Sen and Sengupta paper is an interesting attempt at providing an empirical basis for the debate about reform and growth.

Tenancy reform offers a more promising explanation of the reform–growth link than land redistribution. Compared with the 3 per cent or so of cropped area that was redistributed under the Left Front land reforms, the proportion of land under share tenancy in West Bengal at the time of Operation Barga was around 7 per cent.[13] A recent study by Banerjee and Ghatak (1995) examines the effects of tenancy reform on productivity. These authors go further than Sen and Sengupta (1995) in spelling out a mechanism through which the dual policy of raising crop shares and increasing tenant security might have affected productivity. Banerjee and Ghatak's theoretical model is based on the idea that while a higher crop share and greater security might improve the incentives for a tenant to work harder and to supply greater amounts of non-marketable inputs, the removal of the threat of eviction can have the opposite effect. The net outcome could not be predicted on the basis of the theoretical model alone, and the authors (like Sen and Sengupta, 1995) use district-wise variations in growth between 1981 and 1993 to test their model empirically.

Banerjee and Ghatak (1995) found that after controlling for factors such as rainfall, public irrigation and district-specific fixed effects, growth in the production of *aman*, *aus* and *boro* rice was positively correlated with the progress of tenancy reform.[14] The relationship was found to have been statistically significant in the cases of *aman* and *aus*, but not in the case of *boro*. We noted before that the main sources of growth in rice production in West Bengal were higher yields all round, and the increase in *boro* acreage. Banerjee and Ghatak found that while inter-district variations in the yield of major crops were indeed positively and significantly correlated with the progress of tenancy reform, the same was not true for the expansion of *boro* acreage. Furthermore, the evidence on the mechanism through which higher yields were associated with tenancy reform was not decisive. While both HYV adoption and the increase in shallow tubewells were positively correlated with tenancy reform, the relationship was not found to have been statistically significant.

All of the studies reviewed here have contributed in advancing our understanding of the possible causes of acceleration of agricultural growth in West Bengal, and the possible connections between agrarian reform and agricultural productivity. While there is some empirical evidence of positive effects of reform on growth, it has been more difficult to find the precise mechanisms through which these effects were realised.

Further Research Issues

A number of the issues summarised above require further investigation. In particular, the idea that growth took off in West Bengal as well as in other states in eastern India as a result of favourable prices and the availability of new HYVs needs to be examined more thoroughly.[15] Similarly, the mechanism through which reforms might have influenced productivity requires further consideration. We hope that others more competent than ourselves might be motivated to take up some of these issues. On the reform–growth relationship in West Bengal, we offer a few thoughts of a speculative nature on potential future lines of enquiry.

One of these lines of enquiry emerges from our reading of the history of agrarian reform in West Bengal. Much of the discussion of the possible effects of reforms on growth begins, understandably, with the reforms themselves. The Left Front reforms are thus understood as marking the beginning of a process that might have sparked off or facilitated growth. However, the history of agrarian politics in West Bengal shows that it is possible to interpret the Left Front reforms as the *end* of a process. As we have shown elsewhere (Sengupta and Gazdar, 1997), the reforms were implemented in a context of heightened class-based political mobilisation in West Bengal. The key episodes in this mobilisation were the peasant movements in the 1960s leading to the formation of the United Front governments in the latter part of that decade. The early land reforms under the UF governments were more or less based upon direct actions by the landless and the land-poor to redistribute rights over land. The dismissal of the United Front governments and the suppression of the class-based movements that had emerged at the time did little to ameliorate the social and political conflict in the countryside.

It was finally with the election of the Left Front government that the changed balance of political power in the countryside was recognised and then institutionalised through agrarian and political reform. As such, the Left Front reforms marked not the beginning of attempts to change the distribution of power, but an end to further conflict, and the formal recognition of the changed balance of power (see Bhattacharyya, this volume). The institutionalisation of this change is likely to have significantly reduced the uncertainty surrounding claims to land. If political conflict over land had had a significant dampening effect on land-related investment (e.g., boring wells), by reducing the uncertainty the Left Front reforms might have removed important constraints on investment.[16]

Opposition to agrarian reform is often based on the claim that upsetting existing property rights arrangements can have a detrimental effect on investment. In a situation of heightened class conflict, however, the implementation of land reform might be an effective way of restoring confidence in property rights. Much depends, therefore, on the political context in which the reforms arise. Some critics have decried the shift in West Bengal politics from activism to institutionalisation as the failure of the Left Front to uphold the interests of the poor (see, for example, Webster, 1992). There is, without doubt, much validity in such criticisms. Our arguments here are, as we have mentioned at the outset, somewhat speculative, but they are based on a reading of West Bengal's recent economic and political history. If they do turn out to carry some weight then it is possible that, at least from the point of view of growth, institutionalisation was a positive strategy.

There are further important insights from other disciplines into the nature of social and political change in West Bengal that are not yet adequately reflected in the economic analysis of the recent agricultural growth. 'Environmental' changes, such as greater social equality, greater self-confidence amongst the poor, the strengthening of their overall political position (in addition to their specific benefits from reform such as assignment of land under land reforms, and improvement in the bargaining position of tenants after Operation Barga), the greater proximity, approachability and responsiveness of local government, and the presence of a disciplined 'social-democratic' political organisation, have been noted in a number sociological and political studies of the state's rural areas.[17]

Admittedly, such changes are difficult to measure and quantify, or to model in a formal economic framework. The real significance of the Left

Front reforms, nevertheless, may well lie in these overall political changes rather than in specific agrarian reform programmes such as redistribution or tenant security. The agrarian reforms affected relatively small proportions of total cropped area, but they did directly affect relatively large proportions of the rural population. The political empowerment of the poor was thus out of proportion to the precise area of land over which claims were renegotiated. These developments are likely to have influenced the working of rural markets, as well as the efficiency of public investment, and thus ultimately, economic growth. The study of the effects of political changes of this nature can be an extremely challenging and fruitful area of further research.

RURAL POVERTY AND WELL-BEING

Conventional measures of poverty, such as mean expenditure and the head-count ratio of poverty (i.e., the proportion of the population with expenditures below a predetermined poverty line), largely reflect the level and distribution of private incomes, or the consumption of private commodities. Indicators of the health status of the population, such as mortality rates, and those concerned with its educational status, such as literacy rates, directly measure outcomes that are in great part determined by public investment and the consumption of public goods. In this section we adopt a broad-based definition of well-being, which includes indicators of private income and consumption as well as outcomes such as rates of mortality and literacy, in order to examine the recent trends in rural poverty in West Bengal in its several dimensions.

Expectation of Change

Given the importance of agriculture in the rural economy of India, growth in agricultural output is rightly considered to be a key factor in reducing poverty and promoting well-being.[18] If growth coincides with an increase in inequality, however, the impact of increases in overall output or average consumption on poverty reduction might be neutralised or negated. This has been a common concern in the debate about the processes that led to agricultural growth in western India in the 1960s and 1970s.[19] The recent growth experience of West Bengal is interesting in this regard, particularly because it has coincided with (and, arguably, been buttressed by) agrarian and political reforms. Agrarian reforms such as land redistribution and security of tenure and higher crop shares for sharecroppers might be expected to allow for more participatory growth.

The combination of political change and agricultural growth could also be expected to have a positive impact on other aspects of well-being such as health, education and the relative position of women. Political radicalism is often associated with advances in the position of women, and in India health and educational backwardness (of the whole population, including males) have been found to be linked to gender disparity.[20] Furthermore, interventions is the field of health and educational development require effective local government institutions, and West Bengal has had a fairly unique record in establishing elected PRIs in this period.

A number of issues raised here concerning the relationship between economic change and poverty require detailed micro-level analysis of power relations in the village economy, gender politics, and the working of local democracy.[21] It is important, nevertheless, to place such micro-level analyses within the context of broader trends in poverty-related variables that are discernible from more aggregated data. Nearly all the evidence cited in this section is from secondary sources aggregated up to the state level, and corresponding figures for rural India and for rural areas of Bihar and Orissa are also reported for comparison. We try, wherever possible, to track the trend of diverse well-being indicators, focusing on the early 1980s as the reference period for making comparisons between regions and over time.

Livelihoods and Poverty Ratios

The extent to which the agricultural growth in West Bengal would have been 'poverty-reducing' and participatory depends to a large extent on the importance of agricultural employment in the economy, and the distribution of gains from growth within the agricultural sector.

Agriculture dominates the economic structure of rural West Bengal. Wage labourers, however, amount to around a third of the total rural workforce, and nearly a half of all workers in agriculture.[22] Agricultural labourer households are among the poorest sections of the rural population.[23] The main factors affecting the livelihoods of these households are real wage rates and employment days.

Secondary evidence on the real wages of agricultural labourers suggests that the trend in wage rates shadowed the trend in agricultural output quite closely throughout this period. This is illustrated in Figure 3.3. Whether these rising wage rates translated into higher earnings depends on what happened to the level of employment. Although we have no firm

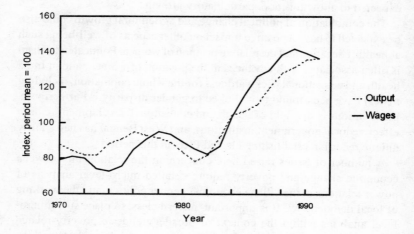

Figure 3.3: Trends in the output of foodgrains and real wages of agricultural labourers in West Bengal, 1970–71 to 1991–92 (three-year moving averages)
Sources: Acharya, 1988 and CMIE, July 1993. Annual wage rates compiled by Bipul Chattopadhyay, IEG, Delhi 1994, based on data from the Agricultural Wages of India series.

evidence on employment trends, it is plausible to argue that the overall effects of growth on employment were positive. The factors which led to a rise in productivity—greater irrigation, higher cropping intensity, greater use of fertilisers, and a switch-over to higher yielding varieties—would have led to increased demand for labour.[24]

The contrast between West Bengal's agricultural take-off and the 'green revolution' in north-western India is noteworthy. In Punjab, Haryana and western Uttar Pradesh, the adoption of new biotechnology coincided with the increasing mechanisation of farm operations. Although the new technologies would have led to a higher demand for labour for some operations, some of this effect was dampened by mechanisation. There was also a corresponding rise in the eviction of tenants and the resumption of land for self-cultivation by landowners.[25] In sum, then, the relative position of the landless with respect to sharing directly in the benefits of agricultural growth deteriorated. In West Bengal, on the other hand, there is little evidence as yet of such labour-displacing mechanisation.[26] Similarly, eviction of tenants has been made much more difficult after Operation Barga. The landless, therefore, through higher wages and employment, land redistribution and security of tenure, have been in a relatively better position to share in the growth process in West Bengal than elsewhere (see also Williams, this volume).

According to National Sample Survey data the proportion of the rural population living below the poverty line declined from around three-fifths in the early 1970s to around a third in the late 1980s (see Table 3.1) (based

Table 3.1
Head-Count Ratio of Rural Poverty in India and the Eastern States

Year	Bihar	Orissa	West Bengal	India
1973.25	69.19	67.03	60.51	55.36
1978.00	66.21	65.52	56.25	50.60
1983.50	69.94	56.76	49.21	45.31
1987.00	56.45	44.95	34.10	38.81
1988.00	58.57	47.86	34.87	39.60
1992.50	67.81	36.57	28.15	43.47

Source: Ozler and Datt, 1996, based on various NSS rounds.

on Ozler and Datt, 1996). Since 1983, the rate of decline in the head-count ratio of rural poverty in West Bengal has been faster than in any other major Indian state. Mean per capita consumption has risen steadily since the early 1970s and the rate of growth was considerably higher after 1983 (see Table 3.2). These trends confirm the view that acceleration in agricultural growth coincided with the acceleration in the decline of rural poverty.

Table 3.2
Estimated Trend Growth Rates (Per Cent Per Year) in West Bengal

	1970 to 1983	1983 to 1992
Foodgrains output	−0.04	4.29
Agricultural wages	1.07	5.68
Mean per capita consumption	1.21	2.11
Inverse head-count ratio of poverty	2.19	3.20

Sources: Foodgrains output: CMIE, 1993; agricultural wages: Bipul Chattopadhyay, IEG, 1994, Delhi, based on Agricultural Wages of India data; mean consumption and head-count ratio of poverty: Ozler and Datt, 1996, based on NSS data.

Longer trends in head-count ratios and mean consumption are also instructive (see Figures 3.4 and 3.5). These trends indicate that for most of the period for which data are available, rural poverty in West Bengal moved in tandem with the rest of the country. After peaking in the late

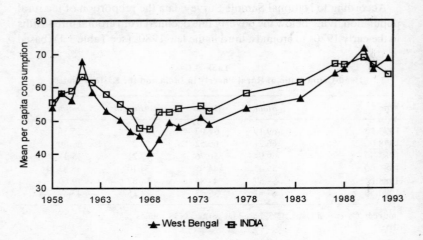

Figure 3.4: Rural mean per capita consumption in West Bengal and India, 1958 to 1992
Source: Ozler and Datt (1996), based on NSS data.

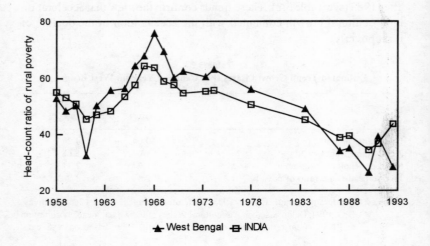

Figure 3.5: Head-count ratio of rural poverty in West Bengal and India, 1958 to 1992
Source: Ozler and Datt (1996), based on NSS data.

1960s, the head-count ratio of rural poverty declined in West Bengal (as it did in India as a whole) throughout the 1970s and 1980s. The head-count ratio in West Bengal remained above that of India for the whole period from the early 1960s until the late 1980s, when it fell below the national average. From around 50 per cent in the late 1950s and early 1960s, the head-count ratio of rural poverty rose to over 70 per cent in the late 1960s and then fell back to around 50 per cent in the mid-1980s. The gains throughout the 1970s and early 1980s simply redressed the impact of the massive increases in poverty in the earlier period. From the mid-1980s onwards, however, new low levels of rural poverty were reached.

Education and Health Outcomes

Health and educational indicators provide an additional dimension to the evaluation of population well-being for at least two important reasons. First, health and educational outcomes can legitimately be regarded as intrinsically valuable goals of economic development.[27] The relationship between private consumption or income and health and educational outcomes is mediated by social factors (e.g., the provisioning of public goods) that vary between locations and over time. Indicators of poverty based on private consumption are therefore poor proxies for the health and educational status of the population, and the latter need to be measured directly. Second, information on consumption and income is generally available only at the household level, while a number of health and education indicators are readily available at the individual level. The latter can therefore provide valuable insights in the event of significant intra-household inequalities in the allocation of resources.

Infant Mortality

State-wise data on infant mortality rates (IMR) in India are available from the Sample Registration System. Table 3.3 presents three-year moving averages for rural areas in West Bengal, India as a whole and in Bihar and Orissa from 1981 onwards. In the early 1990s, West Bengal's IMR of 72 infant deaths per 1,000 live births was the fifth lowest after Kerala (17), Punjab (62), Maharashtra (67) and Tamil Nadu (67), among the major Indian states. Another four states—Andhra Pradesh (76), Bihar (74), Gujarat (75) and Haryana (75)—had rural IMRs that were very close to West Bengal's (*Sample Registration Bulletin*, 28[2]: Table 12, July 1994). West Bengal's ranking in the early 1990s had improved from its.

Table 3.3
Rural Infant Mortality Rate (Number of Deaths Under the Age of 1 Per 1,000 Live Births) (Three-Year Moving Averages)

	India	Bihar	Orissa	West Bengal
1981–83	116	114	136	95
1982–84	114	105	135	91
1983–85	111	102	134	87
1984–86	108	103	133	81
1985–87	105	106	131	78
1986–88	104	102	128	76
1987–89	102	99	127	78
1988–90	96	90	126	75
1989–91	90	81	127	75
1990–92	86	74	124	72

Source: *Sample Registration Bulletin*, July 1994, Vol. 28, No. 2, Table 12.

position in the early 1980s when six states had lower rural IMRs. It is worth noting, however, that this improvement in rank over the decade was due mainly to the particularly slow IMR decline in Andhra Pradesh, and an actual increase in the IMR in Karnataka.

The trend in West Bengal's proportional change in rural IMR over this period can be seen in Figure 3.6. It shows that though the rural IMR

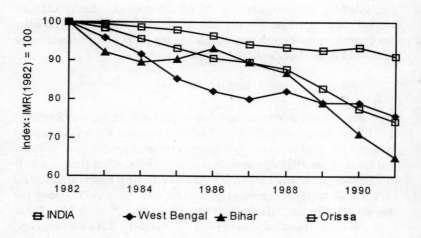

Figure 3.6: Trends in rural Infant Mortality Rates, 1982 to 1991
Source: *Sample Registration Bulletin,* July 1994, 28 (2): Table 12.

declined relatively faster in West Bengal than in Bihar and Orissa as well as in India as a whole in the early 1980s, the pace of decline slowed down from around 1988. In the entire period from 1982 to 1991, the proportionate decline of the rural IMR in West Bengal was identical to that of India as a whole, and slower than that of Bihar. Bihar in fact experienced fairly rapid progress from 1988 onwards in spite of the fact that it had relatively little success in reducing the head-count ratio of rural poverty over this period.

Literacy and School Participation

According to the 1991 Census, West Bengal's rural literacy rate for the 7+ male population was 68 per cent compared to the Indian average of 64 per cent, and the corresponding literacy rate for females was 47 per cent compared to the Indian average of 39 per cent. Figure 3.7 illustrates

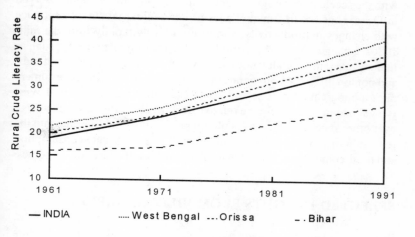

Figure 3.7: Trends in rural literacy rates, 1961 to 1991
Sources: Sharma and Retherford (1987); Nanda (1992).

inter-decadal trends in the crude rural literacy rate for West Bengal, Bihar, Orissa and India as a whole on the basis of the population census data from 1961 to 1991.[28] Of the three states, West Bengal started out with the highest level of rural literacy in 1961 and maintained its rank throughout this period. In the 1980s, the rate of increase in rural literacy was somewhat higher in West Bengal than in Bihar and Orissa. It was very close to the national average.

Age-specific rates of literacy and school participation show that rural West Bengal's progress in this field ought to give cause for alarm. According to NSS data, in 1986–87 nearly half (46 per cent) the children in the 6–14 age group had never enrolled in a school. The states that had worse records than West Bengal were Bihar (62 per cent), Rajasthan (54 per cent), Madhya Pradesh (53 per cent), Uttar Pradesh (52 per cent) and Orissa (47 per cent) (Visaria et al., 1993: Table 15). More recent data suggest that these trends continued into the 1990s, and that in fact rural West Bengal's relative position declined. In 1994, the proportion of rural children in the 6–14 age group who had never enrolled in school was 34 per cent in West Bengal, compared with an all-India average of 29 per cent. Only Bihar (41 per cent), Rajasthan (39), Madhya Pradesh (37), and Uttar Pradesh (36) had worse records (NCAER, 1996: Table 4.3).

To sum up, rural poverty measured in terms of consumption and head-count ratios declined sharply in West Bengal in the 1980s, coinciding with the acceleration in agricultural output. Similar trends were observed in Orissa, but not in Bihar. Wages of agricultural labourers kept pace with changes in productivity, and rose relatively rapidly from the mid-1980s. Changes during this period in other measures of poverty, such as literacy rates and mortality rates, however, were in line with longer trends, and not too different from trends in other states. In the case of enrolment rates among children of school-going ages, the situation in rural West Bengal had in fact undergone relative decline. West Bengal's advantage over other states in raising agricultural productivity did translate into the reduction of poverty measured in terms of wages, private consumption and head-count ratios, but not in terms of health and education outcomes.

SELECTED INSIGHTS FROM VILLAGE SURVEYS

Patterns of change and stagnation in educational and health indicators are explored further here, with the help of micro-level data available from intensive socio-economic and anthropometric surveys carried out in six villages in West Bengal between 1987 and 1989. The villages and the surveys have been described in greater detail elsewhere (see Sengupta and Gazdar, 1997; Sengupta and associates, 1991; Gazdar, 1992). It is sufficient to observe here that the survey villages represented diverse agro-climatic as well as socio-economic conditions. All resident households in these villages were canvassed for the demographic, educational and economic characteristics of individual members, as well as for their joint economic activities. Anthropometric measurement was carried out for all children aged 5 years or under.

The village survey data used here are over ten years old. There are two ways, however, in which they continue to be relevant to the purposes of this chapter. First, they allow us to explore patterns of deprivation and the dynamics of change. As we have already shown using secondary state-level data, these patterns have shown a great deal of resilience. There is much to be gained from analysis of micro-level data, even if they are somewhat dated. Second, these data continue to be of relevance because they give a snapshot view of socio-economic conditions in a set of villages at a crucial period, namely, at a time when there were clear indications of an agrarian take-off.

The following section investigates the nature of the literacy problem by reconstructing the gender-wise time profile for literacy among people from various caste groups using literacy rates of different age cohorts in the survey year. Next, the relationship between household income and the nutritional status of children is examined, and the 'income elasticity' of malnourishment is compared with the effect of other factors such as gender inequality. These various exercises using micro-level data cannot, of course, be interpreted as statistical inferences for the whole of rural West Bengal. They nevertheless do provide insights that might be of some value in understanding the processes by which broad aggregated poverty indicators might change over time in response to economic and political changes.

Dynamics of Literacy

It has already been noted that changes in literacy rates in rural West Bengal in the 1980s were in line with earlier trends and also with trends in neighbouring states. In order to understand the dynamics of aggregate literacy rates it is important to analyse the structure of educational achievement by social grouping. In particular, caste and gender are important correlates of educational status in India.

Table 3.4 shows the gender- and caste-wise breakdown in literacy rates in the WIDER village surveys. The female and male literacy rates for the WIDER villages taken together correspond quite closely with the average for rural West Bengal from the 1991 Census. Caste-wise contrasts are extremely instructive. While the general caste Hindu males had nearly total

Table 3.4
Literacy Rates (Percentage of Population Aged 7 and Above) by Gender and Caste

	Female	Male	Both
Rural West Bengal	38	62	50
WIDER sample	40	60	51
Caste Hindus	71	92	82
Scheduled Castes	22	43	33
Scheduled Tribes	5	24	15
Muslims	15	30	22

Sources: Census 1991, WIDER village surveys, 1987–89.

literacy, the Scheduled Tribe females in the very same villages were almost entirely illiterate. Gender and caste inequalities are thus key characteristics of the literacy situation.

Further insights on this issue may be obtained from a consideration of changes in literacy rates over time. These can be gauged from the current literacy rates of different age cohorts. Assuming that people become literate if at all by the age of 15, and that the chances of acquiring literacy thereafter are low, literacy rates of different age cohorts can tell us something about patterns of change in literacy.[29] Table 3.5 gives literacy rates by gender and caste for people who reached the age of 15 in different periods.

Table 3.5
Literacy Rates (Percentage of Population) by Caste and Age Cohort

Aged 15 in	Caste Hindus		Scheduled Castes		Scheduled Tribes	
	Males	Females	Males	Females	Males	Females
1983–87	95	88	58	30	41	0
1978–82	95	82	38	26	30	2
1973–77	95	76	40	19	19	5
1963–72	96	77	44	11	12	0
1953–62	91	53	31	7	30	0

Source: WIDER village survey, 1987–89.

It is striking that, at the two extremes of the literacy scale, there is no discernible trend over time; for general caste Hindu males the 1953–62 cohort was already very close to full literacy, while for Scheduled Tribe females there has been no significant departure from total illiteracy during the entire reference period. The groups in between accounted for the changes in aggregate literacy rates between the 1950s and the late 1980s.

During the most recent years, or the period corresponding with important institutional and productivity changes, there is evidence of significant improvement in literacy among Scheduled Caste and Scheduled Tribe males.[30] Among Scheduled Caste females, however, progress was slower, and Scheduled Tribe females registered no progress at all.

Social Inequalities and Nutrition

It is not only in the case of basic education that gender- and caste-based inequalities play a role in determining final well-being outcomes. Gender and caste are also known to be important correlates of other measures of poverty such as mortality rates and malnutrition. While the household is the main locus of gender disparities in the allocation of resources, caste-wise disparities are, obviously, associated with a wider set of social and economic relations. If land-ownership is closely correlated with caste for historical reasons, and if income is closely associated with land-ownership, it might be argued that caste shows up as an important correlate of poverty not because it is important in its own right, but because of its correlation with historical patterns of asset ownership.

The WIDER village surveys included an anthropometric survey of all children under the age of 5 in six villages.[31] It is therefore possible to carry out statistical tests of the relevance and weight of various economic and social determinants of child nutrition using multivariate analysis. The anthropometric survey was used to construct a weight-for-age index along the lines of the index widely used in rural health centres in India.[32] Besides the classification of children into the well-nourished and the under-nourished, four categories of undernourishment: 'slight', 'moderate', 'severe' and 'disastrous' were used. If a child is found to be in the latter two categories, health centre guidelines call for clinical intervention including hospitalisation if necessary.

Only sixty-eight out of the sample of 436 children, or 16 per cent of the total, were nutritionally normal. About one-third were slightly malnourished, another third moderately so, and 15 per cent suffered from severe or disastrous malnutrition.[33] For the purposes of our multivariate analysis we reclassified as undernourished all children anthropometrically classified as moderately malnourished or worse.[34] Explanatory variables included per capita household income, the gender of the child, the child's age (as well as the square of age in order to allow for non-linearity in the relationship), dummy variables for Scheduled Caste and Scheduled Tribe children, and a dummy variable for whether any female member of the household had received secondary education. Logit analysis was used to estimate the probability of a child being well-nourished; the results are reported in Table 3.6.

Table 3.6
Predicted Probability of a Child (Under 5 Years) Being Well-Nourished

Dependent Variable	Logit Maximum Likelihood Estimates		
	Parameter Estimates	Standard Error	Pr>Chi-Square
Intercept	−3.8553	1.7233	0.0253
Log per capita income	0.7426	0.2395	0.0019
Age in months	−0.0513	0.0262	0.0506
Square age	0.0005	0.0004	0.2112
Sex: Male = 0, Female = 1	−0.7709	0.2048	0.0002
Female secondary schooling, No = 0, Yes = 1	0.4613	0.2948	0.1176
Scheduled Caste	0.0094	0.2508	0.9702
Scheduled Tribe	−0.3099	0.2888	0.2832
Chi-square for covariates	−2 LOG L	57.787 with 6 DF (p = 0.0001)	
	Score	48.904 with 6 DF (p = 0.0001)	

Sample: 436 observations
Response variable Nutritional Status (NSTAT)
NSTAT = 0 if well-nourished, 225 observations
NSTAT = 1 if malnouished, 211 observations

Source: Calculated by authors from WIDER village data.

Household per capita income was found, as expected, to be positively correlated with the probability of a child being well-nourished. Likewise, gender had the predictable effect: girls were less likely to be well-nourished than boys. In both cases the estimated coefficients were statistically significant. Caste dummy variables were not statistically significant. The dummy variable for female education, although of the expected sign (i.e., higher female education meant improved nutritional status), was statistically significant only at the 12 per cent level. It might be added that when the model was tested without the female education dummy, the coefficient for the dummy variable for Scheduled Tribe children was found to have been significant at the 10 per cent level. These results suggest that caste does not independently affect nutritional status if the effects on other variables such as income, but also, importantly, female education, are taken into account.

In order to simplify the interpretation of these results, Table 3.7 reports the estimated probability predictions of male and female children aged 30 months (which was close to the average age in the sample) being well-nourished, given the income and educational characteristics of their households. It is possible to assess the respective effects of income, gender and female education on the nutritional status of children using different values for the explanatory variables.

Table 3.7
Predicted Probability of a Child Aged 30 Months Being Well-Nourished for Various Household Characteristics

	Male	Female
Income = Sample average; Female secondary schooling = 0	0.605	0.415
Income = Sample average; Female secondary schooling = 1	0.708	0.529
Income = Sample average 2; Female secondary schooling = 0	0.718	0.542
Income = Sample average 3; Female secondary schooling = 0	0.776	0.616

Source: Calculated by authors from WIDER village data.

According to our estimates a male child from a household with mean sample income (Rs 1,367 per capita per year), and in which no females had received secondary schooling, had a 61 per cent predicted probability of being well-nourished. For a female child from a similar home the predicted probability was much lower at around 42 per cent (Table 3.7, first row). The predicted probability of being well-nourished for a male child from an average-income home, but with at least one female who had received secondary schooling, was 71 per cent, and that for a female child from such a home it was 53 per cent (Table 3.7, second row). The presence of an educated female, other things being equal, improves the chances of nutritional well-being by around 10 percentage points.[35]

The relative effects of household income and its female educational status in determining the probability of avoiding undernourishment can be seen by considering the following question: at what level of household income would children from a household without educated females face similar predicted probabilities of undernourishment as those from an average income household where there was an educated female? The third row in Table 3.7 gives the predicted probabilities of being well-nourished for children from households where there are no educated females but the household income is *twice* as high as the sample average. The predicted probabilities are similar to those for children from average income households where there is an educated female. The presence of an educated female, therefore, has the same effect on child nutrition as the *doubling* of household income.

The relative effects of income and gender disparities on nutritional status can be gauged in a similar manner. With twice the average household income (third row) a female child is still less likely to be well-nourished compared to a male child in a household with average income (first row). It is only when income is raised to *three* times the average (fourth row) that the nutritional status of a female child begins to resemble that of a male child in a household with average income. The complete elimination of gender disparities, while keeping income constant, would have a similar impact on the nutritional status of a female child as the raising of income by a factor of three while retaining the level of gender disparity.

Our results give some indication of the respective roles of various social factors in the intermediation between household income and an outcome-based indicator of poverty: in this case, nutritional status. Similar observations apply to other indicators of poverty such as illiteracy. While our results confirm that income (and therefore income growth) has a positive effect on poverty reduction, they also strongly indicate that this effect can be seriously blunted by the persistence of gender- and caste-based social inequalities. This, arguably, was the case in rural West Bengal in recent years where, in spite of the rapid decline in consumption-based indices of poverty, there was little change in the rates of progress of health and educational indicators.

CONCLUSION

The sustained period of growth in agricultural productivity in West Bengal following a long period of stagnation and decline may be viewed as a turning point in the state's rural economy. Even in the context of longer-term trends, and in relation to developments elsewhere in India, including in its neighbouring states, the agricultural take-off in West Bengal appears to be impressive, though somewhat less dramatic than originally expected, and certainly not unique.

The debate about the causes of West Bengal's growth acceleration in the 1980s and early 1990s has been quite instructive. To sum up briefly, the paper by Saha and Swaminathan (1994) statistically established the trend break not only in the production of rice or foodgrains, but across a range of crops. Given the coincidence of growth with reform, the authors express a justifiable interest in the possible connection between the two processes, but do not pursue the point further (though see Rawal and Swaminathan, 1998). Harriss (1993), mainly on the basis of evidence from village surveys and case studies, points out problems with the two versions of the reform–growth relationship that have been widely postulated, namely, the size–productivity effect, and the cooperative public irrigation thesis. He points, instead, to changes in factor price ratios, and the availability of new technology as possible explanations of the take-off.

Since the agricultural take-off also occurred in West Bengal's neighbouring states, it would seem natural that this issue ought to be investigated more thoroughly with reference to all three states. Sen and Sengupta (1995) show statistically that though the take-off also occurred elsewhere, quantitative as well as qualitative differences in West Bengal's experience justify the search for state-specific explanations of growth. Even if one accepts Harriss' analysis of the causes of take-off, it is not necessary to abandon the pursuit of the study of the reform–growth relationship. Both Sen and Sengupta (1995) and Banerjee and Ghatak (1995) continue with this pursuit and find some evidence of a positive correlation between tenancy reform and growth.

It has been more difficult, however, to find the precise mechanism through which reforms translated into better productivity. Economic analysis of the reform–growth relationship might gain something from paying more attention to insights gained from other disciplines into the recent political and social changes in rural West Bengal. In particular, the Left Front reforms might be interpreted as the institutionalisation of a changed balance of class power in the countryside. The effects of this changed balance of power on economic behaviour provide a useful area of enquiry.

The aim of this chapter has been to draw attention to some other equally instructive aspects of West Bengal's recent development experience. Agricultural growth, whether induced by public policy or market response, or indeed by a combination of the two, did have a positive impact on the rate of decline of important aspects of rural poverty. Factors that led to growth also contributed towards greater participation in the growth process by the poor. Wages of agricultural labourers closely shadowed changes in output, and redistributive land reforms further widened the base over which the benefits of growth were shared. As a result of these factors, West Bengal saw sharp declines in the head-count ratio of poverty and increases in mean consumption.

By contrast, improvement in other aspects of well-being was less marked and did not deviate significantly from longer trends or from the experience of other parts of rural India where agricultural growth had been less impressive. The resilience of longer trends in aspects of well-being such as education and health can in some measure be seen as a consequence of social conditions, especially inequalities due to gender and caste, and the absence of major public intervention in these areas. Income growth in itself is unlikely to lead to a dramatic reduction in these forms of deprivation. In our opinion, therefore, the debate about the relationship between socio-economic conditions and population well-being needs to engage with a range of issues far wider than those specifically concerned with agrarian structure and agricultural growth.

NOTES

1. For a fuller description of these studies see Sengupta and associates (1991), Gazdar (1992) and Sengupta and Gazdar (1997).
2. In much of the discussion that follows we refer to the growth in agricultural output while presenting statistical evidence for foodgrains only. Subsequent discussion of patterns of growth narrows the focus down still further to rice production. The official definition of foodgrains includes cereals (rice, wheat, bajra, jowar, barley, maize, millet, etc.) and pulses. Our focus on foodgrains in general and rice in particular is in line with the concerns of the literature, and not misleading, given the importance of rice in the state's agrarian economy. Throughout the 1980s, foodgrains accounted for over 80 per cent of total cropped area, and rice nearly 90 per cent of total area under foodgrains in West Bengal (Chandhok et al., 1990).
3. Sen and Sengupta (1995: Table 2). It should be added that precise estimates of rates of growth are particularly sensitive to the choice of base year as the harvests in 1981–82 and 1982–83 were uncharacteristically poor. Thus the growth rate figure reached by using 1983 as a base year is likely to be an overestimate (see Sengupta and Gazdar, 1997). Nevertheless, estimates of trend changes in growth in the early 1980s are robust.
4. The reliability and consistency of agricultural data from West Bengal have been called into question from time to time (Ray, 1994). See also James Boyce's objections to Government of West Bengal data on agricultural growth reported in Rogaly, Harriss-White and Bose (this volume). While different sources of data might yield different magnitudes of growth, there can be little doubt that historically high rates of growth have sustained for over ten years.
5. Figure 3.2 plots the index of three-year moving averages of foodgrains output with period means for each state set at 100.
6. The problem of base-year sensitivity mentioned in note 3 also applies to these estimates. It should be noted, however, that for all choices of base year the state rankings in the growth of foodgrains output in the 1980s are preserved.
7. Although the CPI(M) has by far the largest representation of all the left parties in the panchayats, some of the latter are controlled by Forward Block (FB), Revolutionary Socialist Party (RSP) and Communist Party of India (CPI) representatives, where they play the same role.

Agricultural Growth and Well-Being 87

8. We have discussed the historical and economic context of the Left Front reforms elsewhere (see Sengupta and Gazdar, 1997).
9. For a recent review and restatement see Dasgupta (1993).
10. Of this relatively small area, only about half was redistributed under the Left Front after 1977. The rest was redistributed under the United Front governments of the late 1960s. See Sengupta and Gazdar (1997) for details of land redistribution in different periods.
11. The exercise was statistical in the sense that it was simply aimed at capturing a statistical relationship between growth and a number of variables without specifying the precise mechanism through which these variables affect growth.
12. It should be noted that Sen and Sengupta's measure of tenancy reform is somewhat problematic. They use the number of sharecroppers recorded per 1,000 hectares of cropped area in the district. Other things being equal, a district with a smaller average farm size would have more sharecroppers recorded per unit of land. Since farm size is known to be inversely related to agricultural productivity in India, Sen and Sengupta's tenancy variable may well be capturing some of this effect, which has nothing to do with the progress of tenancy reform.
13. Bhaumik (1993: Table 3.7, based on National Sample Survey data for 1982). With under 8 per cent of land under registered share tenancy, exercises to measure the direct influence of Operation Barga on agricultural productivity are limited. Even if registered share tenants increased their output significantly, this increase would make a much smaller proportional dent in the total output figures for the state.
14. Banerjee and Ghatak use the proportion of sharecroppers registered in a district as the measure for the progress of tenancy reform. This is not prone to the difficulty pointed out above with Sen and Sengupta's (1995) measure of tenancy reform.
15. The examination could usefully differentiate between districts of Bihar and Orissa which border West Bengal and other districts in those states for a tighter control of agro-ecological variables.
16. A similar effect of reforms of reducing uncertainty and thus allowing profitable investment to take place relates to tenancy. Many observers point out that there is now the widespread practice in the state of farmers with shallow tubewells on their plots leasing in plots of neighbouring farmers in order to optimise their use of their groundwater resource. Seasonal leasing is outside the purview of tenancy regulation and owners of land leased out seasonally are assured of its reversion to them at the end of the lease period. If prior to the reforms there was uncertainty about the status of seasonally leased land—terms of tenancy, after all, have been highly politicised issues in Bengal at least since the 1930s (see for example, Dhanagare, 1983)—the establishment of tenancy reform, by clarifying the legal status of seasonal leasing, would have improved the profitability of investment in tubewells. On the distributional effects of seasonal leasing see Sengupta and Gazdar (1997) and Webster (this volume).
17. See Sengupta and Gazdar (1997) and the references cited there; also see Bhattacharyya (this volume) and Williams (this volume).
18. For a recent empirical confirmation of the strong inverse relationship between agricultural productivity and a range of rural poverty indices see Datt and Ravallion (1996).
19. See Singh (1990) for a review of this debate.
20. See, for example, Basu (1992) on the former point, and Murthi et al. (1997) on the latter. Nevertheless, in making the link between political radicalism and advances in the position of women, Basu draws on a comparative study of West Bengal and Maharashtra, which shows that the redistribution of land titles to men in West Bengal

may have worsened the position of women there. See also Sengupta and Gazdar (1997).
21. See, for example, Harriss (1993) for evidence on access to irrigation and perceptions of relative gains from growth, Basu (1992) on the role of gender relations in political mobilisation, and Williams' chapter on local democracy in this volume.
22. According to the population census of 1991 the proportion of male main workers who reported their primary occupation as agricultural wage labour in rural West Bengal was 32 per cent in 1991 (Nanda, 1992: Table 4.1).
23. See, for example, Gazdar (1992) for evidence from village-level data. See also Rogaly (this volume).
24. There is also some evidence that the rate of employment in West Bengal was higher than in most other Indian states. According to a recent national survey, the average number of person days a year for which an adult agricultural labourer in West Bengal was employed was the second highest among the major states at 178, compared to the national average of 137 days (NCAER, 1996).
25. See Singh (1990) for a review of the evidence on these issues.
26. This difference in the pattern of mechanisation between north-western Indian states and West Bengal is largely due to the different agronomic conditions in the two regions. Small plot sizes and the high degree of land fragmentation in West Bengal makes tractorisation less viable than, say in Punjab and Haryana, although ploughing by rented tractor is a common and growing practice even in the intensively cultivated areas of West Bengal. Furthermore, paddy farming in the state involves a number of specialised labour-intensive operations, such as transplanting, that cannot be easily mechanised.
27. See Drèze and Sen (1995) for a recent exposition of this view with reference to India.
28. Figure 3.7 uses crude literacy rates (i.e., the total number of literates as a proportion of the entire population) rather than age-specific literacy rates. The reason is that consistent time series were not available for age-specific literacy rates.
29. Strictly speaking, this 'backward projection' method also assumes that mortality rates are similar for literate and illiterate people within the relevant groups; see Drèze and Loh (1995) for further discussion.
30. Similar patterns of change were found in primary school enrolment in the Muhammad Bazar block in Birbhum district over the mid-1970s onwards by Lieten (1992).
31. Two of these six villages (Kuchly and Sahajapur in Birbhum district) had been first surveyed for child nutrition in 1983. The results of that study had provided the material for one of the earlier contributions to the debate on gender biases within households in the allocation of nutritional and health inputs (Sen and Sengupta, 1983).
32. Sen and Sengupta (1983) classified their data according to the 'Harvard' standards of weight-for-age. We followed their classification.
33. It should be noted that in the two villages for which benchmark data from 1983 were available, there had been some improvement in nutritional conditions over the intervening period (for details see Sengupta and Gazdar, 1997).
34. In an index of this nature, any cut-off point is somewhat arbitrary. The reason for choosing the cut-off used here is that 'mildly malnourished' or 'slightly malnourished' might mean those suffering from temporary ailments, whereas the category 'moderately malnourished' is likely to capture those suffering permanent damage due to poor nutrition and health.
35. The positive relationship between female education and the health status of children has been observed for various kinds of data in India; see, for example, Murthi et al.

(1995), and the literature cited there. In a society where child care is mainly regarded as a female concern, the education of women is likely to promote greater effective awareness of health issues, and better access to health facilities.

REFERENCES AND SELECT BIBLIOGRAPHY

Acharya, S. 1988. 'Agricultural Wages in India: A Disaggregated Analysis,' *Indian Journal of Agricultural Economics*, 44(2): 121–39.
Banerjee, Abhijit V., and **Maitreesh Ghatak.** 1995. 'Empowerment and Efficiency: The Economics of Tenancy Reform.' Mimeograph, Department of Economics, Harvard University.
Basu, Amrita. 1992. *Two Faces of Protest: Contrasting Modes of Women's Activism in India.* Berkeley: University of California Press.
Bhaumik, Sankar Kumar. 1993. *Tenancy Relations and Agrarian Development: A Study of West Bengal.* New Delhi: Sage.
Boyce, James K. 1987. *Agrarian Impasse in Bengal: Institutional Constraints to Technological Change.* Oxford: Oxford University Press.
Centre for the Monitoring of the Indian Economy (CMIE). 1993. Performance of Agriculture in Major States (1967–68 to 1991–92), July 1993.
Chandhok, H.L., and the **Policy Group.** 1990. *India Database—The Economy, Annual Time Series Data*, Volume 2. New Delhi: Living Media Books.
Dasgupta, Partha. 1993. *An Inquiry into Well-Being and Destitution.* Oxford: Clarendon.
Datt, Gaurav, and **Martin Ravallion.** 1996. 'Why Have Some Indian States Done Better Than Others at Reducing Rural Poverty?' Mimeograph, World Bank.
Dhanagare, D.N. 1983. *Peasant Movements in India 1920–1950.* Delhi: Oxford University Press.
Drèze, Jean, and **Amartya Sen.** 1995. *India: Economic Development and Social Opportunity.* Oxford and Delhi: Oxford University Press.
———. (eds). 1997. *Indian Development: Selected Regional Perspectives.* Oxford and Delhi: Oxford University Press.
Drèze, Jean, and **Jackie Loh.** 1995. 'Literacy in India and China.' Working Paper No. 29, Centre for Development Economics, Delhi School of Economics.
Gazdar, Haris. 1992. 'Rural Poverty, Public Policy and Social Change: Some Findings from Surveys of Six Villages.' WIDER Working Paper WP 98, May 1992, World Institute for Development Economics Research, Helsinki.
Government of West Bengal (GoWB). 1993. *Economic Review 1992–93.* Calcutta.
Harriss, John. 1993. 'What is Happening in Rural West Bengal? Agrarian Reform, Growth and Distribution,' *Economic and Political Weekly*, 28(24): 1237–47.
Ishikawa, Shigeru. 1974. *Economic Development in Asian Perspective.* Tokyo: Kinokuniya Bookstore.
Jose, A.V. 1984. 'Poverty and Income Distribution—The Case of West Bengal,' A.R. Khan and Eddy Lee (eds) *Poverty in Rural Asia.* Bangkok, International Labour Organisation, Asian Employment Programme, ARTEP.
Lieten, G.K. 1992. *Continuity and Change in Rural West Bengal.* New Delhi: Sage.
———. 1994. 'For a New Debate on West Bengal,' *Economic and Political Weekly*, 29(29): 1835–8

Majumder, Debanshu, and Sunil Sengupta. 1990. 'Level of Living Among the Rural Poor in West Bengal (1987–89)—A Study of Six Villages, Part I.' Working Paper, WIDER Project on Rural Poverty, Social Change and Public Policy, Santiniketen, 1990.

Minhas, B.S., L.R. Jain and S.D. Tendulkar. 1991. 'Declining Incidence of Poverty in 1980s: Evidence Versus Artefacts,' *Economic and Political Weekly*, 26(27–28): 1673–83.

Misra, Anshuman, and Sunil Sengupta. 1990. Food Consumption in Selected Poor Households (1987–89)—A Study of Six Villages, Part I.' Working Paper, WIDER Project on Rural Poverty, Social Change and Public Policy, Santiniketan, 1990.

Murthi, Mamta, Anne-Catherine Guio and Jean Drèze. 1995. 'Mortality, Fertility and Gender Bias in India.' Discussion Paper No. 61, Development Economics Research Programme, Suntory and Toyota International Centres for Economics and Related Disciplines (STICERD), London School of Economics.

———. 1997. 'Mortality, Fertility and Gender Bias in India,' Jean Drèze and Amartya Sen (eds) *Indian Development: Selected Regional Perspectives*. Oxford and Delhi: Oxford University Press.

Nag, Moni. 1989. 'Political Awareness as a Factor in Accessibility of Health Services: A Case Study of Rural Kerala and West Bengal,' *Economic and Political Weekly*, 24(8): 417–26.

Nanda, Amulya Ratna. 1992. *Census of India 1991, Series–1, Final Population Totals*. Paper–2 of 1992, Registrar General and Census Commissioner, New Delhi.

National Council of Applied Economic Research (NCAER). 1996. *Human Development Profile of India: Inter-State and Inter-Group Differentials*. Report, NCAER, New Delhi.

National Sample Survey Organisation (NSSO). *Sarvekshana*, various, including 12(3), January–March 1989 and 14(3), January–March 1991.

Ozler, Berk, and Gaurav Datt. 1996. *A Database on Poverty and Growth in India*, Policy Research Department, The World Bank, Washington DC, January 1996.

Rawal, Vikas, and Madhura Swaminathan. 1998. 'Changing Trajectories: Agricultural Growth in West Bengal 1950–1996,' *Economic and Political Weekly*, 33(40): 2595–602.

Ray, S. Datta. 1994. 'Agricultural Growth in West Bengal,' *Economic and Political Weekly*, 29(29): 1883–84.

Rogaly, Ben. 1994. 'Rural Labour Arrangements in West Bengal, India.' D.Phil thesis, University of Oxford.

Saha, Anamitra, and Madhura Swaminathan. 1994. 'Agricultural Growth in West Bengal in the 1980s: A Disaggregation by Districts and Crops,' *Economic and Political Weekly*, 29(13): A–2 to A–11.

Sample Registration Survey. 1994. *Sample Registration Bulletin*, July 1994, 28(2).

Sen, Abhijit, and Ranja Sengupta. 1995. 'The Recent Growth in Agricultural Output in Eastern India, with Special Reference to the Case of West Bengal.' Paper presented at the Workshop on Agricultural Growth and Agrarian Structure in Contemporary West Bengal and Bangladesh, Centre for Studies in Social Sciences, Calcutta, 9–12 January 1995.

Sen, Amartya, and Sunil Sengupta. 1983. 'Malnutrition of Rural Indian Children and the Sex Bias,' *Economic and Political Weekly*, 18(19–21): 855–64.

Sengupta, Sunil. 1981. 'West Bengal Land Reforms and the Agrarian Scene,' *Economic and Political Weekly*, 16(25–26): A–69 to A–75.

Sengupta, Sunil (and associates). 1991. 'Rural Economy and Poverty in West Bengal and Public Policy Interventions.' Final Report of the WIDER Project on Rural Poverty, Social Change and Public Policy, Santiniketan, September 1991.

Sengupta, Sunil, and **Haris Gazdar.** 1997. 'Agrarian Politics and Rural Development in West Bengal,' Jean Drèze and Amartya Sen (eds) *Indian Development: Selected Regional Perspectives.* Oxford and Delhi: Oxford University Press.

Sharma, O.P., and **Robert D. Retherford.** 1987. *Recent Literacy Trends in India.* Occasional Paper No. 1 of 1987, Office of the Registrar General and Census Commissioner, New Delhi.

Singh, Inderjit. 1990. *The Great Ascent: The Rural Poor in South Asia.* London and Baltimore: John Hopkins University Press for the World Bank.

Sinha, A., D. Majumder, D. Mondal and **K. Chattopadhyay.** 1993. 'Glimpses of Literacy Profile: An Evaluation Study of the Mass Literacy Programme in Birbhum.' Report, Agro-Economic Research Centre, Visva-Bharati, Santiniketan.

Visaria, Pravin, Anil Gumber and **Leela Visaria.** 1993. 'Literacy and Primary Education in India, 1980–81 to 1991: Differentials and Determinants,' Journal of Educational Planning and Administration, 7(1), January 1993.

Webster, Neil. 1992. *Panchayati Raj and the Decentralisation of Development Planning in West Bengal: A Case Study.* Calcutta: K.P. Bagchi.

4

Slowdown in Agricultural Growth in Bangladesh: Neither a Good Description Nor a Description Good to Give

RICHARD PALMER-JONES

Few could doubt the importance of finding ways to improve the well-being of people in Bengal,[1] but it is not at all clear that public policy has been very successful in doing this. In part this is because, I suggest, of the unsatisfactory nature of much of the policy discourse in which there has been much noise and not very much light. Again, few would disagree with the proposition that public policy should depend on the evidence.[2] The evidence however is often contested, for example in two areas critical to policy in Bangladesh, namely, trends in agricultural growth and poverty. Participants in policy debates use different evidence, and interpret the

---

I am happy to acknowledge the indulgence of the editors for letting me make a late submission to this book, and to Ben Rogaly and Barbara Harriss-White for editorial comments. Quazi Shahabuddin helped with data and so did Sattar Mandal who also put me to work on the written version of these ideas. Participants in a BIDS seminar and Sam Jackson engaged me in many critical arguments, and made useful suggestions. I take the usual responsibilities for errors of fact and interpretation.

same evidence differently; it has different meanings for different people. Much of this is unavoidable, but the point reminds us that we do choose what evidence to amass and put forward according to the interests that we have, a point most eloquently put by an eminent son of the Bengali soil, Amartya Sen (1980: 355).

Debates about the course, pace and implications of agricultural growth for the well-being of the masses in Bengal are a good example of the way the choice of evidence that is put forward, and the interpretation of that evidence, affect the policy prescriptions that are drawn from it. Since the independence of Bangladesh more than a quarter of a century ago, the best prognoses from a wide range of perspectives, though optimistic about the potential, have been pessimistic about the achievements in terms of agricultural growth and poverty alleviation in the absence of land reforms. These diagnoses have shared the view that the key underlying factors behind the slow rate of growth of agriculture and continuing poverty lay in the impediments posed by the agrarian class structure. Two major features of the agrarian structure argument are: first, that agricultural growth is hindered; and second, that the pattern of agricultural growth impoverishes the majority; together, these two ideas lead to the expectations that agricultural growth will be slow and that poverty will increase, or at least not be rapidly reduced.[3]

Evidence has been put forward that the recent performance of public policy in West Bengal has exceeded these expectations, with respect to both agricultural growth and poverty alleviation, and that these successes have largely been due to policies based on the diagnosis that agrarian structure posed the main constraint to agricultural growth and poverty alleviation (Lieten, 1990; Saha and Swaminathan, 1994: A–10; Bandyopadhyaya, 1995; Dasgupta, 1995). In Bangladesh, so the argument goes, public policy has not been so centrally influenced by the agrarian structure argument; perhaps there was unexpectedly rapid agricultural growth in the 1980s, but this was much less than in West Bengal (Bose, 1995).[4] However, it often appears in policy discourse that, rather than national policy, it is the innovative and radical activities of NGOs in Bangladesh— which are strongly influenced by the agrarian structure argument (BRAC, 1979; 1980)—that have played the most important role in Bangladesh's development achievements. (For example, World Bank, 1990a: 71, specifically refers to the poverty-reducing effects of the promotion of groundwater irrigation in Bangladesh by NGOs;[5] see also World Bank, 1990b.[6]) At the Workshop on Agricultural Growth and Agrarian Structure in Contemporary West Bengal and Bangladesh held in Calcutta in January 1995, in several of the papers included in this volume, and in the recent independent reviews of the Bangladesh economy (Sobhan, 1995; 1997), the position generally adopted has been that, after quite rapid growth in agricultural production towards the end of the 1980s in Bangladesh, there has been a slowdown in growth or even stagnation of agricultural output in the 1990s.[7]

Together these observations may lead the unwary to the conclusion that the original diagnosis was largely correct—that the agrarian structure often presented as typical of eastern India[8] poses the main obstacle to development in the absence of policies specifically aimed at overcoming these constraints. Here, I present evidence that this interpretation of the trends in agricultural production in Bangladesh is mistaken. Belief in the existence of a recent downturn in the growth of agricultural production reflects on the one hand poor policy analysis, the main fault of which has been to rely excessively on short term trends, and, on other, a particular political economy model of agrarian society. I put forward an interpretation of the evidence that should lead to a clearer understanding of patterns of agricultural growth and perhaps rather more emphasis on some of the beneficial impacts of the pattern of growth and commercialisation of agriculture, and also to higher priority for policies which sustain, enhance, and ameliorate the social and environmental impacts of agricultural commercialisation, rather than resisting and contesting it. Key to this is an undersanding of the agricultural economy of Bangladesh in terms of its supply response, and the determinants of income, consumption and well-being and their distribution among the population.

This paper has four parts. Following this introduction the next part describes the trends in agricultural growth and their determinants, and the policy debates surrounding these. Having established that agricultural growth in Bangladesh did not demonstrably slow down when looked at from the longer run perspective, I discuss briefly whether there has been a deterioration in poverty and real wage rates associated with these patterns of agricultural growth. Finally, I draw some conclusions as to the implications of the trends I put forward as the facts:[9] agriculture grew; agricultural wage rates did not fall; poverty did not deteriorate and may have been alleviated. This pattern needs to be sustained and built upon. This is not the place to spell out exactly what this would entail that is different from policies that are presently being pursued, or from others proposed by those who give most weight to agrarian structure. This is the case not least because the logic of present policies is not always easy to discern, and the structuralists seldom do more than to criticise the policy stance of others and exhort us 'to confront the private sector interest groups (classes and fractions) which have persistently extracted surplus from producers through control over price formation and access to resources' (Adnan, this volume). Like much policy prescription, including that to 'get prices right', this is rather voluntaristic in that it does not spell out the political practicalities; while the structuralists do have arguments about the determination of policy practice (often attributed to aid institutions), I do not attempt such an account, but merely assert that if agricultural growth has been sustained, and has been more beneficial for the poor than other diagnoses suggest, it is likely that more could be done if policy discourses had explicitly acknowledged these 'facts' and used them in policy design and implementation, as a consequence of which beneficial agricultural growth and commercialisation might be sustained. There are good historically based arguments for thinking that the (perhaps) unintended effects of commercialisation could actually contribute to bringing about the demise of 'vested interests' and the throwing off of (semi-) feudal shackles; after all, even Marx and Engels thought that![10] Participants in policy discourse must choose how to portray and interpret events.

TRENDS IN AGRICULTURAL GROWTH IN BANGLADESH

Figures 4.1a and 4.1b show the trends in foodgrains production by fiscal year (FY) in Bangladesh from 1990 to 1997.[11] Overall, the growth of rice and foodgrains production was marginal, and up to FY 1995 aggregate rice and grains production had fallen. Only wheat production had shown any increase, but this crop is of minor significance in overall production; similarly the decline in *aus* production has little aggregate effect because it is also a relatively minor contributor to the overall position.

The current emphasis on agricultural stagnation in Bangladesh comes from similar sources and repeats arguments put forward in the mid-1980s about an apparent slowdown in agricultural growth at that time (Like-Minded Group [LMG], 1985, published 1990;[12] Osmani and Quasem, 1984, published 1990; see also Boyce, 1987, for the longer term analysis). As in the mid-1980s, in 1995 the argument pointed to a decline in the growth of the numbers of shallow tubewells (STW) as the immediate cause (and indicator) of the slowdown in growth of agricultural production (Khan, 1990; Sobhan, 1995 and 1997; Adnan, this volume). Understanding the verdict of 'agricultural slowdown' in the mid-1990s requires a review of the situation in the mid-1990s in conjunction with the similar 'panic'[13] in the mid-1980s.

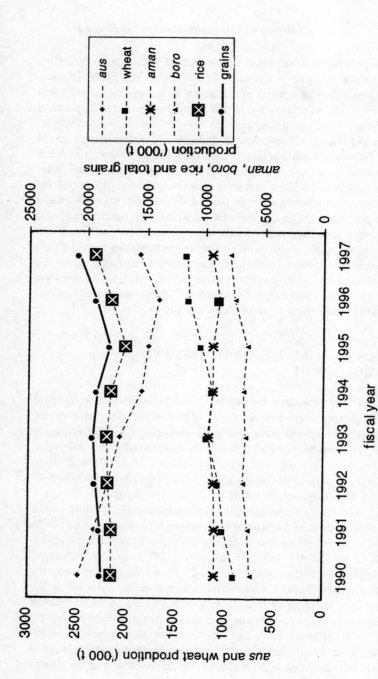

Figure 4.1a: Production of major foodgrains in Bangladesh, 1990–97

Sources: 1960–94: BBS various dates, *Yearbook of Agricultural Statistics of Bangladesh*, Dhaka, Bangladesh Bureau of Statistics; 1995–97: World Food Programme, *Bangladesh Foodgrain Digest*, Dhaka, various dates.

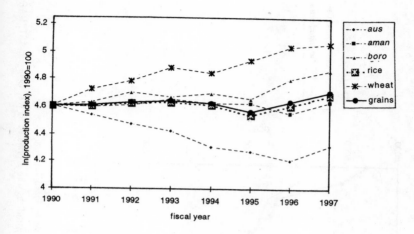

Figure 4.1b: Indexes of major foodgrains production in Bangladesh, 1990–97
Sources: As for Figure 4.1a.

Growth in the number of STWs was crucial[14] because the major source of the growth of agriculture in Bangladesh was the spread of the use of STW irrigation equipment in irrigating HYV *boro* rice. In 1995 it appeared that, following a period of rapid growth in STW numbers due to the liberalisation of import and spacing regulations in 1986 and 1987, the rise in STW numbers had slackened considerably. The figures of STW numbers contributed in a large way to the strength of the belief that a slowdown in agricultural growth was occurring. Figure 4.2 shows the trends in numbers of STW and estimates of the areas irrigated by STWs. There certainly seems to have been a slowdown[15] in the rate of increase of numbers of installed STW. However, the interpretation that this marked a change to a more long-lasting reduction in the growth rate depended on an all too ready willingness to believe in this possibility—rather than, for example, looking carefully for transient features of the agricultural economy that might account for this phenomenon, or examining the unexploited potential for further growth of STW-based irrigation.

Since the extensive margin of cultivation in what is now Bangladesh was reached by the late 1950s if not before, the growth of agriculture has had to come from intensification.[16] The two main forms of intensification are irrigation and flood control (and drainage without or with irrigation—

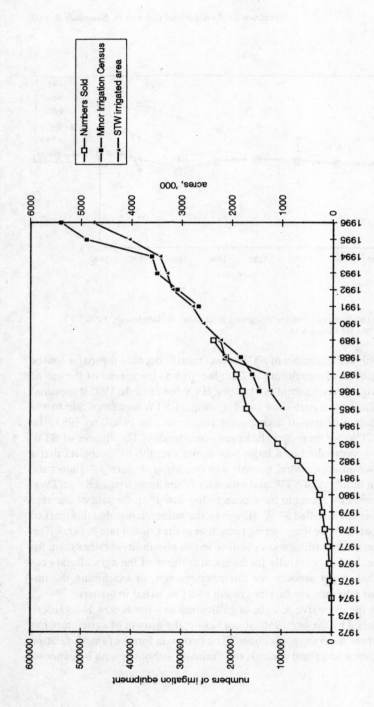

Figure 4.2: Numbers of STW and areas irrigated by STW in Bangladesh, 1972–96
Sources: 1974–89: Numbers sold are calculated from the records of various projects and institutions who distributed STWs; 1990–96: AST, DAE and NMIDP.

Flood Control and Drainage [FCD] and Flood Control, Drainage and Irrigation [FCDI] respectively). In both cases intervention in the environment facilitates the cultivation of high yielding varieties (HYV) with fertilisers and other agro-chemicals. Increases in cropping intensity could be achieved by extending cultivation in the dry *rabi* and *boro* seasons (November through May or June), or through the substitution of higher productivity crops or varieties for lower productivity crops traditionally grown in the monsoon or *rabi* seasons. Rescheduling of crops is often required along with, in some cases, a reduction in time between harvesting and desirable planting times of the next crop, putting pressure on tillage and power resources. These latter forms of intensification (FCD and FCDI) had been limited by the capital intensity and technical and other limitations of these schemes (Hughes et al., 1994), but from the mid-1960s the spread of small scale mechanically powered irrigation equipment had produced the rapid growth of agricultural production of *boro* rice and wheat. In the late 1960s low lift pumps (LLP) had been rapidly deployed under rental arrangements by the East Pakistan Agricultural Development Corporation (EPADC), making use of readily available surface water sources to spread irrigated HYV *boro* rice (see Figure 4.3 for the spread of different types of minor irrigation equipment). Deep tubewells (DTW) were also promoted under rental arrangements by EPADC (and subsequently by the Bangladesh Agricultural Development Corporation [BADC]), but spread more slowly from the late 1960s to the present. From the mid-1970s, privately owned shallow tubewells spread rapidly, irrigating HYV *boro* rice from groundwater. These were sponsored initially by the BADC, but later through credit schemes of the nationalised banks (Palmer-Jones, 1992, gives the sales of STW by different programmes in the early 1980s).

While the growth of *boro* rice production has provided much of the structural change in agriculture over the last three decades, the overall level of agricultural production in Bangladesh is still dominated by the *aman* rice crop which also has a profound effect on year-to-year fluctuations in agricultural growth and consequently on its short term trend. This rice crop occupies most of the cultivable land in the *kharif* season, and still accounts for more than half of foodgrains production. Apart from the catastrophic fall in *aman* production during the war of independence in 1971 from which it took until about 1977 for production to recover to where it might have been, and the two successive years of flood in 1987 and 1988, the *aman* crop has grown fairly steadily over the three decades up to the mid-1990s. Figure 4.4 shows this longer run trend

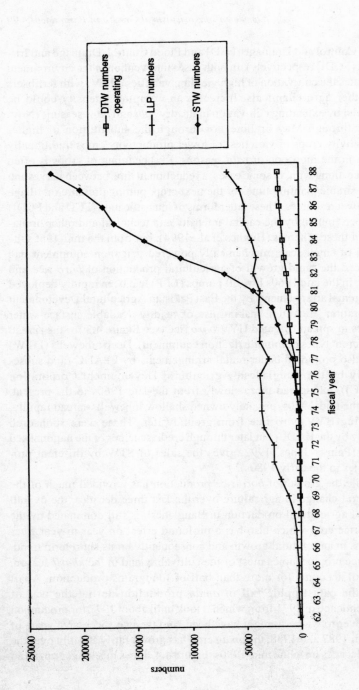

Figure 4.3: Distribution of minor irrigation in Bangladesh, 1962–88
Sources: BBS (1996); David (1994); National Minor Irrigation Programme (personal communication, 1997).

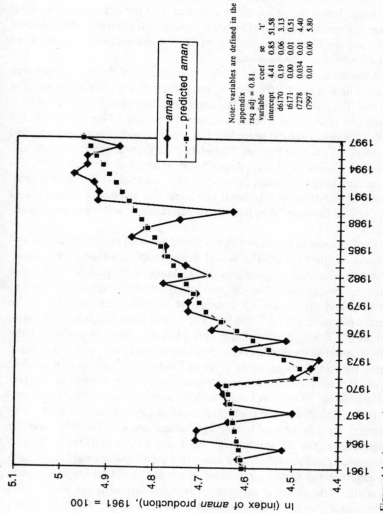

Figure 4.4: *Aman* production in Bangladesh, 1960–96
Sources: As for Figure 4.1a.

and indicates quite clearly how production in 1989–90 was above the long run trend.[17] It may be that production of *aman* will stagnate through the late 1990s, in which case there may be a break in the long run trend in the early 1990s, but it would be speculative to predict this now without other information which leads to this conclusion. My own view is that we will see the further penetration of HYVs and continuing growth of production in the *aman* season.

After the initial boom in LLP-based irrigation of the *boro* crop in the second half of the 1960s, production of *boro* rice stagnated through the 1970s. The initial rise in STW-based irrigation was associated with a rapid rise in irrigated wheat production in the late 1970s, but this tailed off in the early 1980s to be replaced by rapid growth in *boro* rice irrigated mainly by STW (see Figures 4.5 and 4.6[18] for growth trends of wheat and *boro*). A slowdown in the sales of STW occurred in the mid-1980s, and there may have been some faltering in the growth of STW irrigated areas under *boro* (although locational shifts and the increasing irrigated area per STW at this time[19] may have partly offset this slowdown in new installations).

Explanations of this slowdown differed between groups of donors and associated intellectuals. The World Bank group[20] attributed it to bureaucratic impediments (explicitly in World Bank, 1989). Another self-styled 'like-minded group' of donors was inclined to explain both the level of poverty and the slowdown in terms of impediments imposed by the agrarian structure (LMG, 1985).[21] The World Bank had been promoting increased deregulation of the agricultural sector, in particular the privatisation of agricultural inputs and output marketing, with some output price support to maintain the level and stability of agricultural product prices (mainly the rice price) so as to maintain incentives to rice production (World Bank, 1979). The Like-Minded Group was more inclined to attribute the slowdown to the unequal and fragmented farm structure preventing the adoption of lumpy agricultural inputs like STW, and the imperfect nature of agricultural input and output markets leading to monopsony and monopoly and associated inefficiencies and exploitation in a market-dominated agricultural sector. It would take too much space to spell out the extent of differences in policy diagnosis and prescription at this time, but the main lines of conflict ran between the World Bank which was pursuing an agenda of structural adjustment, and the LMG and Bangladeshi 'nationalists' who favoured a more interventionist strategy. In relation to the minor irrigation sector the World Bank was pressing for removal of subsidies on minor irrigation equipment (and

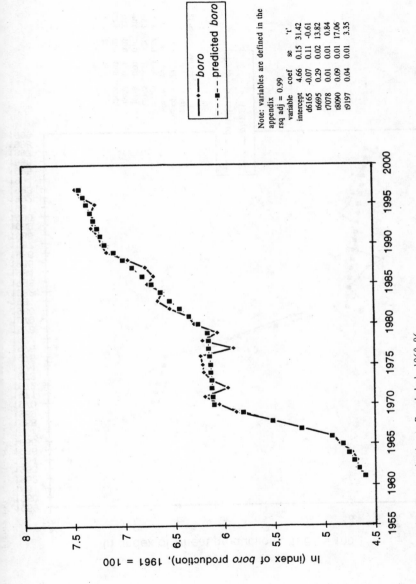

Figure 4.5: *Boro* production in Bangladesh, 1960–96
Sources: As for Figure 4.1a.

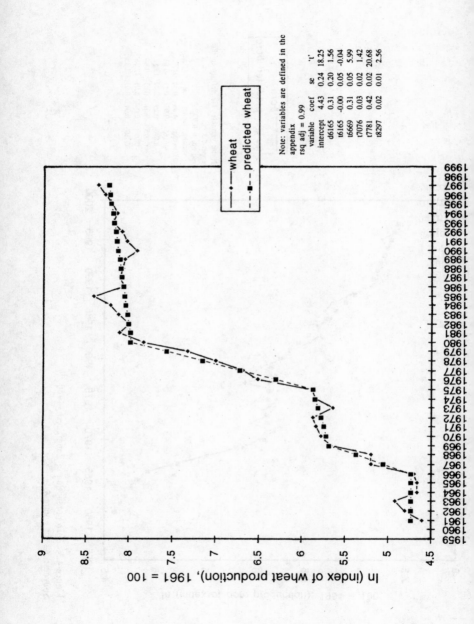

fertilisers), for private sector import and distribution of STW and LLP, and sale of DTW to cooperatives. The critics of the World Bank pressed for the reintroduction of controls on distribution and spacing of STW to be implemented by BADC, for a continuing role for BADC in renting out and maintaining DTW and LLP, and for continuing subsidies for DTW— which we may note are still being actively considered (Sobhan, 1997: 38). The history of DTW in South Asia is discussed at length in Palmer-Jones (1993b, 1994b).

This conflict of policy analysis was in large part what prompted the Government of Bangladesh in 1986 to request the United Nations Development Programme (UNDP), as a neutral agency, to institute an Agricultural Sector Review (ASR) which would authoritatively assess the current state of agriculture and recommend appropriate agricultural policies (see Norbye, 1990, for an account and assessment of the ASR).[22] As it happened the ASR concluded largely along the lines of the World Bank and supported the deregulation of agricultural input and output markets (ASR, 1989). The Government of Bangladesh (GoB) removed restrictions on imports and on the spacing of irrigation equipment, and there followed a number of years of rapid increase in installations of STW in particular. *Boro* output growth resumed immediately, with dramatic increases in 1988 and 1989, and with further but lower increases up to 1992.

Then, just as in the mid-1980s, it appeared that there was a slowdown in new installations of STW in the 1992–93 *boro* season,[23] and that there was also a slight fall in *boro* production in this year (the first reported absolute fall in *boro* production since 1986). This fall was associated with an increase in the diesel price enforced by the GoB in the aftermath of the Gulf War. Other factors were physical shortages of diesel and fertiliser, lower input levels used by farmers, and drought (David, 1994).[24] Sales of STW did not pick up in 1993–94 and the total number of installed STW hardly increased at all, according to the Census of Minor Irrigation (see Figure 4.2). The production of *boro* increased slightly (Figure 4.5). The average size of new STW may well have fallen as smaller scale STWs were becoming more popular (IIMI, 1996). These would have irrigated smaller areas per tubewell. Also, there appears to have been a secular trend towards smaller command areas per installed horsepower (ibid.), which would have had similar effects. It is not clear what accounts for this trend, but mining of groundwater is not a factor since this has not generally occurred (IIMI, 1996: 161–86). Increase in the seasonal maximum draw-down may have been a factor, but it is more likely that ensuring a margin of irrigation capacity in the face of increased (threat of) competition, together with unpredictable irrigation demand due to very variable rainfall in the crucial months of March, April and May, the unreliability of diesel supplies and risks of breakdown are among the key factors.

The trends of the relatively less significant *aus* crop are shown in Figure 4.7. As noted by many authors, the rapid growth of *boro* has been to some extent at the expense of *aus* (which declined in absolute terms especially in the 1990s [up to 1996]), and of wheat. Wheat grew rapidly towards the end of the 1970s but there has since been only a slight upward trend, rather more persistent in the 1990s.

While trends in each crop have their own periodicities and interpretations (as shown here), it is the aggregate trend that affects the overall food security position, and captures the attention of policy analysts. Aggregate rice production fell in 1987–88 and 1988–89 as a result of flood damage to the *aman* crop. In 1989–90 *aman* surpassed its previous trend level and with continuing growth of *boro* in this year, aggregate output rose to record levels. However, since then aggregate rice production has increased only slightly. It fell slightly in 1993–94 largely as a result of a fall in *aman* production not offset by a significant rise in *boro* production (which in 1993 had fallen below its 1992 peak). It appeared that aggregate production would fall again rather more dramatically in 1994–95 because *aman* and *aus* production had fallen compared to 1993, and it was anticipated that *boro* production would not rise much given pervasive stagnationist preconceptions. It has been argued that this again represented a serious slowdown in the rate of growth of agriculture, and it was expected that production would continue to stagnate (see Adnan and Shahabuddin's contributions to this volume).

It was against this background that the 1995 Calcutta Workshop was presented with the existence of a slowdown in agricultural production, the likelihood of which was all too plausible for those inclined towards explanations of trends of agricultural production in terms of the obstacles posed by agrarian structure. The reported dramatic increases in agricultural production in West Bengal (Saha and Swaminathan, 1994) too readily lent themselves to support of the 'agrarian structure as obstacle' view. This follows from the argument that the agrarian reforms in West Bengal would have removed much of the obstruction to agricultural growth posed by these impediments, although it emerged through the Workshop that there were a number of problems with this argument.[25] My own view, articulated at the Workshop, based in part on experience of the earlier

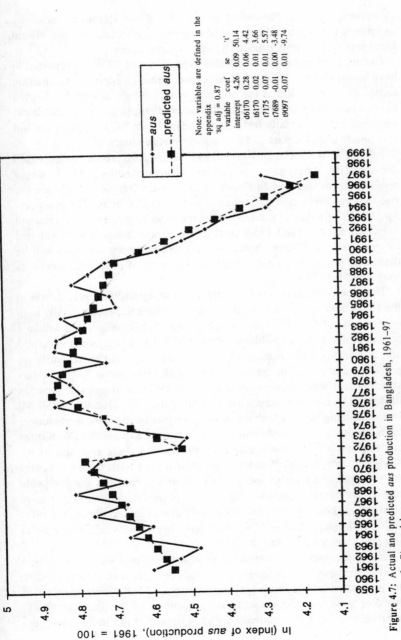

Figure 4.7: Actual and predicted *aus* production in Bangladesh, 1961–97
Sources: As for Figure 4.1a.

slowdown, was that the causes of any slowdown should be carefully examined, particularly the trends of rice prices in relation to agricultural input prices including, especially, agricultural wages.

Events seem to have born out this view. Sales of STW are reported to have boomed in the 1994–95 *boro* season, although *boro* rice production turned out to be less than the previous year. Drought adversely affected *boro* as well as *aman*, and politically-based disruption in the agricultural input markets, especially fertiliser (Sobhan, 1995: 142), led to lower and less timely use of inputs and consequent yield reductions.

Sales of STW were again high in 1995–96 and *boro* production increased substantially. Aggregate rice production rose despite a fall in *aman* production (in part due to floods in north-west Bangladesh), since the rise in *boro* production more than offset the fall in *aman*. The prospects for 1996–97 appeared excellent at the time of writing as *aman* has returned to its long run trend and it appears that *boro* production will be maintained or even increase (factors behind the present prospects for *boro* will be discussed later). Aggregate production appears to have returned to its long trend line.[26]

The gloomy predictions of a slowdown in agricultural production in the 1990s have been based on analyses which took the exceptionally high production in 1989–90 as their baseline (see Shahabuddin's and Adnan's papers in this volume; Sobhan, 1995; 1997). These prognoses were and are, it is argued here, profoundly misleading, and were based partly on prejudice and partly on the use of a particular base year. Proper understanding and better predictions come from taking a longer run view, as illustrated by the trend graphs given in this paper. The crop year 1989–90 was misleading because both *aman* and *boro* production were above longer term trends and remained so in 1990–91, mainly as a result of fortuitous circumstances.[27] Placing agricultural production (here rice and major foodgrains production)[28] in the context of longer term trends indicates the misleading nature of using 1989–90 as the baseline for analysing subsequent trends. Figure 4.8 shows how aggregate production may have been above the long run trend for a number of years in the early 1990s, but over the period up to 1997 there have been four years above trend and four years on or below it.

Of course, since growth was not accelerating, there was bound to be a short term slowdown as longer run trends reasserted themselves. Also, the prolonged and dramatic growth of irrigated HYV *boro* meant that this crop now covered a significant proportion of the area potentially suitable for it; as a consequence the growth rate (if not immediately the

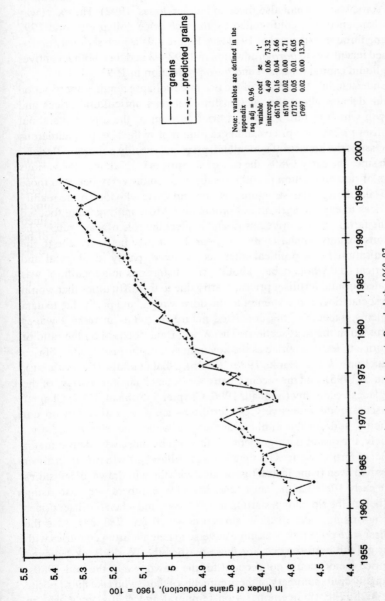

Figure 4.8: Actual and predicted total grains production in Bangladesh, 1960–97
Sources: As for Figure 4.1a.

absolute increase in production) of the *boro* crop would be expected to fall somewhat (a point also made in Palmer-Jones, 1992). The exact way this happened was confounded by the rice price collapse in mid-1992 lasting through to the end of 1994 (see Figure 4.9), which seriously undermined incentives to *boro* production in 1992–93 and then both incentives and liquid capital for *aman* and *boro* production in 1993–94.[29]

Understanding the impact of the rice price collapse requires appreciation of the debates about the relationships between agricultural prices and growth which in turn requires a brief account of the arguments about fertiliser prices, rice prices and agricultural growth.[30] Despite numerous analyses of the impacts of agricultural price and subsidy policies in Bangladesh since the early 1980s, the role of rice prices in relation to the various costs of rice production is still the subject of controversy between those who take the agrarian structure position and those who look more readily to prices to explain agricultural production. Most analysts have focused on the fertiliser–rice price ratios, and neglect the role of wage rates. The reasons for this emphasis are not clear, but it may be partly due to the organisational and political interests involved, partly ideological and historical (the whole debate about 'price distortions in agriculture' was initiated around fertiliser prices), partly due to the difficulties that would be associated with any attempt to subsidise wages, and partly due to data problems, especially the contested nature of trends in money wages (discussed in the next section). The World Bank focused on the ratio of rice price to fertiliser price as the key indicator of the profitability of rice production (World Bank, 1989, 1990b); Shahabuddin (this volume), Hossain (1995), and the recent Centre for Policy Dialogue reviews of the Bangladesh economy (Sobhan, 1996: Chapter 5; Sobhan, 1997: Chapter 7) discuss quite extensively the fertiliser–rice price ratios, but do not explicitly address the implications of the wage rate–rice price ratio. Crudely, the gist of the argument put forward by those who deny a major impact of fertiliser prices on trends in agricultural growth is that fertiliser prices had been rising fastest (associated with the withdrawal of subsidies in the early 1980s) at the same time that STW numbers were also rising fast (before the slowdown starting in 1983–84), and when fertiliser prices were not rising, sales of STW slowed down (from 1983–84). Also the fertiliser–rice price ratio was more adverse in neighbouring countries with better growth performance; for these reasons the World Bank believed that price factors could not account for the slowdown of the mid-1980s. An equally crude summary of the arguments of those who take the opposite view highlights their emphasis on temporal and spatial increases in

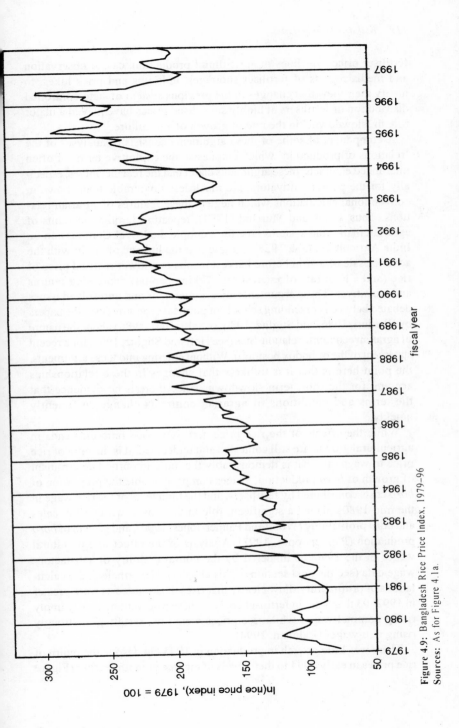

Figure 4.9: Bangladesh Rice Price Index, 1979–96
Sources: As for Figure 4.1a.

fertiliser prices, declines in agricultural production, casual observation and media reports of fertiliser shortages, hold-ups and price hikes, to justify their view that changes in the previous system of state-controlled distribution of fertilisers at highly subsidised prices have played a major role in 'slowdowns' in the rate of growth of agriculture.[31]

A component of some of these arguments consists of analyses of the 'relations of production' which emphasise the exploitive terms of often interlinked contracts; these arguments claim that the real incentives, especially for the poorer cultivators, are much less favourable than shown in conventional calculations which neglect these features of agrarian relations. Thus, Crow and Murshid (1994), repeating familiar accounts of agrarian marketing systems (see for example the Royal Commission on India Agriculture, GoI, 1928),[32] argue that the linkage of credit with the advance sale of a crop before harvest results in a lower realised price of rice (and a high rate of interest on working capital) depressing returns and incentives. These arguments raise contentious and unresolved issues debated between contending schools of political economy (viz., the papers by Bhaduri, 1986, and Stiglitz, 1986; see also Hart, 1986) about the nature of agrarian economic relationships (see Hoff and Stiglitz, 1993, for a recent statement of the orthodox position). Without entering into these arguments, the point here is that it is unlikely that changes in these relationships account for the short term slowdowns since there is no evidence that the terms and conditions of agrarian contracts change sufficiently quickly.[33]

While the effects of the rice price–fertiliser price ratio on trends in agricultural growth are still contested and undecided, it is the ratio of rice price to wage rate that is demonstrably the more important determinant of growth of *boro* production. Wages compose double the proportion of *boro* costs contributed by fertilisers, and the high level of the rice wage in the mid-1980s played a significant role in the slowdown in STW sales and *boro* profitability (Wood and Palmer-Jones, 1990), and growth of *boro* production (Palmer-Jones, 1992). Analysis of the effect of agricultural wages in the 1990s is hindered by the non-availability of satisfactory wage data (see the next section of this chapter). Nevertheless, the calculations of profitability show clearly that rises in labour costs were greater in 1992–93 than rises in fertiliser costs.[34] These calculations clearly imply rising nominal wages while rice prices were low, resulting in strongly rising rice-wages (Rahman, 1994).

There was little growth in production in 1993–94 despite the return of rice prices in early 1994 to their levels of eighteen months earlier (Figure

4.9). New installations of STWs seem to have been quite high for the 1992–93 season despite the collapse in rice prices at the harvest of *aman* in 1992. However, the low returns to *boro* production in 1993 due to the low price at harvest in May 1993, and the continuing low price of rice at the harvest of *aman* in late 1993, led to a collapse in purchases of new STWs and low expenditure on agricultural inputs for the 1993–94 *boro* crop. Cultivators would have had relatively little cash left over from the sale of the *aman* crop of 1993, when prices were still low, to invest in *boro* production in 1993–94, and informal agricultural credit would also have been scarce following a year of low rice prices. Both farmers' and credit suppliers' liquidity and profit expectations from the 1993–94 *boro* crop would have been low;[35] hence the collapse in sales of STW.

While the role played by the fall in rice prices in the slowdown in *boro* production has been noted, one line of explanation attributes the low rice price itself to agrarian structures (Adnan, this volume). According to this view (which is very similar to the classical neo-Marxist 'underconsumptionist' model—see below), while the agrarian structure may not have posed an obstacle to the development of a market for irrigation water and some spread of this technology (in contrast to what others had argued [Boyce, 1987]), the unequal distribution of benefits from this technology would result in a failure of demand by the poor, who benefited little from the new technology. This failure of effective demand for rice could account in part for the failure to maintain rice prices at incentive levels.[36] Furthermore, collaboration between the rural elite and members of state bureaucracies would undermine attempts to implement output price support policies.

However, it is clear that at least in the short run (and, I would expect, also in the longer run subject to the caveat that rice price collapses such as that which started in mid-1996 can be largely avoided), neither explanation seems very plausible in the light of actual outcomes. Neither the revival in STW sales in 1994–95 nor the renewed growth of *boro* production in 1995–96 fit well with the idea of structural or environmental obstacles (whether from limits to groundwater availability or from land degradation due to prolonged intensive mono-crop rice cultivation) imposing a limit at this time on the spread of groundwater-based irrigation. The sudden nature of the collapse in the rice price[37] does not fit very well with the underconsumptionist explanation of the apparent slowdown;[38] this explanation should manifest itself in longer term trends rather than in sudden changes, whether in rice prices or in agricultural growth. Failure to implement output price support (or rather the continuing political economy of periodic subsidised rice imports) is not necessarily related to the agrarian structure either. Other explanations, such as short term pursuit of political and/or financial advantage by the party in power, have some plausibility even though interventions may be explained and legitimised by reference to 'agrarian structure as obstacle' arguments.

After the apparent hiatus in 1992–93, growth in numbers of STW was resumed and rice prices returned (even if only for a couple of years) to their previous trend. Agricultural growth seems to be not far from its long run trend. Structural obstacles to further groundwater-based irrigated agriculture have not emerged and the failure to maintain rice prices seems to be better explained in terms of a simple rent-seeking model (Bardhan, 1984) rather than the grand theory of underconsumption.

POVERTY AND AGRICULTURAL WAGE RATES

The idea that agricultural growth, when associated with commercialisation of agriculture, is 'immiserising'[39] is very widespread and has historically deep roots.[40] There are many variants, but perhaps the most compelling metaphor is the parable of the talents (Pearse, 1980), which is often interpreted as 'the rich getting richer and the poor poorer'. Detailed discussion of this approach here would be redundant in view of its more than adequate exposition by Adnan (this volume).[41] Many explanations of varying complexity and levels of abstraction can be given for the impoverishing effects of growth and commercialisation of agriculture, but here we are concerned with the prediction rather than the process.

In addition to the test of predictions of slow growth of agricultural output, a further test of the agrarian structure model is its expectation of increasing impoverishment. If poverty has increased greatly, it would tend to be interpreted as support for this position, while if poverty has decreased then this would be evidence against it.[42] Perhaps, more realistically, one could argue the falsificationist view that the failure of poverty to rise dramatically would raise questions about the argument which holds agrarian structure as an immiserising obstacle to agricultural growth.[43]

While the focus of those who espouse such a position is on unequal agrarian relationships, and on the increasing inequality of these relationships, it is perhaps not unreasonable to investigate this using concept of poverty.[44] At several points Adnan draws attention to the worsening income distribution, implying at least increasing relative poverty. This, together with his account of the failure of rice prices to rise, implies absolute impoverishment, since it is unlikely, to say the least, given the probable high income elasticity of demand of the poor for foodgrains and rice in particular,[45] that a fall in relative rice prices due to deficiency of effective demand would occur if absolute poverty had not also increased.[46]

Some effort has been devoted to estimating the trend in poverty in Bangladesh. The main sources of information have been the various household expenditure surveys (HES) and the trend in real wage rates (especially of agricultural labourers). Neither source is entirely satisfactory—there are both conceptual and operational lacunae. While there are a variety of views on the trend in poverty towards the end of the 1980s (Rahman and Hossain, 1995; Ravallion and Sen [R&S], 1996; B. Sen, 1995; 1997), the most recent and quantitatively authoritative work on both poverty and real wages (R&S, 1996; Sen, 1995; 1997) argues that poverty increased from the mid-1980s to the early 1990s and failed to fall up to the mid-1990s, during the period of the rise in *boro* rice production based on STW groundwater irrigation. It is further argued in these works that this conclusion is supported by the trend in real wages of agricultural labourers, which, according to them, was downwards.

Table 4.1 and Figure 4.10 report head-count measures of poverty in

Table 4.1
Estimates of Population in Poverty in Bangladesh from Household Economic Surveys, 1973–74 to 1991–92

Year	BBS (FEI Method)[1]					R&S (CBN Method)[2]		
	Rural		Urban		National	Rural	Urban	National
	Graph	Direct	Graph	Direct	Direct			
1973–74	82.9		81.4					
1981–82	73.8		63.0	66.0	73.0			
1983–84	57.0	61.0	66.0	67.7	62.61	53.8	40.9	52.3
1985–86	51.0	54.7	56.0	62.6	55.65	45.9	30.8	43.9
1988–89	48.0	47.8	44.0	47.6	47.75	49.4	35.9	47.8
1991–92	50.0	47.6	46.8	46.7	47.52	52.9	33.6	49.7

Sources: BBS, 1991: 33, 1995: 31; R&S, 1996: 773.
[1]The Food Energy Intake (FEI) method is based on the observed household income level at which per capita calorie consumption reaches a standard level required to escape poverty. There are minor discrepancies between the two BBS sources, which account for differences between these figures and some of those reported by R&S (1996: 762).
[2]The Cost of Basic Needs (CBN) method is based on the cost of a normative bundle of foods required to escape poverty, valued at prices recovered from the household expenditure survey, to which an allowance for non-food consumption at the poverty line is added (R&S provide further details).
[3]Graph refers to a graphical method of interpolating the data using a regression to estimate the calorie-expenditure relationship. Direct refers to linear interpolation.

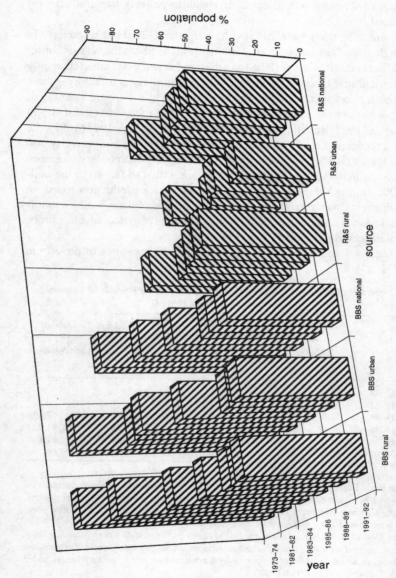

Figure 4.10: Rural and urban poverty in Bangladesh, 1973–74 to 1991–92
Sources: BBS and R&S estimates.

Bangladesh based on the household expenditure surveys by the Bangladesh Bureau of Statistics (BBS) and by R&S. The main discussion centres on trends after the 1981–82 HES, mainly because of the break in methods which render incomparable the earlier HES and the more recent surveys.

While the BBS figures show a decline in poverty throughout the 1980s and a relatively small rise in 1991–92 in rural poverty, R&S show a sharp fall in 1985–86 followed by rises in 1988–89 and 1991–92. A further difference is that BBS shows a higher incidence of poverty in urban than rural areas in 1983–84 and 1985–86, and a higher incidence in rural than in urban areas otherwise. R&S have more rural than urban poverty in all the years for which they present results.

This is not the place for a detailed consideration of all the arguments put forward by R&S in their various papers; suffice it to say that there are substantive problems with their preferred method of estimating poverty from the different HES, and their arguments about the trends in real wages are also misleading. Details of these weaknesses can be obtained from the author of this chapter.

Table 4.2 shows calculations of the rice wage, and compares these with the food and non-food Consumer Price Indices (CPIs) used by R&S; the rice wage and the food wage were clearly higher at the end of the 1980s than at the beginning, even though they were at their 1980s peak in 1985–86 (Figure 4.11). It is the rise of the non-food CPI that drives the real wage calculated by R&S down from its peak in the mid-1980s. Even so it remains the case that the real wages as calculated by R&S were higher at the beginning of the 1990s than at the beginning of the 1980s (Figure 4.12).

The basis of my criticism is not controversial; as is well known, and as R&S (and also Wodon, 1997)[47] acknowledge, the method they use is based on a Laspeyres index which exaggerates the inflation of a price index when it is rising and underestimates the reduction when it is falling. While the FEI method may have significant disadvantages for comparing poverty over time or space, the CBN method also has disadvantages which should be allowed for. The most obvious way to do this in the present case is to use the coarse rice price as the yardstick for inflation of the poverty line, mainly because rice is such an important component of consumption of the poor. It might be realistic to add some inflation above that of the rice price to allow for faster inflation of non-staple commodities, not offset by quality or other factors, and to perform a sensitivity analysis of poverty comparisons.

Table 4.2
Rice Wages of Agricultural Labourers and Food and Non-Food Rural Cost of Living Indices, Bangladesh, 1981-95

FY	Coarse Rice Price (tk/kg)	Agricultural Wage (BBS) (tk/day)	Rice Wage (kg/day)	Period of Wage Data	Rural Food Poverty Line (Food CPI)	Rural Non-Food CPI	Index of Ratio of Non-Food to Food CPI	Food Wage Index[1]	Real Wage Index[2]
1981	4.84	13.52	2.79	July 1980–June 1981					
1982	5.61	15.47	2.76	—					
1983	6.23	17.06	2.74	—					
1984	6.7	19.38	2.89	—	100	100	100	100	100
1985	7.5	24.69	3.29	—					
1986	7.25	29.71	4.10	—	117.47	122.44	104.23	130.50	129.09
1987	8.54	32.6	3.82	—					
1988	9.19	30.3	3.30	—					
1989	9.36	32.7	3.49	—	137.80	150.40	109.14	122.44	119.61
1990	9.34	33.6	3.60	July 1989–Dec 1989					
1991	10.096	42.2	4.18	Dec 1990–June 1991					
1992	10.721	41.5	3.87	July 1991–Dec 1991	164.46	203.54	123.76	130.21	122.65
1993	9.195			na					
1994	9.597	39.5	4.12	July 1993–June 1994					
1995	12.273	40.8	3.32	July–Aug, Oct–Dec 1994					

Sources: Coarse rice price: 1981-84 United States Agency for International Development; 1985-95 World Food Programme Open Market Price. Wages: BBS, Agricultural Year Books, up to 1994. Rural food CPI: R&S, 1996: 770. Rural non-food CPI: R&S, 1996: 770.

[1](index of agricultural wage)/(index of rural food poverty line) (from R&S)

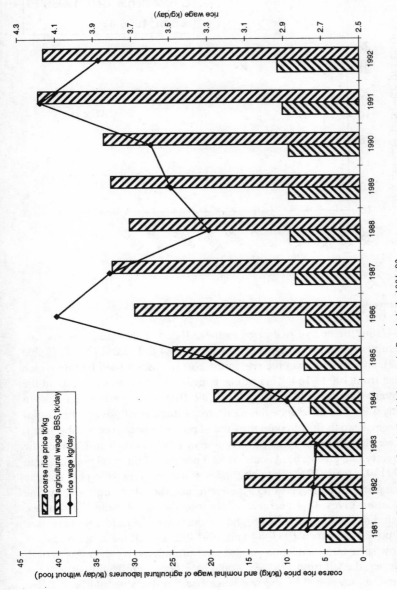

Figure 4.11: Rice wages of agricultural labourers in Bangladesh, 1981–92
Sources: As for Figure 4.1a.

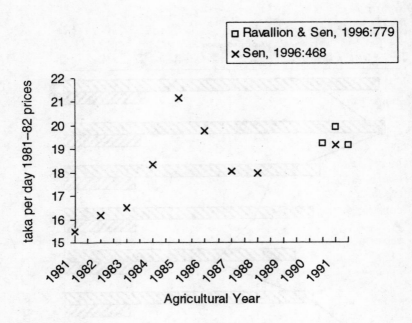

Figure 4.12: Real wage rate of agricultural labourers in Bangladesh, 1981-91

Using the coarse rice price as the inflator of the rural poverty line results in an index rising from 100 in 1983–84 to 160 in 1991–92 (Table 4.2), slightly less than the rise in the cost of R&S's food bundle which rose from 100 to 164. This can be compared with the rise to 203 in the poverty line actually used by R&S; the BBS rural poverty line rose to only 143. The difference these different estimates of poverty line inflation make to the head-count measure of poverty are of course substantial; if we start from the R&S poverty line in 1983–84 their inflation of the poverty line gives a head count of 52.3 per cent of the rural population in 1991–92; their CBN food price index inflation gives 46.4 per cent, the rice price inflation gives 43.7 per cent, and the BBS rural poverty line inflation gives 34.4 per cent of the population. All these estimates, including that chosen by R&S, show reduction in the head-count measure of poverty between 1983–84 and 1991–92, and all the other measures show substantial reductions in this measure of poverty over this period. Further, the fall from 1981–82 would have been significantly greater because, as suggested by the real wage rise,[48] poverty should have fallen from 1981–82 to 1983–84. Furthermore, we should not be over-impressed by the sharp reduction in poverty in the mid-1980s, which was in part due to transitory factors. The mid-1980s were characterised by rapid (for Bangladesh) growth of both GDP and agriculture, with improved seasonality of employment, a one-off rural construction programme associated with the Upazila decentralisation programme from 1983 to 1985, and expansion of the Food For Works and Vulnerable Groups Feeding programmes (as well as expansion of the poverty-oriented NGO and Grameen Bank programmes for which Bangladesh has become famous). The Upazila decentralisation programme initiated by the Ershad regime in 1982 was associated with an approximately 25 per cent increase in total allocations to decentralised institutions in 1983–84 and 1984–85, especially the Upazila Development Grant, which was mainly spent on construction of Upazila headquarters and housing (ASR, 1989, 5: 68). Wood and Palmer-Jones (1990: 156), with very detailed but localised data, show rising rural money wages and employment from mid-1983 to the end of 1985. I suggest on the basis of the evidence presented here, and also my casual observations in rural Bangladesh over this entire period (see also Rahman and Hossain, 1995), that a more balanced assessment of trends in poverty since the late 1970s would be that there has been a steady decline, with fluctuations; there was an above trend period in the mid-1980s due to the factors mentioned above, and a below trend period in the mid-1980s associated with the two years of heavy flooding in 1987 and 1988. The generally downward trend in poverty has been largely due to the growth of agriculture, which brought substantial benefits to many of the poor through its employment-intensive nature. The steady spread of Government- and NGO-supported poverty alleviation programmes played a lesser role. Note that I am not arguing here that inequality has decreased, or that social relations have not deteriorated. The evidence on rising income and expenditure inequality from the HES is not being contested, but this is not conclusive for absolute poverty, which is what I am concerned with. However, it should be noted that my views would not be so firm were it not that the conclusion that poverty has been almost certainly decreasing coincides with my casual observations in rural Bangladesh since the early 1980s.

CONCLUSION

This paper argues that the policy discourses which hold that agricultural production in Bangladesh experienced a significant slowdown or even stagnation in the mid-1990s are erroneously based on the interpretation of short term trends in foodgrain production using the baseline of the crop year 1989–90, together with a conjunctural situation brought about by the rice price collapse of mid-1992. Placing the figures of production in their longer term context, and examining the short term fluctuations of each particular crop, especially the quantitatively more important *aman* and *boro* crops, shows that so far there is no good reason to think that growth along the longer run trends of aggregate grains production is unlikely to be sustained in the immediate future. On the basis of general knowledge of rural Bangladesh and its agro-ecology, as well as analysis of past trends, the predictions that there are further opportunities to expand the area irrigated by STW, and that there may be efficiency gains as alternative crop technologies and cropping patterns become better understood by cultivators and others, can be entertained. This will keep irrigated *boro* production rising, providing price incentives do not suffer many shocks such as those of mid-1992 (which were repeated in mid-1996). *Aman* production may also continue to rise as there appear to be considerable further opportunities to adopt HYV *aman*, especially as more HYVs adapted to the particular conditions of Bangladesh become available and their use spreads. The further growth of irrigation may promote intensification of *aman*, and perhaps other crops as well (viz., the apparent turnaround in the decline in *aus* production in 1996), as this development provides increasing levels of well-being, and some insurance against natural hazards in other crop seasons.

While the majority of rural Bangladeshis are still very poor, it is important not to conclude that the situation has deteriorated if it has not, or indeed that it has improved if it has not. Rural poverty will decline in relative and perhaps absolute numbers provided growth of the type described is maintained, especially if Bangladesh also has success in industrial growth and population does not grow rapidly.[49] A type 2 error (accepting as true a false conclusion) could have unfortunate results, if for example it led to the rejection of one growth strategy that was relatively successful, or to accepting another that did not adopt crucial aspects of a 'successful' strategy, as happened briefly, I believe, in the mid-1980s, when civil servants in the Ministry of Agriculture obstructed the allocation of licenses to private traders for the import and distribution of STW equipment (Palmer-Jones, 1992).

Perhaps it is the case that rural poverty has decreased and real wages have not fallen or may even have risen during a period of agricultural growth and commercialisation, led by the spread of irrigation services selling STWs mainly used to irrigate HYV rice, also for sale. I believe that the balance of argument supports this view. The policy implications cannot of course be read off from evidence of this type. It is always possible that greater advances in human well-being could have been achieved in some other way, whether by some modification of the strategy which prevailed during this period, or by a completely different strategy. But such a discussion is not relevant here; what is relevant is that a powerful policy model—agrarian structure as obstacle to agricultural growth, and as conducive to impoverishment—has twice prematurely and mistakenly heralded the arrival of a slowdown in agricultural production and increasing impoverishment, when in fact neither has occurred. I am not sure what this implies for the agrarian structuralists, since I do not propound that model. The main point of this paper is that this erroneous model should be, at least, largely rethought. Loose empirical work can do serious damage.

To return to the sub-title of this paper: Sen (1980) points out that one might well wish to choose the description to give according to the circumstances and the purposes one has, so that a good (accurate) description may not always be a good (in terms of achieving one's objectives) description to give. Unfortunately, he does not provide any guidance on how to choose the description to give. We make choices about the evidence and how to interpret and present it, and these choices often have consequences for public policy (and, perhaps, for the well-being of people). Interventions in policy debates obviously create 'noise' which distracts attention from other tasks, and they have influence on the course of events. There are also silences in respect of factors they do not address. The perception of a crisis was and is supported by various preconceptions and theoretical perspectives—the agrarian structure model—and finds support in the unsound interpretation of the figures on poverty and the wage rates of agricultural labourers (even though these are not always attached to the agrarian structure argument). The apparent slowdown in agricultural production in the early to mid-1990s coincided with the election to power of the Bangladesh National Party in Bangladesh, and the discourse may in part have been motivated by political considerations. Nevertheless, the policy implications of these positions may actively undermine the processes of growth that have in fact been observed (as noted above), and that, I argue, have played a large role in preventing deterioration in the position of the poor, or even brought many of them significant benefit. Those who hold contrary positions, who put them forward in the mid-1980s and were found wrong, and who have again put them forward and, I have suggested, been found wrong, should, to the extent that they have individual agency, perhaps consider (changing) their position.

APPENDIX

Variables used in Time Trend Regressions[1]

Time trend variables are similar to those in Palmer-Jones, 1992. The following conventions are used:

Dependent Variables

Dependent Variables are the natural logarithm of the index of production with 1960–61 as the base year

(i.e., $y_i = \ln \{100 \cdot$ (production of crop y in year i)/(production of crop y in 1960–61)$\}$

e.g., $\ln$ (index of grains production)$i = \ln \{100 \cdot (aus_i + aman_i + wheat_i + boro_i)/(aus_{61} + aman_{61} + wheat_{61} + boro_{61})\}$

Independent Variables

d6070 takes value 1 in FY 1960–70 and 0 otherwise
t6070 takes the value 0 before FY 1960, 1 in FY 1960, 2 in 1961 ... 11 in 1970 and in all subsequent years.
Other variables are defined in a similar way.

[1] Regressions were estimated in Microsoft Excel Data Analysis module by OLS.

NOTES

1. See Ravallion (1990) for an assessment of the size of the problem.
2. Even Chairman Mao apparently came to believe, perhaps rather belatedly, in the importance of information for policy, even if he did, again slightly belatedly, suggest the need to rely on democracy rather than the cadres for it (Sen, 1997: 17–8).
3. Classical statements are those of Boyce (1987), Jannuzi and Peach (1980), de Vylder and Asplund (1982); see various chapters in Robinson and Griffen (1974) for arguments for immediate land reform or even collectivisation as necessary prerequisites for agricultural growth and poverty reduction. Section 2.3 in Like-Minded Group (1985) (published 1990) contains a thumbnail sketch of the agrarian structure argument. The following is a far from exhaustive list of intellectuals who have been active participants,

besides those mentioned already, and whose works contain large parts of the agrarian structure argument: van Schendel (1981), Adnan (1985; this volume), Rahman (1986), Jansen (1986) (see also Jansen in Norbye, 1990, for unchanged views), Hartmann and Boyce (1983), Hossain (1987), Crow and Murshid (1994).

4. 'I think it can be fairly said that there has been a significant acceleration of the agricultural growth rate in West Bengal. What I know from my colleagues in Bangladesh leads me to believe that any improvement in agricultural performance there has been quite modest by comparison' (Bose, 1995: 14).

5. See also Kahnert and Levine (1993); Wood and Palmer-Jones (1990) give a more qualified view of the relationship between NGO-promoted groundwater developments and those of the private sector.

6. There is a huge literature reporting successful 'evaluation' of the activities of NGOs in Bangladesh (including the Grameen Bank), most of which mentions possibilities of large scale implementation: e.g., with reference to Bangladesh, Fugelsang and Chandler (1988: 192–4); Lovell (1992: 167–9); Kramsjo and Wood (1992). While much of this literature deals only with benefits to those directly involved with the NGOs and the desirability of replication or scaling up for others presumably to be directly involved with them, without dealing with the macro-level social and economic impacts, there can be little doubt that recent crazes, e.g., for micro-credit, are based on the assumption that these institutional innovations do play a major role in overall improvements in well-being in Bangladesh. Of course, such assumptions sit somewhat uneasily with discourses which see little improvement in Bangladesh itself.

7. The World Bank, which was noticeably bullish in the early 1990s about agricultural growth (World Bank, 1992), notes a slowdown in agricultural production in 1995 and an apparent recovery in 1996 (World Bank, 1996). It does not debate the contention that there has been a more sustained slowdown, and this omission implicitly concedes the field in Bangladesh to those who hold to the stagnationist view.

8. Although there is a widely acknowledged 'eastern India problem' (Reserve Bank of India, 1984; Stokes, 1978), there is great variety in the agrarian structure in Bengal, as Bose (1986) most strongly makes us aware (see also Wood, 1981), which we must assume extends throughout eastern India; there is a social variety which is no less than the agro-ecological and climatic variety within the region. Hence it is unlikely that there is just one problem. Nevertheless there is some utility in discussing the similarities as well as the particularities within the region. Three broad types of explanation of the 'eastern India problem' have been given. They differ in the factors given central importance which are, respectively, related to characteristics of the agroecology, agrarian structure, and policy in the region. Very crudely, these accounts can be paraphrased as 'floods, feudals, or Fabians', due to the central role they ascribe to floods (and poor drainage and winters not cold or long enough for wheat), the semi-feudal agrarian structure, and (the failure of) social-democratic (Fabian) policies. My work falls in the last category, in that I attribute quite a larger part of the (recent) slow growth of agriculture and continuing poverty in the region to inappropriate 'Fabian' policies.

9. As noted again later, there are many objections to any 'facts', and anyway these facts are not 'found' but 'produced', in part by the very theories they are supposed to be 'testing'. I do not discuss these issues here, in the interest of brevity.

10. Or so I interpret the famous passage: 'in one word, the feudal relations of property became no longer compatible with the already developed productive forces; they became so many fetters They had to be burst asunder; they were burst asunder' (Marx and Engels, 1973: 72, original 1848); see Hirschman (1982) for the contradictory

relation between the 'douce commerce' and 'feudal shackles' theses; see also Platteau (1994a and b).

11. There are broadly four major foodgrains crops—*aus, aman* and *boro* rice and wheat—and three agricultural seasons—*kharif* 1, *kharif* 2 and *rabi*. I would now add a fourth season by dividing the *rabi* season into two—*rabi* proper and *boro*. *Aus* rice is grown in *kharif* 1, *aman* rice in *kharif* 2 and *boro* rice and wheat in the *rabi* season; it is conventional for the crops to be recorded in the period in which they are harvested, so the FY 1990, for example, which runs from July 1989 to June 1990, includes the *kharif* 1 and *kharif* 2 crops of 1989 and the *rabi* crops of 1990. Figures for production used in this paper come from a variety of sources, mainly the Bangladesh Bureau of Statistics, *Statistical Yearbook of Bangladesh, Yearbook of Agricultural Statistics of Bangladesh*, and the World Food Programme, *Bangladesh Foodgrain Digest*. Agricultural statistics in Bangladesh, are not produced by formal statistical procedures and are subject to considerable margins of error. There are often inconsistencies in the same statistics between different publications within the BBS and other organisations.

12. To maintain the appropriate sense of timing, the original dates of publications which appeared in print somewhat later will be used. The fact that these documents were published with such delays is testimony to their enduring significance to discourse.

13. The term panic refers to a 'folly of collective action' (Goode and Ben-Yahuda, 1994). It is used to describe the way in which an issue becomes publicised and politicised, and invokes a massive, disproportionate and often inappropriate response from the state and other institutions, which may have long-lasting effects. I have described briefly such a panic in relation to the apprehension that the groundwater table was being rapidly depleted due to unregulated expansion of private sector exploitation of groundwater following the failure of the end-of-monsoon rains in north-west Bangladesh in 1982 (Palmer-Jones, 1992: A-131; see Palmer-Jones, 1989: 14–23, for more details). While the agrarian structure model clearly derives from a long enduring political ideology, it can be heavily implicated in more transient panics.

14. Numbers of STW are used as the key indicator because it is generally accepted that growth in STW numbers has been the main contributor to growth in the irrigated area and increase in *boro* production. Up to the early 1980s it was reasonable to cumulate annual new STW installations to measure total installed STW, but by the late 1980s this was no longer reasonable as many of the earlier STW would have been fully broken down, or transferred permanently to other uses. Biannual surveys of minor irrigation were initiated in 1985. These report the total numbers of minor irrigation installations throughout the country. These censuses have been conducted by a variety of organisations with a number of different formats. The Department of Agricultural Extension Block Supervisors have been the main enumerators. In some years validation surveys have been conducted. From 1992–93 these censuses have been conducted annually, and presently a database of all the installed minor irrigation is kept by the Department of Agricultural Extension (DAE). A comparison of the numbers found in a small area of south-east Ghatail corresponds quite well with those found by a map-based local census. There is no simple relationship between STW numbers and irrigated areas. There is uncertainty about how many of the STWs sold were actually in operation during the 1980s. More recently the estimates of area per STW have either been based on standard figures, or on sample survey estimates with quite high standard errors. Also, there is limited recognition of the variable relationship between characteristics of STWs such as horsepower of engine, power source—diesel or electric—

or geographical location and its agro-ecological features, and irrigated area per TW (see Wood and Palmer-Jones, 1990).

15. One feature of the data series might warrant caution in interpreting the slowdown; prior to 1992–93 the area per TW was estimated from the main census reports, while from this year estimates from the validation survey were used. The validation surveys found lower irrigated areas per TW and hence in 1992–93 there may have been a once-and-for-all downward adjustment in areas irrigated. In 1994–95 there were further changes in the census and survey instruments; there appears to have been some stability in 1995–96 and 1996–97.

16. There was much concern about post-harvest losses in the late 1980s; it seems likely that the gains from reducing these losses are limited (Greeley, 1982).

17. Definitions of variables and methods of estimating the time trends are given in the appendix to this chapter.

18. Figure 4.5 shows the longer run trends of *boro* rice production with trend kinks in 1970, 1979 and 1991; Figure 4.6 shows wheat production with trend kinks in 1966, 1969, and 1981; see Boyce, 1987: 267–71, for explanations of kinked exponential models.

19. In the early years STW entrepreneurs (and those to whom they might sell irrigation services) may have been unduly cautious in their determination of the appropriate command area per STW of a given capacity, due to inexperience with the technology. Later they may have expanded the command areas, but pressure of competition and ageing of their equipment might then have led to contractions. Appropriate command areas also depend on soils and other local conditions (Wood and Palmer-Jones, 1990: Chapters 4–7).

20. Including, most importantly, USAID.

21. Their argument holds dependence on foreign assistance from the major multilateral aid agencies as the prime mover in the impoverishment of the masses in Bangladesh; but this works through the policies imposed by, typically, the World Bank, which are expected to have this effect (Adnan, this volume).

22. Norbye's gloss is as follows: 'There were other reasons as well [besides the "slowdown"]: different donor agencies exerted pressures on the Government in order to make it change its policies, and these pressures some times were inconsistent with each other' (Norbye, 1990: 110). The author of this paper was a member of the team which formulated the ASR in 1986, and a member of the ASR during 1987 and 1988.

23. But recall the caveat that the methodology of the Census of Minor Irrigation changed in this year.

24. Understanding the behaviour of farmers and water sellers in evolving economic circumstances is hindered by the lack of suitable studies of the water selling industry. Empirical work on tubewells generally takes the form of 'cost of production' studies which relate to only one year and do not throw much light on responses over time. Insight is further hindered by the failure to see beyond the 'waterlord' model or monopolistic model of groundwater-based irrigation water markets presented most forcefully by Shah (1993), and widely used in Bangladesh. I have criticised Shah's work in detail (Palmer-Jones, 1994c). Some comments are made in the present paper on the logic of farmers and water services sellers to events in the last few years; see also Palmer-Jones (1997).

25. Whatever qualifications were added subsequently (Rogaly et al., 1995), at the Workshop, Dasgupta (1995) presented the most straightforward case for the association between the apparently exceptional rate of growth of agriculture in West Bengal and the agrarian reforms of the Left Front government; Bose (1995; this volume) and

Bandyopadhyay (1995) have also attributed the reforms some role. Sen and Sengupta (1995: 2) find that 'there appears to be a strong inter-district difference in output growth rates and the impact of Operation Barga.' Rogaly et al. (1995) accept that agricultural growth in the 1980s in West Bengal experienced 'a lead over the respectable performance of other states', and concede that '[w]hatever the actual rate of growth in agricultural production in West Bengal, the Left Front government has gained credit from the very high levels implied by the official data' (1995: 1863). One feature to emerge strongly from the Workshop was the unsound basis of the official statistics on agricultural production in West Bengal, and the possibility that the growth rate was inflated. Boyce (1987: 56–67) discusses the biases that could have occurred in West Bengal's agricultural statistics up to 1980, and shows how plausible revisions can significantly alter the size and even the sign of estimated growth trends. An alternative explanation of some advantage in agricultural growth for West Bengal compared to Bangladesh that might be pursued in further research, might be found in their different agro-ecologies. Quite significant areas of West Bengal lend themselves to HYV cultivation in both *aman* and *boro* seasons. The spread of HYV in Bangladesh has been most significant in the *boro* season since there is still a lack of varieties suitable for the more cloudy and flood prone agro-ecology of the *aman* season in eastern Bengal.

26. At the time this was drafted (January 1997) only preliminary estimates of the 1996–97 crops in Bangladesh were available; at the time of revision (July 1997) the expectations in this (unaltered) paragraph have apparently been confirmed by the official acceptance that foodgrains production in Bangladesh in 1996–97 exceeded 20 million tonnes.

27. The very high level of aggregate foodgrains production in 1989–90 results from the coincidence of high *aman* production following two years of floods with high *boro* production associated with the continuing effects of liberalisation of groundwater development and favourable groundwater and economic conditions. The exceptionally high *aman* production following floods may owe something to the fertility effects of flooding, although other factors may well have been involved as well. In general the factors contributing to the growth of *aman* production are not well understood.

28. We have noted earlier the lack of an objective basis for agricultural production statistics in Bangladesh. Foodgrains production is the headline item and much attention is paid to it. Other crops are generally considered minor and less attention is devoted to their figures, and they may be treated as something of a residual after foodgrains production has been estimated. This may lead to considerable bias, as, for example, when *rabi* crops were thought to have fallen in the late 1970s due to the growth of wheat and *boro* rice production. The agricultural census of 1983–84 showed significantly more minor *rabi* crops compared to the BBS series (Boyce, 1987). We await the outcome of the 1997 agricultural census.

29. This account is informed by conversations with farmers in Tangail district in early 1997, when there had been a rice price collapse some six months earlier yielding relatively low returns to *aman* cultivation despite good yields, and expectations were for a low harvest price of *boro*. Much has been written about rural credit markets in Bangladesh, most recently by Crow and Murshid (1994). Unfortunately, for all its documentary interest in showing continuity of forms of market structures in the region over a long period of time, this work produces and reads 'the facts' through a model that may not be very helpful (and consistently fails to account for the full transaction costs of lenders). In a situation of underdeveloped agricultural credit and insurance markets and institutions, farmers' cash and credit situation is very important in determining their ability to invest in agricultural production, and depends crucially

on the outcome of their recent crop (and other) enterprises. Creditworthiness and credit availability are both likely to be favourably affected by high yields and good price outcomes. High yields will tend to ensure subsistence and hence underwrite risk taking; profitability will increase both supply and demand for production credit.

30. These debates have been long, enduring and rancorous at times, and have major organisational and political aspects; only an outline of their character can be given here. Differences over the role of fertiliser subsidies and privatisation of fertiliser distribution have been far more definitive in the conflict between the World Bank and the LMG and their respective associates over agricultural policy analysis than the debate over minor irrigation (viz., LMG, 1985; World Bank, 1990b). Some key references in the fertiliser debate include: Osmani and Quasem, 1984; various articles in Abdullah, 1985; Hossain, 1987; ASR, 1989; LMG, 1985; World Bank, 1990b, 1992; Sobhan, 1995; 1997). While the academic debate in Bangladesh is largely resolved in favour of those who argue that the withdrawal of subsidies and privatisation of fertiliser distribution has not been crucial in the slowing down of agricultural growth (Abu Abdullah, personal communication, 1997), the political debate continues and fertiliser prices and subsidies, imports, exports and in-country distribution remain subject to major political intervention (most recently in the control exerted by the Awami League government over distribution of fertiliser in January 1997), and the issue continues to absorb large amounts of policy analysts' time. The organisational interests of the Bangladesh Agricultural Development Corporation, which has been largely responsible for the distribution of subsidised fertiliser, and more recently the nationalised fertiliser production corporations, have been important in lobbying for interventions in the fertiliser business.

31. They could also draw on empirical analyses which show significant fertiliser price responsiveness (see Hossain, 1985; 1995; Osmani, 1985, among many others). However, it should be noted that these are mainly partial-equilibrium analyses and ignore the implications of rationing that is associated with subsidies, and the fiscal and general equilibrium implications of input subsidies (de Janvry and Sadoulet, 1987; Sadoulet and de Janvry, 1992).

32. By drawing attention to this continuity of form of agrarian marketing systems one should not conclude that their functioning has not changed (Bardhan, 1980).

33. The main point at issue seems to be the question of power in these transactions. Orthodox economists often model them by assuming that the less powerful parties to the transaction are driven to their reservation price, in a way that reflects the market power of the more powerful party. It is not clear to what extent other forms of power are justified, although there are no doubt cases where trickery, force, and access to the organs of the state result in further domination of the less powerful, but there is a question of whether this is a good (quantitatively accurate) description of local society in much of Bangladesh. Even Crow and Murshid admit to spatial variations in the forms, terms and meanings of agrarian transactions.

34. Sobhan, 1995: 164 and 1997: 230. He shows that the dramatic fall in profitability of *boro* rice production in 1992–93 and 1993–94, estimated from crop budget data, is more due to rising labour costs than to fertiliser costs. Labour costs for *boro* rose in 1992–93 and 1993–94 by 496 and 253 tk/ha while fertiliser costs rose by 62 and 2 tk/ha respectively. While unit prices cannot be recovered from these data it is unlikely that increased volumes account for these increases in costs. The role of the rise in labour costs passes without comment in these texts.

35. As noted before, and also later in this chapter, it is not clear what was happening to agricultural wages, which are crucial to calculations of net returns from *boro* production,

at this time; but the labour cost calculations reported here show their level rising again in 1993–94, although not quite so much as in 1992–93.
36. Rising inequality is reported in the various household expenditure surveys; but whether poverty was increasing is not so clear.
37. Sobhan (1995: 136) attributes the rice price collapse to 'off-loading of excess private stocks built up over the previous three years of large harvests, high wheat imports, and declining real prices'. To this account should be added the facts that the public foodgrain stocks were at their highest recent level and that the *aman* crop was expected to be excellent, and indeed the 1992–93 *aman* harvest was the largest ever. The government would be no more able to implement effective price support than previously, so that rice price collapsed beginning around the time that expectations of a good *aman* crop were assured—towards the end of August 1992.
38. This explanation is based on the Marxist theory of the crises to which capitalism is liable; such models were very prominent in the 1970s as they played a major role in theories of underdevelopment which drew on Lenin's theory of imperialism. Adnan (this volume) presents such an account of the slowdown in agricultural production. The general decline in profitability of *boro* rice production to the extent that it occurred is not conclusive, since it is to be expected as a technology matures; farmers will be willing to invest more in such a maturing technology because their knowledge of likely outcomes (and their ability to manage unfavourable circumstances) will improve.
39. Certainly relative poverty and inequality are expected to increase, and in many versions absolute poverty also (although we may note the view of another participant in the Calcutta Workshop that 'respect and social prestige were more important than material factors such as food' [Beck, 1994: 194]). This suggests that a situation of deepening but shared poverty might be described as having less absolute and relative poverty.
40. We can sketch a line of descent from the Schoolmen of Greece and their hostility to the commercial economy (Parry and Bloch, 1989: 2; Parry, 1989: 82–5), through Marx's views on the development of capitalism in England in Part 8 of Volume 1 of Capital, via European Marxists, especially Kautsky and Lenin, involved in political writings in late 19th and early 20th century Europe, English socialists and historians such as R.H. Tawney commenting on the agrarian disturbances of the 16th century, through Maurice Dobb in Cambridge writing on the transition from feudalism to capitalism, to the modern generation of Indian and foreign intellectuals who have contributed to the debates about the mode of production in Indian agriculture, the nature and effects of the Green Revolution, and so on. Among other examples of this line, we note Daniel Thorner's important role in popularising similar ideas in India through the 1950s and 1960s (Harriss, 1992). Amit Bhaduri's influential model of semi-feudalism has passages that are almost identical to passages in Dobb's 'Studies in the Development of Capitalism' (Dobb, 1946; Bhaduri, 1973). In 1968 Bhaduri already knew that the Green Revolution would polarise and impoverish the rural masses (Bhaduri, 1968).
41. Adnan holds the demand constraint/underconsumption argument leading to low rice price as the key constraint and the underlying cause of the slowdown; he also puts forward a classic account of the 'agrarian structure as obstacle' arguments.
42. It is always hard to hit a target especially when it moves, and there is a danger of constructing straw bodies to criticise. The reader must decide whether the positions I am resisting did predict that STW and associated irrigation service markets would not spread as they have (which Adnan, this volume, denies), and whether they entail immiseration which, I will argue, has not occurred. And the reader must also decide what the implications of the evidence arranged here for these theories are—to what extent they should be modified or rejected.

43. I do not commit myself on the question of how agricultural growth will affect poverty, since I am concerned only to critically discuss the argument that poverty has risen while agriculture has grown; however, I believe that the evidence supports the view that poverty has fallen substantially, but is far from being eliminated (see also Rahman and Hossain, 1995).
44. It is enormously difficult to agree upon an unproblematic measure of well-being (Nussbaum and Sen, 1993; Lipton and Ravallion, 1995); none of the following plausible concepts or methods—anthropometry, health, income, expenditure, consumption—is without criticism. There are pragmatic reasons for working with poverty as assessed by a head count of those who live in households whose per capita expenditure falls below a 'poverty line', and for believing that real wage rates would not be negatively associated with well-being. It is conceivable that real wage rates could stagnate but increases in the number of days of employment result in increases in per capita expenditure, and reduction in head-count measures of poverty. And it is also possible that real wage rates could rise or stagnate but that a reduction in days of employment results in rising poverty. However, it is more likely that either constant or rising real wages are associated with falling poverty (as a proportion of the population at least), while only falling real wages would be likely to be associated with increasing poverty. Note that while money wages might be sticky downwards, real wages can fall because prices rise faster than wages. Short run fluctuations in real wage rates are heavily influenced by the volatility of rice prices.
45. Not that such high elasticities are always found in empirical work.
46. The underconsumption argument is not explicit about poverty in part because the argument is more usually couched in terms of rising inequality and more adverse social relations. However, there is a clear implication that poverty was increasing, both by the authors discussed here and among many who espouse the agrarian structure argument. Adnan is explicit about growing inequality, and comments on the absence of measures 'ensuring minimum wages or maintaining the level of real wages' (Adnan, this volume).
47. Wodon's method suffers similar problems; again, details can be obtained from the author of this chapter.
48. See Palmer-Jones (1993a; 1994c) and Palmer-Jones and Parikh (1998) for further details of real wages of agricultural labourers in Bangladesh.
49. Also, it should be understood that the success of a number of poverty alleviation enterprises probably depended crucially on increased agricultural production and associated commercialisation. This is perhaps especially true for the fashionable microenterprise and NGO projects for which Bangladesh is well known, as indicated in Wood and Palmer-Jones (1990). These insights are also mentioned in Wood and Sharif, 1997, although not elaborated in any way, perhaps in the erroneous belief that proper policy analysis to maintain and sustain agricultural growth, and so on, is going on elsewhere, for which I see little evidence.

REFERENCES AND SELECT BIBLIOGRAPHY

Abdullah, Abu Ahmed (ed.). 1985. 'Agricultural Inputs in Bangladesh,' *Bangladesh Development Studies*, 13(3–4): 141–46.
Abdullah, Abu Ahmed, M. Hassanullah and **Quazi Shahabuddin.** 1995. 'Bangladesh Agriculture in the Nineties: Some Selected Issues,' R. Sobhan (ed.) *Experiences*

with *Economic Reform: A Review of Bangladesh's Development.* Dhaka: University Press for Centre for Policy Dialogue.
Abdullah, Abu Ahmed, and **Quazi Shahabuddin.** 1997. 'Recent Developments in Bangladesh Agriculture,' R. Sobhan (ed.) *Growth or Stagnation? A Review of Bangladesh's Development.* Dhaka: University Press for Centre for Policy Dialogue.
Adnan, Shapan. 1985. 'Classical and Contemporary Approaches to Agrarian Capitalism,' *Economic and Political Weekly,* 20(30): PE-53 to PE-64.
Agricultural Sector Review (ASR). 1989. *Bangladesh Agriculture: Performance and Policies.* Dhaka: United Nations Deveolpment Programme (UNDP).
Atkinson, A.B. 1987. 'On the Measurement of Poverty,' *Econometrica,* 55: 749-64.
Bandyopadhyay, Nripen. 1995. 'Agrarian Reforms in West Bengal—An Enquiry into their Impact and Some Problems.' Paper presented at the Workshop on Agricultural Growth and Agrarian Structure in Contemporary West Bengal and Bangladesh, Centre for Studies in Social Sciences, Calcutta, 9-12 January 1995.
Bangladesh Rural Advancement Committee (BRAC). 1979 (published 1983). *Who Gets What and Why: Resource Allocation in a Bangladesh Village.* Dhaka: BRAC Prokashana.
———. 1980 (published 1983). *The Net: Power Structure in Ten Villages.* Dhaka: BRAC Prokashana.
Bardhan, Pranab K. 1980. 'Interlocking Factor Markets and Agrarian Development: A Review of Issues,' *Oxford Economic Papers,* 32: 82-98.
———. 1984. *The Political Economy of Development in India.* Oxford: Basil Blackwell.
Beck, T. 1994. *The Experience of Poverty: Fighting for Respect and Resources in Village India.* London: Intermediate Technology Publications.
Bhaduri, Amit. 1968. 'New Agricultural Policy,' *Frontier,* 1(13): 12-3.
———. 1973. 'A Study in Agricultural Backwardness under Semi-Feudalism,' *Economic Journal,* 83: 120-37.
———. 1986. 'Forced Commerce and Agrarian Growth,' *World Development,* 14(2): 267-72.
Bose Sugata. 1986. *Agrarian Bengal: Economy, Social Structure and Politics, 1919-1947.* Cambridge: Cambridge University Press.
———. 1995. 'Agricultural Growth and Agrarian Structure in Bengal: An Historical Overview.' Paper presented at the Workshop on Agricultural Growth and Agrarian Structure in Contemporary West Bengal and Bangladesh, Centre for Studies in Social Sciences, Calcutta, 9-12 January 1995.
Boyce, James K. 1987. *Agrarian Impasse in Bengal: Institutional Constraints to Technological Change.* Oxford: Oxford University Press.
Crow, Ben, and **K.A.S. Murshid.** 1994. 'Economic Returns to Social Power: Merchants' Finance and Interlinkage in the Grain Markets of Bangladesh,' *World Development,* 22(7): 1011-30.
Dasgupta, Biplab. 1995. 'West Bengal's Agriculture since 1977.' Paper presented at the Workshop on Agricultural Growth and Agrarian Structure in Contemporary West Bengal and Bangladesh, Centre for Studies in Social Sciences, Calcutta, 9-12 January 1995.
David, W.P. 1994. 'Minor Irrigation Development in Bangladesh.' Background Paper No. 4. Paper presented at the DAE/ATIA Workshop on Research to Promote Intensive Irrigated Agriculture, Khamarbari, Dhaka.
de Janvry, A., and **E. Sadoulet.** 1987. 'Agricultural Price Policy in General Equilibrium Models: Results and Comparisons,' *American Journal of Agricultural Economics,* 69: 230-46.

de Vylder, S., and D. Asplund. 1982. *Agriculture in Chains: Contradictions and Distortions in a Rural Economy, the Case of Bangladesh.* New Delhi: Vikas.

Dobb, Maurice. 1946. *Studies in the Development of Capitalism.* London: Routledge and Kegan Paul.

Fugelsang, A., and D. Chandler. 1988. 'Participation as Process—What We can Learn from the Grameen Bank, Bangladesh. Dhaka: Grameen Bank.

Goode, E., and N. Ben-Yahuda. 1994. *Moral Panics: The Social Construction of Deviance.* Oxford: Blackwell.

Government of India. 1928. *Royal Commission on Indian Agriculture, 1928.*

Greeley, M. (ed.). 1982. 'Feeding the Hungry: A Role for Post Harvest Technology?' *Bulletin of the Institute of Development Studies,* 13(3): 51–60. Brighton, Sussex.

Harriss, John. 1992. 'Does the "Depressor" Still Work? Agrarian Structure and Development in India: A Review of Evidence and Argument,' *Journal of Peasant Studies,* 19(2): 189–227.

Hart, G. 1986. 'Interlocking Transactions: Obstacles, Precursors, of Instruments of Agrarian Capitalism,' *Journal of Development Economics,* 23: 177–203.

Hartmann, B., and James K. Boyce. 1983. *A Quiet Violence: View from a Bangladesh Village.* London: Zed Books.

Hirschman, A.O. 1982. 'Rival Interpretations of Market Society: Civilising, Destructive, or Feeble?' *Journal of Economic Literature,* 20(4): 1463–84.

Hoff, K., A. Braverman and J.E. Stiglitz (eds). 1993. *The Economics of Rural Organisation.* Oxford: Oxford University Press for the World Bank.

Hossain, Mahbub. 1985. 'Price Responsiveness of Fertiliser Demand in Bangladesh,' *Bangladesh Development Studies,* 13(3–4): 41–66.

———. 1987. *The Assault that Failed: A Profile of Absolute Poverty in Six Villages of Bangladesh.* Geneva: United Nations Research Institute for Social Development (UNRISD).

———. 1995. 'Agricultural Policy Reforms in Bangladesh: An Overview.' Paper presented at the Workshop on Agricultural Growth and Agrarian Structure in Contemporary West Bengal and Bangladesh, Centre for Studies in Social Sciences, Calcutta, 9–12 January 1995.

Hughes, R., Shapan Adnan and B. Dalal-Clayton. 1994. *Floodplains or Flood Plans? A Review of Approaches to Water Management in Bangladesh.* London: International Institute for Environment and Development (IIED).

International Irrigation Management Institute (IIMI) (with Bureau of Socio-Economic Research and Training of Bangladesh Agricultural University). 1996. 'Study on Privatisation of Minor Irrigation in Bangladesh.' Final Report, IIMI, Colombo.

Jannuzi, F., and J.T. Peach. 1980. *The Agrarian Structure of Bangladesh: An Impediment to Development.* Colorado: Westview Press.

Jansen, Eirik G. 1986. *Rural Bangladesh: Competition for Scarce Resources.* Oslo: Norwegian University Press.

———. 1990. 'Processes of Polarisation and the Breaking of Patron–Client Relationships in Rural Bangladesh,' O.D.K. Norbye (ed.) *Bangladesh Faces the Future.* Dhaka: University Press.

Kahnert, F., and G. Levine (eds). 1993. *Groundwater Irrigation and the Rural Poor: Options for Development in the Gangetic Basin.* Washington: World Bank.

Khan, A.R. 1990. 'Poverty in Bangladesh: A Consequence of and a Constraint on Growth,' *Bangladesh Development Studies,* 18(3): 19–34.

Kramsjo, B., and Geoffrey D. Wood. 1992. *Breaking the Chains: Collective Action for Social Justice among the Rural Poor.* London: Intermediate Technology Press.

Lieten, G.K. 1990. 'Depeasantisation Discontinued: Land Reforms in West Bengal,' *Economic and Political Weekly*, 25(40): 2265–71.
Like-Minded Group (LMG). 1985 (published 1990). *Rural Poverty in Bangladesh: A Report to the Like-Minded Group.* Dhaka: University Press.
Lipton, M., and M. Ravallion. 1995. 'Poverty and Policy,' J. Behrman and T.N. Srinivasan (eds) *Handbook of Development Economics*, Volume 3B. Amsterdam: Elsevier.
Lovell, C.H. 1992. *Breaking the Cycle of Poverty: The BRAC Strategy.* Dhaka: University Press.
Marx, Karl, and Friedrich Engels. 1973 (originally published in 1848). *Manifesto of the Communist Party.* Harmondsworth: Penguin Books.
Moulton, B.R. 1996. 'Bias in the Consumer Price Index: What is the Evidence?' *Journal of Economic Perspectives*, 10(4): 159–78.
Norbye, O.D.K. (ed.). 1990. *Bangladesh Faces the Future.* Dhaka: University Press.
Nussbaum, M.C., and A. Sen (eds). 1993. *The Quality of Life.* Oxford: Clarendon Press.
Osmani, S.R. 1985. 'Pricing and Distribution Policies for Agricultural Development in Bangladesh,' *Bangladesh Development Studies*, 13(3–4): 1–40.
Osmani, S.R., and M.A. Quasem. 1984 (published 1990). *Pricing and Subsidy Policy for Bangladesh Agriculture.* Dhaka: Bangladesh Institute of Development Studies.
Palmer-Jones, Richard. 1989. 'Groundwater Management in Bangladesh: Review, Issues, and Implications for the Poor.' Paper presented at the Workshop on Groundwater Management, Institute of Rural Management, Anand (IRMA), January 1989.
———. 1992. 'Sustaining Serendipity? Groundwater Irrigation, Growth of Agricultural Production and Poverty in Bangladesh,' *Economic and Political Weekly*, 27(39): A–128 to A–140.
———. 1993a. 'Agricultural Wages in Bangladesh: What the Figures Really Show,' *Journal of Development Studies*, 29(2): 277–300.
———. 1993b. 'Deep Tubewells for Irrigation, Drainage and Salinity Control: The Life and Times of an Inappropriate Technology.' Mimeograph, School of Development Studies, University of East Anglia.
———. 1994a. 'An Error Corrected? And What the Figures Really Show,' *Journal of Development Studies*, 30(2): 346–51.
———. 1994b. 'The Hand Over of Deep Tubewells.' Paper presented at the International Conference on Irrigation Management Handover, Wuhan, 20–24 September 1994.
———. 1994c. 'Water Markets in South Asia: A Discussion of Theory and Evidence,' M. Moench (ed.) *Selling Water: Conceptual and Policy Debates over Groundwater Markets in India.* Ahmedabad: Vikram Sarabhai Centre for Development Interaction (VIKSAT).
———. 1997. 'Groundwater Management in South Asia: What Role for the Market,' M. Kay (ed.) *Water: Economics, Management, Demand.* Proceedings of the 18th European Conference on Irrigation and Drainage, Oxford, September 1997, E&FN Spon., London.
Palmer-Jones, Richard, and A. Parikh. 1998. 'The Determination of Agricultural Wage Rates in Bangladesh,' *Journal of Agricultural Economics*, 49(1): 111–33.
Parry, Jonathan. 1989. 'On the Moral Perils of Exchange,' Jonathan Parry and Maurice Bloch (eds) *Money and the Morality of Exchange.* Cambridge: Cambridge University Press.
Parry, Jonathan, and Maurice Bloch (eds). 1989. *Money and the Morality of Exchange.* Cambridge: Cambridge University Press.
Pearse, A. 1980. *Seeds of Plenty, Seeds of Want: A Critical Analysis of the Green Revolution.* Oxford: Oxford University Press.

Platteau, J-P. 1994a. 'Behind the Market Stage where Real Societies Exist: 1. The Role of Public and Private Institutions,' *Journal of Development Studies*, 30(3): 533–77.
———. 1994b. 'Behind the Market Stage where Real Societies Exist: 1. The Role of Public and Private Institutions,' *Journal of Development Studies*, 30(4): 753–817.
Rahman, A. 1986. *Peasants and Classes: A Study in Differentiation in Bangladesh*. Dhaka: University Press.
Rahman, Hossain Zillur. 1994. 'Low Price of Rice: Who Loses, Who Gains? Findings from a Recent Survey of Rural Bangladesh.' Report for Analysis of Poverty Trends Project, Working Paper, New Series No. 5. Dhaka: Bangladesh Institute of Development Studies.
Rahman, Hossian Zillur, and M. Hossain (eds). 1995. *Re-Thinking Rural Poverty: A Case for Bangladesh*. New Delhi: Sage.
Ravallion, M. 1990. 'The Challenging Arithmetic of Poverty in Bangladesh,' *Bangladesh Development Studies*, 18(3): 35–54.
Ravallion, M., and B. Sen. 1996. 'When Method Matters: Monitoring Poverty in Bangladesh,' *Economic Development and Cultural Change*, 44(4): 761–92.
Reserve Bank of India. 1984. *Agricultural Productivity in Eastern India* (Volume 2). S.R. Sen Report, Bombay, Reserve Bank of India.
Robinson, E.A.G., and K. Griffen. 1974. *The Economic Development of Bangladesh in a Socialist Framework*. London: Macmillan.
Rogaly, Ben, Barbara Harriss-White and Sugata Bose. 1995. '*Sonar Bangla*? Agricultural Growth and Agrarian Change in West Bengal and Bangladesh,' *Economic and Political Weekly*, 30(29): 1862–8.
Sadoulet, E., and A. de Janvry. 1992. 'Agricultural Trade Liberalisation and the Low Income Countries: A General Equilibrium-Multimarket Approach,' *American Journal of Agricultural Economics*, 74: 268–80.
Saha, Anamitra and Madhura Swaminathan. 1994. 'Agricultural Growth in West Bengal in the 1980s: A Disaggregation by Districts and Crops,' *Economic and Political Weekly*, 29(13): A–2 to A–11.
Sen, Amartya. 1980. 'Description as Choice,' *Oxford Economic Papers*, 32(3): 333–70.
———. 1997. 'Development Thinking at the Beginning of the 21st Century.' DERP No. 2, Suntory and Toyota International Centres for Economics and Related Disciplines (STICERD). London School of Economics, London.
Sen, Abhijit, and Ranja Sengupta. 1995. 'The Recent Growth in Agricultural Output in Eastern India, with Special Reference to the Case of West Bengal.' Paper presented at the Workshop on Agricultural Growth and Agrarian Structure in Contemporary West Bengal and Bangladesh, Centre for Studies in Social Sciences, Calcutta, 9–12 January 1995.
Sen, B. 1995. 'Recent Trends in Poverty and its Dynamics,' R. Sobhan (ed.) *Experiences with Economic Reform: A Review of Bangladesh's Development*. Dhaka: University Press for Centre for Policy Dialogue.
———. 1997. 'Poverty and Policy,' R. Sobhan (ed.) *Growth or Stagnation? A Review of Bangladesh's Development*. Dhaka: University Press for Centre for Policy Dialogue.
Shah, T. 1993. *Groundwater Markets and Irrigation Development: Political Economy and Political Policy*. New Delhi: Oxford University Press.
Sobhan, R. (ed.) 1995. *Experiences with Economic Reform: A Review of Bangladesh's Development*. Dhaka: University Press for Centre for Policy Dialogue.
———. (ed.). 1997 *Growth or Stagnation? A Review of Bangladesh's Development*. Dhaka: University Press for Centre for Policy Dialogue.

Stiglitz, J.E. 1986. 'The New Development Economics,' *World Development*, 14(2): 257–65.
Stokes, E. 1978. 'Dynamism and Enervation in North Indian Agriculture: The Historical Dimension,' E. Stokes (ed.) *The Peasant and the Raj*. Cambridge: Cambridge University Press.
van Schendel, Willem. 1981. *Peasant Mobility: The Odds of Life in Rural Bangladesh*. Assen, van Gorcum.
Wodon, Q.T. 1997. 'Food Energy Intake and Cost of Basic Needs: Measuring Poverty in Bangladesh,' *Journal of Development Studies*, 34(6): 66–101.
Wood, Geoffrey D. 1978. 'Class Differentiation and Power in Bandokgram: The Minifundist Case,' M.A. Huq (ed.) *Exploitation and the Rural Poor—A Working Paper on the Rural Power Structure in Bangladesh*, Bangladesh Academy for Rural Development (BARD), Comilla. Reproduced as Chapter 1 in Wood, 1994.
———. 1981. 'Class Formation in Bangladesh, 1940–1980,' *Bulletin of Concerned Asian Scholars*, 13(4): 2–15.
———. 1994. *Bangladesh: Whose Ideas, Whose Interests?* Dhaka: University Press.
Wood, Geoffrey D., and **Richard Palmer-Jones.** 1990. *The Water-Sellers: A Co-operative Venture by the Rural Poor*. West Hartford: Kumarian.
Wood, Geoffrey D., and **I.A. Sharif** (eds). 1997. *Who Needs Credit? Poverty and Finance in Bangladesh*. Dhaka: University Press.
World Bank. 1979. *Bangladesh: Food Policy Issues*, Washington D.C., World Bank, Report No. 2761–BD.
———. 1989. 'Selected Issues in Agricultural Devleopment.' Mimeograph, World Bank, Dhaka.
———. 1990a. *World Development Report, 1990: Poverty*. Oxford: Oxford University Press for the World Bank.
———. 1990b. 'Bangladesh: Poverty and Public Expenditures: An Evaluation of the Impact of Selected Government Programs.' World Bank, Asia Country Department 1, Washington.
———. 1992. *Bangladesh Food Policy Review: Adjusting to the Green Revolution* (2 volumes). Washington: World Bank.
———. 1996. *Bangladesh Annual Economic Update: Recent Economic Development and Medium Term Reform Agenda*. Dhaka: World Bank and University Press.

5

Agricultural Growth Performance in Bangladesh: A Note on the Recent Slowdown

QUAZI SHAHABUDDIN

INTRODUCTION

The early 1990s witnessed a drastic slowdown in rice production growth in Bangladesh, in spite of arguments to the contrary by Palmer-Jones (this volume). After a big jump in production in 1989–90 (due to increase in *aman* rice production by more than 2 million tonnes), growth seems to have decelerated considerably in subsequent years. The trend growth rate for the 1990–91 to 1993–94 period was as low as 0.37 per cent, as compared to 2.79 per cent during the 1970s and 2.32 per cent during the 1980s.[1]

This slowdown in production growth occurred in the absence of any major natural disasters (such as flood, drought or cyclone) in the early 1990s (except for the drought in 1994–95). One can, therefore, legitimately ask whether such a drastic slowdown in growth is a transitory phenomenon

---

The author is grateful to Abu Abdullah and Rushidan Islam Rahman for their valuable comments on an earlier draft of the paper. However, the usual disclaimer applies.

(part of the regular production cycle), or is it threatening to become a permanent feature of the rice production system as technological frontiers are gradually approached? No serious investigation into the causes of this slowdown seem to have been conducted so far, although Abdullah et al. (1995) identify some proximate determinants of the slowdown in terms of the use of chemical fertilisers and irrigation expansion, as well as the trend in land productivity and farmers' production incentives in recent years.

TRENDS IN INPUT USE AND FACTOR PRODUCTIVITY

Fertiliser and irrigation constitute the two critical inputs in the dissemination of modern rice technology. Chemical fertilisers were introduced in Bangladesh in the late 1950s. However, their importance has increased with the diffusion of 'seed–fertiliser–water' technology since the mid-1960s. In fact, use of chemical fertilisers increased by about twenty times from 100 thousand tonnes in 1965–66 to 2 million tonnes by 1989–90 (Chowdhury and Shahabuddin, 1992). Urea accounts for about 71 per cent of total fertiliser consumption, while the shares of Triple Super Phosphate (TSP) and Murate of Potash (MP) are roughly 23 and 6 per cent respectively (Shahabuddin and Zohir, 1995).

Total fertiliser consumption increased at an annual rate of 9.36 per cent over the two decades covering the 1972–73 to 1993–94 period (Table 5.1). The rate of growth of fertiliser consumption declined in the 1980s

Table 5.1
Trend Growth Rate of Fertiliser Use by Type in Bangladesh

Period	Urea	TSP	MP	SSP	Total
1972–73 to 1993–94	9.46	7.89	9.43	–	9.36
1972–73 to 1979–80	12.63	14.32	16.39	–	13.25
1980–81 to 1989–90	9.88	9.38	9.80	–	9.74
1990–91 to 1993–94	5.41	–24.72	–11.18	91.74	1.54

Sources: Statistical Yearbooks, BBS and authors' calculations.

compared to its growth rate during the earlier decade. This remains true for each type of fertiliser used. However, the decline has become more pronounced in recent years, with an annual growth of only 1.54 per cent during the 1990–91 to 1993–94 period. In fact, total fertiliser consumption declined in absolute terms (by over 100,000 tonnes) in 1993–94. A slight increase in Urea and Single Super Phosphate (SSP) consumption could not compensate for the large decrease in TSP and MP use in that year (Abdullah et al., 1995).

Since its introduction in the 1960s, modern irrigation in Bangladesh was initially almost entirely dependent on utilisation of surface water. The extraction of groundwater assumed greater importance over time. In fact, low lift pumps drawing upon surface water were followed by suction-mode tubewells (shallow tubewells) drawing from near-surface aquifers, and finally, force-mode tubewells exploiting deep aquifers were used for expanding irrigation in the country (Shahabuddin and Zohir, 1995). Irrigated area increased at a rate of 4.90 per cent per annum over the 1972–73 to 1993–94 period (Table 5.2). As a result, the share of irrigated

Table 5.2
Trend in Irrigation Coverage and Growth Rates

	1972–73 to 1993–94	1972–73 to 1979–80	1980–81 to 1989–90	1985–86 to 1989–90	1990–91 to 1993–94
		(Annual Average for the Period)			
Area under irrigation (as % of Net Cropped Area [NCA])	24.17	16.50	25.27	29.17	36.78
Area under surface water irrigation (as % of NCA)	14.86	15.12	14.59	13.49	15.02
Area under ground water irrigation (as % of NCA)	9.31	1.39	10.68	15.67	21.75
	(Annual Growth Rates)				
Modern irrigation	7.92	4.34	9.67	9.01	0.22
Traditional irrigation	–3.13	1.07	–6.15	3.16	–0.15
Total irrigation	4.90	2.85	6.12	8.04	0.17

Sources: Statistical Yearbooks, BBS and authors' calculations.

area (as percentage of net cropped area) increased from about 17 per cent during the 1970s to about 37 per cent in the early 1990s. The growth rate of irrigated area was only 2.85 per cent during the 1970s, increasing to 6.12 per cent during the following decade. This growth accelerated further to 8.04 per cent during the late 1980s following certain liberalisation policies (e.g., withdrawal of all restrictions on import of irrigation equipment by the private sector, removal of restrictions on standardisation and siting etc.) that have been pursued by the government since 1987–88. Most of this growth was accounted for by the adoption of modern irrigation methods (Table 5.2). However, the growth in irrigated area declined perceptibly in the early 1990s averaging only 0.17 per cent during the 1990–91 to 1993–94 period.

While returns from fertiliser use and irrigation are still positive, time series evidence on productivity generally suggests declining incremental returns with respect to both (Mahmud et al., 1993; Abdullah et al., 1995). The same trends may also signal the contribution of other variables. Incremental fertiliser productivity declined sharply from the beginning of the 1980s; such declines may be explained in terms of declining soil fertility due to the intensification of rice monoculture, and expansion of HYV boro into less suitable lands.[2] The estimated output and input indices which formed the basis of the total factor produtivity exercise by Dey and Evenson (1991) also suggest that the incremental gains from aggregate input use declined during the 1980s compared to the earlier decade.

To assess the sustainability of growth in crop agriculture through intensified rice cultivation, Mahmud et al. (1993) look at the past pattern of growth in rice production, disaggregated by region. In Table 5.3, the different regions (former districts) of the country are ranked according to the rate of growth of rice production during three overlapping ten-year periods beginning with the late 1960s when HYV rice was first introduced. It can be seen that the growth rate of rice production at the national level conceals strikingly large variations across regions (see also Crow, this volume). More importantly, the growth points have shifted between periods. While all the regions have had at least medium growth in one period or another, the early starters have generally lagged behind other regions in the later periods. The exhaustion of easy sources of irrigation is a likely reason why production growth at the regional level has not been sustained over prolonged periods. But the explanation of this may also partly lie in the hypotheses mentioned before regarding the agronomic constraints to intensified rice cultivation.

MAJOR POLICY REFORMS IN THE 1980s

The 1980s have witnessed major changes in agricultural policies in Bangladesh. Significant growth performance was also achieved during this period. Whether the two are causally linked is a hotly contested issue (see Adnan, Palmer-Jones, this volume). In our view, the burden of evidence seems to be overwhelmingly in favour of such a causal link

Table 5.3
Ranking of Regions According to Trend Rate of Growth of Rice Production
(per cent annual growth rate within parentheses)

Annual Growth Rate Ranking	Time Periods		
	1967–68 to 1977–78	1973–74 to 1983–84	1979–80 to 1989–90
High (above 4%)	Tangail (10.56*) Noakhali (5.12*) Chittagong (4.12*)	Tangail (5.77*) Pabna (5.66*) Bogra (5.08*) Patuakhali (4.85*)	Kushtia (8.18*) Jessore (6.66*) Faridpur (6.37*) Bogra (4.71*) Rajshahi (4.09*)
Medium (between 2% and 4%)	Jessore (3.55*) Mymensingh (3.52*) Dhaka (2.27*) Pabna (2.06*)	Mymensingh (2.95*) Dhaka (2.86*) Barisal (2.59*) Khulna (2.46*) Rangpur (2.15*) Sylhet (2.00)	Dinajpur (3.61*) Rangpur (3.18*) Comilla (2.83*) Noakhali (2.47*)
Low (below 2%)	Bogra (1.92) Barisal (1.81) Kushtia (1.60) Rajshahi (1.18) Rangpur (1.08) Dinajpur (0.84) Comilla (0.82) Khulna (−0.36) Faridpur (−0.79) Sylhet (−3.43)	Chittagong (1.83*) Rajshahi (1.80*) Comilla (1.34) Noakhali (1.04) Dinajpur (0.77) Faridpur (0.63) Jessore (0.22) Kushtia (−0.19) Tangail (−1.12)	Barisal (1.77*) Pabna (1.67) Khulna (1.66) Patuakhali (1.41) Mymensingh (0.87) Chittagong (0.51) Sylhet (−0.35) Dhaka (−0.72)

Source: Mahmud et al., 1993.
*Growth rates are estimated by fitting semi-logarithmic trend lines. The asterisk sign indicates that the estimated growth rate is statistically significant at the level of 5 per cent or less.

(Abdullah and Shahabuddin, 1993).[3] Two major areas where such policy changes exerted a significant influence related to:

1. Privatisation of fertiliser distribution and elimination of subsidies;
2. Evolution of minor irrigation policy towards increased private sector participation in procurement and distribution, and reduction in subsidies.

Changes in the fertiliser marketing system were initiated in the late 1970s. Significant policy changes during the late 1980s and early 1990s have completely privatised the fertiliser market in the country. Beginning July 1987 private dealers were allowed to procure fertiliser in bulk quantities at a higher discount rate from the factories as well as from four larger supply centres known as Transport Discount Points (TDP). Since March 1989, lifting prices were equalised across private traders and the Bangladesh Agricultural Development Corporation (BADC), the parastatal organisation which had monopolised the procurement and distribution of fertiliser before the privatisation process was initiated. Beginning in July 1991, private traders have engaged in direct import of fertiliser. An International Fertiliser Development Corporation (IFDC) estimate suggests that the share of the private sector in the fertiliser market increased from less than 5 per cent to more than 90 per cent over the 1987–92 period. Available evidence supports the contention that such policies have elicited favourable responses from farmers in terms of growth performance. Abdullah and Shahabuddin (1993) provide direct evidence to show that fertiliser markets have been functioning competitively in Bangladesh following their privatisation in the early 1980s.[4]

Fertiliser prices at the farm level had been deregulated throughout the country by 1983. There has not been any direct subsidy on Urea since the early 1980s. However, the practice of influencing market prices by administering issue prices continued. This was especially true in the cases of TSP and MP, and was also applicable to private imports until recently. Thus, subsidies continued during the second half of the 1980s. All subsidies have, however, been withdrawn since the end of 1991. There are two major implications of policies related to fertiliser pricing (Shahabuddin and Zohir, 1995). The withdrawal of subsidies on fertiliser raised the cost of fertilisers for farmers, which in turn increased the fertiliser–paddy price ratio. The latter was further affected adversely due to a decline in rice prices in the mid-1990s (see Palmer-Jones, this volume). Second, since the withdrawal of subsidy has primarily raised the prices of TSP and MP, another implication is the possibility of use of undesirable mixes of various types of fertilisers. At a more general level, an increase in the fertiliser–paddy price ratio is likely to reduce the intensity of fertiliser use and thereby adversely affect crop yields.

Private sector participation in the market for irrigation equipment also commenced during the late 1970s. Private import and sale of shallow tubewells were allowed in 1978–79. However, such imports were subject to standardisation requirements and a Groundwater Ordinance was introduced in 1987 to control the siting of shallow and deep tubewells.

Since 1988, the government has withdrawn all restrictions on import of irrigation equipment by the private sector, eliminated import duty on agricultural machinery and removed restrictions on standardisation and siting. These policy changes have been accompanied by the elimination of subsidies on minor irrigation equipment. More importantly, irrigation management has gone through a gradual metamorphosis from public ownership with bureaucratic management to public ownership with cooperative management and, finally, to private ownership with private management. Eventually, the irrigation equipment which had been owned by the public sector was privatised.

The policy of removing restrictions on standardisation and siting of tubewells had a positive impact on minor irrigation expansion.[5] The spectacular spurt in minor irrigation development, particularly shallow tubewells, between 1987–88 and 1990–91 was undoubtedly caused by the increased availability of cheaper Chinese and Korean engines. Not only did such policy changes make available to farmers cheaper (if less durable) brands of engines, but the resulting competition along with the elimination of duty caused a fall in the prices of the standardised brands as well (Abdullah and Shahabuddin, 1993). Interestingly, however, the high pace of growth during the first few years of policy reforms has turned out to be a short-lived one. The rates of increase in the number of tubewells fielded and in the area under groundwater irrigation have declined after 1990–91 (Shahabuddin and Zohir, 1995).

AGRARIAN STRUCTURE AND AGRICULTURAL GROWTH

The two major elements of agrarian structure which are conjectured to influence growth are: (a) the concentration of landownership, and (b) the prevalence of share tenancy. Nearly 70 per cent of the farmers in Bangladesh have holdings of less than 1 hectare, more than 40 per cent are tenants or partial tenants, and about 23 per cent of the total cultivated area is farmed under tenancy (BBS, 1986). Thus, one would expect a priori the agrarian structure in Bangladesh to constrain the diffusion of modern varieties.[6] Hossain et al. (1994) attempted to empirically verify this hypothesis using farm level data from sixty-two villages in Bangladesh. Their multivariate analysis, which they employed to identify the determinants of adoption of modern varieties, shows that while the coefficient of farm size turned out to be positive and significant for the dry season suggesting a higher rate of adoption on large farms, the tenancy variable was insignificant for both dry and wet seasons indicating that the tenurial status of the farm does not affect the adoption rate when the effects of all other variables are controlled.[7]

Abdullah (1995) has attempted using district level data to ascertain how much of the variation in growth rates of rice output (which largely reflects the rapidity of shifting from traditional to modern varieties) is explained by factors related to the agrarian structure. His multivariate analysis involved regressing the rate of growth of rice output (during the 1980–81 to 1993–94 period) on four 'structural' variables: percentage of cultivated area operated by large (over 7.5 acres) farmers, percentage of tenant-operated area, percentage of landless households and the average number of fragments per holding. None of the variables came out statistically significant in this empirical exercise, thereby providing evidence against any significant role of the 'agrarian structure' (narrowly defined in terms of distribution of landholdings and incidence of tenancy) in determining growth performance in Bangladesh.[8]

CONCLUDING OBSERVATIONS

As Palmer-Jones argues at length (this volume), long term growth rates often conceal more recent trends. A closer look at the recent growth of the crop economy suggests that annual growth rates in rice production slowed down considerably in the early 1990s. So did the growth in area under mechanised irrigation and in the use of fertiliser. Both price and non-price factors may have contributed to this, as reflected in declining land productivity and adverse movements in the fertiliser–paddy price ratio. However, the evidence available is insufficient to draw concrete conclusions about whether the recent slowdown in production is a temporary phenomenon, or it is threatening to become a permanent feature of the rice production system as technological frontiers are gradually approached. In either event, while growth of crop production in the foreseeable future is likely to remain largely dependent on expansion of HYV rice, there is, as Mahmud et al. (1993) rightly emphasise, a need for crop diversification as a means of sustaining agricultural growth and productivity.

NOTES

1. For the most up-to-date production figures at the time of going to press, see Palmer-Jones (this volume).
2. In fact, the declining trend in fertiliser productivity tends to support two hypotheses relating to the agronomic constraints to the growth of rice based crop agriculture. First, the intensification of rice monoculture is liable to be detrimental to soil fertility which is one reason why agricultural scientists advocate crop diversification. There is an increasing concern in Bangladesh about the likely adverse effect on crop yields resulting from the depletion of micro-nutrients and organic matters in soil. Second, the rapid expansion of the area under HYV boro may have increasingly led to its cultivation on relatively less suitable lands.
3. An International Food Policy Research Institute (IFPRI) study (Ahmed, 1994) using multivariate analysis estimated that the production of rice could have been 35 per cent lower (than the level attained in 1992–93) had the input market reform measures not been put in place. The impact on production was achieved through the effects of reforms in the increased use of fertiliser (by 75 per cent) and expansion in irrigated area (by 67 per cent).
4. The mark-up of prices paid by farmers over the issue prices was fairly modest from 1991 to 1993 fluctuating between 21.45 and 26.98 per cent for Urea, and between 13.62 and 26.86 per cent for TSP. Given that these margins must cover transportation costs from factories and ports to the farmers, and yield 'normal' profits to wholesalers and retailers, they should be considered quite moderate (but for a contrary view, see Adnan, this volume).
5. Between 1987–88 and 1988–89, the number of STWs (and private force-mode tubewells) fielded increased by 40 per cent from 183,000 to 223,000. Between 1988–89 and 1990–91, this further increased to 276,000—an increase of 53 per cent over two years (Abdullah and Shahabuddin, 1993).
6. Much of the literature on adoption of modern varieties argues that large farmers and owner-cultivators are in a more favourable position to adopt them than small and tenant farmers. For an excellent survey of that literature and a review of empirical studies, see Griffin (1974), Feder et al. (1985), and Lipton and Longhurst (1989).
7. In this empirical exercise, the tenancy variable was defined in terms of percentage of cultivated land rented. In reality, however, the terms and conditions of tenancy may have a greater bearing on production incentives than the incidence of tenancy per se (for similar hypotheses regarding West Bengal, see Gazdar and Sengupta, this volume).
8. While this is very far from the kind of complete specification attempted by Hossain et al. (1994), one would still expect any significant effects of the structural variables on growth performance to show up in the form of significant coefficients.

REFERENCES

Abdullah, Abu Ahmed. 1995. 'Agrarian Structure in the Study of Underdevelopment.' Mimeograph, Bangladesh Institute of Development Studies.

Abdullah, Abu Ahmed, Quazi Shahabuddin and **M. Hassanullah.** 1995. 'Bangladesh Agriculture in the Nineties: Some Selected Issues,' R. Sobhan (ed.) *Experiences with Economic Reform: A Review of Bangladesh's Development.* Independent

Review of Bangladesh Development, Dhaka: University Press for Centre for Policy Dialogue.

Abdullah, Abu Ahmed, and **Quazi Shahabuddin.** 1993. 'Critical Issues in Bangladesh Agriculture: Policy Response and Unfinished Agenda.' Mimeograph, Bangladesh Institute of Development Studies.

Ahmed, R. 1994. 'Liberalisation of Agricultural Input Markets in Bangladesh: Process, Impact, Lessons.' Mimeograph, International Food Policy Research Institute, Washington D.C.

Bangladesh Bureau of Statistics (BBS). 1986. *The Bangladesh Census of Agriculture and Livestock, 1983–84*, Vols 1 and 4, Dhaka.

Chowdhury, N. 1992. 'Rice Markets in Bangladesh: A Study in Structure, Conduct and Performance.' IFPRI Report, Mimeograph, International Food Policy Research Institute, Washington D.C.

Chowdhury, O.H., and **Quazi Shahabuddin.** 1992. 'A Study of Food Situation and Outlook of Asia: Country Report for Bangladesh.' IFPRI Study, Mimeograph, International Food Policy Research Institute, Washington D.C.

Dey, M.M., and R.E. Evenson. 1991. 'The Economic Impact of Rice Research in Bangladesh.' Mimeograph, International Rice Research Institute/Bangladesh Agricultural Research Council.

Feder, G., R.E. Just and D. Zilberman. 1985. 'Adoption of Agricultural Innovations in Developing Countries: A Survey,' *Economic Development and Cultural Change*, 33(2): 255–98.

Goletti, F. 1993. 'The Changing Public Role in a Rice Economy Moving Towards Self-Sufficiency: The Case of Bangladesh.' IFPRI Report, Mimeograph, International Food Policy Research Institute, Washington D.C.

Griffin, K. 1974. *The Political Economy of Agrarian Change: An Essay on the Green Revolution.* Cambridge: Harvard University Press.

Haggblade, S. 1994. 'Evolving Food Markets and Food Policy.' Keynote paper presented at the IFPRI seminar held in Dhaka, 2–4 May 1994.

Hossain, Mahbub, M. Abul Quasem, M.A. Jabber and M.M. Akash. 1994. 'Production Environments, Modern Variety Adoption, and Income Distribution in Bangladesh,' Christina David and Keijiro Otsuka (ed.) *Modern Rice Technology and Income Distribution in Asia.* Boulder and London: Lynne Rienner Publishers/Manila: IRRI.

Hossain, Mahbub, and M.R. Dhaly. 1991. 'Tracer Study on Recent Agricultural Policies in Bangladesh.' A report prepared by the Bangladesh Institute for Development Studies for Asian Development Bank, Manila.

Lipton, M., and R. Longhurst. 1989. *New Seeds and Poor People.* London: Unwin Hyman.

Mahmud, W., S.H. Rahman and S. Zohir. 1993. 'Agricultural Growth through Crop Diversification in Bangladesh.' BIDS–IFPRI Study, Mimeograph, Bangladesh Institute of Development Studies/International Food Policy Research Institute.

Shahabuddin, Quazi. 1995. 'Assessment of Domestic Production, Consumption and Trade Prospects of Foodgrains in Bangladesh.' Mimeograph, Bangladesh Institute of Development Studies.

Shahabuddin, Quazi and S. Zohir. 1995. 'Projections and Policy Implications of Medium and Long Term Rice Supply and Demand.' Country Report for Bangladesh, IRRI BIDS Study, Mimeograph, Bangladesh Institute of Development Studies.

6

Why is Agricultural Growth Uneven? Class and the Agrarian Surplus in Bangladesh

BEN CROW

INTRODUCTION

This paper tells an intriguing story about contrasts between two areas. In an 'advanced' area, in the north-west of Bangladesh, investment in irrigation is associated with accumulation by rich peasants. This is an area of rapid agricultural growth and local prosperity. In a 'backward' area, in the south-east of the country, there is little accumulation by cultivating classes. The grain surplus is appropriated by traders and landlords, and this is an area with little investment in irrigation and little growth in agrarian output. Nevertheless, at least one of the prominent traders from this area has become a partner in two Dhaka garment factories employing part of Bangladesh's newly-emerging female proletariat and supplying clothes to the global economy.

The paper hints at connections between backward agriculture, poverty and non-agrarian investment. This part of the story, however, remains tentative. With more substantial foundations, the paper provides a clear contrast between two local regimes of accumulation—one connected with agrarian investment and increased agricultural productivity, the other generating an agrarian squeeze and agricultural stagnation. These contrasting regimes of accumulation have been observed for almost ten years, intensively from 1987 to 1989, and then periodically till 1997. Improved roads and telephones, and the increasing density of commodity relations, are changing economic relationships in both areas. The contrast between the two areas has been reduced in the last ten years, but it has not been erased.

The data presented in this paper describe marked contrasts in the scale of surplus product and the forms of surplus appropriation in two different rural areas of Bangladesh. It is argued that these two areas, with contrasting social relations, are broadly representative of the most advanced and most backward areas of the countryside, each covering approximately one-third of the land area of Bangladesh.

The paper is organised as follows: the following section discusses the concept of surplus and the different forms which surplus can take. The next section describes the contrasts between the two areas for which data are presented in this paper. Then, a brief summary is provided of the method used to place households in categories of rural social class. Using these categories, the scale of paddy output by class and how it is distributed in sales and kind payments is described. These data are then examined in relation to the surplus or deficit of the household. Next, the question of who accumulates grain is explored. The final section summarises the conclusions of this paper.

FORMS OF SURPLUS PRODUCTION

In Marxian social theory, surplus appropriation provides the primary explanation for the making of rich and poor. What is surplus? Heilbroner (1985: 33–4) explains the idea as follows:

> Surplus itself, in all societies, refers to the difference between the volume of production needed to maintain the work force and the volume of production the work force produces. It is not always easy to measure this difference with exactitude, or to compare one surplus with another when the two are embodied in different kinds of goods. But the general notion of a margin over and above that required for the maintenance—the 'reproduction'—of society is a basic concept of classical political economy that offers no stumbling block (see, for example, Adam Smith . . .).

The production of a surplus over the needs of daily maintenance allows both the emergence of class society and social progress. The support of non-productive workers, including those employed in the state and in religion, depends upon the production and appropriation of the surplus of direct producers. Historically, the emergence of a surplus has enabled a wider division of labour and of social classes with distinct economic roles. The character of social progress, however, depends in large part upon the ways surplus is appropriated and the ends for which it is used.

Surplus has been appropriated in many different ways. *Jajmani* exchanges in the classical South Asian caste system laid down rules for the distribution of surplus, so that artisans were paid an amount of grain for their goods and services, and temples were allocated another amount. These *jajmani* rules codified established practice with respect to the distribution (or appropriation) of surplus in kind, that is, primarily grain.

Such rules introduce a useful distinction between surplus *product* and surplus *labour time*. *Jajmani* practices concerned the distribution of surplus product. In capitalist society it is not, in general, products in kind which are appropriated, but the labour time of employees. Under capitalism, the products of labour are deemed the property of the employer (the owner of the means of production), rather than the property of the producer. The employer owns all production, and pays the workers a wage out of the returns from the sale of those goods or services. In this case, the appropriation of surplus can be measured by the proportion of labour time generating the profits accruing to the employer relative to the labour time generating the wage paid to the worker. The employers of labour appropriate the surplus labour power of their employees. The scale of appropriation can be measured in labour time, or as the appropriation of the value of the surplus product. The two concepts of surplus labour power and surplus value are often used interchangeably (see, for example, Shaikh, 1987: 348).

The form of surplus appropriation, and the ends for which it is used, have consequences not only for the making of rich and poor, but for the direction of social change, and particularly for the pace and direction of economic change. For example, surplus appropriation in feudal Europe established an economic system which was 'largely incompatible with, if not positively antithetical to, specialization, productive investment and innovation in agriculture' (Brenner, 1987). The kind payments made by serfs to lords in exchange for land and 'protection' established an economic system prone to stagnation and periodic crises.

The differences between regimes of accumulation in Bangladesh are, of course, less stark, less extensive and less comprehensive than the differences between modern capitalism and Middle Ages feudalism in Europe. The parallels, in the forms of surplus appropriation and relations of production, nevertheless raise important questions.

In rural Bangladesh, surplus is appropriated as product, labour time and value. Surplus product is appropriated through sharecropping and other kind payments for loans, labour power and commodities. The hiring of labourers appropriates surplus time. Trade and lending of various kinds may appropriate or redistribute surplus in the forms of product and/or value.

Contrasts in rural Bangladesh in the form of surplus appropriation in kind rather than labour time, for example, arise from significant differences in relations of production. Where these differences in relations of production are extensive, they may indicate differences in class domination, and the aggregate use of appropriated surplus. The contrasts this paper describes in Bangladesh raise questions about the appropriation of surplus product by large traders and landlords, and its consequences for investment by cultivators. Is this form of surplus appropriation antithetical to 'specialization, productive investment and innovation in agriculture'?

In a number of publications, Wood has suggested that rural class relations in Bangladesh may be characterised by the use of capital in exchange rather than in production (1984), that entrepreneurialism is required to shift investment away from rent and usury into forms which bring increased productivity (1991), that monetary wealth is not invested in productive agricultural capital. Thus, he argues that the labour process in agriculture is not transformed:

> ... this wealth stimulates the attempt among tenants and labourers to achieve a rise in the level of absolute surplus-value of their labour, where they try to work longer or harder without any capital-assisted increase in their productivity to meet their debt, rental and purchase obligations (Wood, 1993: 107)

Wood concludes that the use of capital in exchange raises the level of absolute surplus value accruing to certain classes without generating productive agricultural capital (see also Wood, this volume).

Previous empirical discussions of the 'marketed surplus' (Quasem, 1987; Chowdhury, 1988) have tended to ignore the possibility that there could be accumulation in the exchange process. These discussions have also ignored non-market transfers, including share-crop payments and kind loans. The discussion between Quasem and Chowdhury, for example, assumes that all grain which is not marketed is retained for household use. This chapter shows that assumption to be wide off the mark.

CONTRASTS BETWEEN 'ADVANCED' AND 'BACKWARD' AREAS

There is considerable regional variation in the countryside of Bangladesh. This variation includes differences in vulnerability to flood, drought and cyclone, as well as differences in social relationships of production, reproduction and exchange. This study examines some of the variation by focusing on areas of agricultural growth and stagnation. We studied an 'advanced' area, where there has been substantial investment in agriculture, and a 'backward' area where there has not (Crow, forthcoming a).

The advanced area is in the north-west of the country, in Bogra district. It is an area of differentiated owner-occupancy, where the expansion of tubewell irrigation in the 1970s allowed several crops to be grown each year. Green revolution technologies, including high yielding seed varieties and heavy application of fertiliser, have been encouraged by the wide availability of irrigation. This is an area, in other words, where there is considerable investment in agriculture. Growth of agricultural output was reported at 6 per cent per annum prior to our fieldwork (Government of Bangladesh, 1985). In the early 1980s, the district was reported to be producing a grain surplus of 82–100 per cent over local needs (Giasuddin and Hamid, 1986: 16). In 1997, this district remained an area of substantial foodgrain export. Our data include four villages in this area.

The 'backward' area, where agricultural growth was stagnant just prior to our fieldwork (0.3 per cent per annum), is in Noakhali in the south-east of the country. In this region we sampled four villages in two contrasting areas: two villages in a peri-urban, 'plains' area where there is some irrigation and many households produce more than one crop per year, and a *char* area, where land had recently been reclaimed and resettled after riverine erosion. This latter area, with a grain deficit of 8–16 per cent in the early 1980s (Giasuddin and Hamid, 1986: 26), drew our attention because it appears to be broadly similar to many *char*, *bhil*, and flood-prone areas constituting a substantial part of the rural area of Bangladesh. These areas have extensive cultivation, very limited irrigation, complex and contested tenurial relations, often with a high proportion of sharecropped land, and low levels of agricultural investment and growth. There is a growing literature describing conditions in these 'backward areas' (Ahmed and Jenkins, 1991; Hanchette and Alam, 1993; M.Q. Zaman, 1991). Frequently these are areas where poor households have gained access to land under the patronage of large landholders, whose holding may derive from forcible 'grabbing' of the land.

In the Noakhali *chars* we studied, land which had been lost to river and sea action was reclaimed by the construction, in the 1970s, of a long 'cross-dam' made of earth. Poor households migrated from other areas where they had been dispossessed (primarily by erosion and social differentiation), and began cultivating the new land. Struggles (see particularly Ahmed and Jenkins, 1991) in the courts between different groups claiming ownership of the land, and on the land between *lathials* (thugs) and settlers, created a relationship between a class of powerful urban 'landowners' which employed the *lathials*, and a class of generally poor cultivating households. This relationship is legitimated and reproduced primarily through sharecropping contracts agreed to under duress (ibid.). Cultivating households have access to relatively extensive areas of land, which are generally not irrigated, and grow a single crop of paddy each year. In this area, finance and grain markets are dominated by commission agents (*aratdars*) in a local town. Levels of investment in new agricultural technologies (irrigation, drainage, fertilisers) are lower than in the advanced area of Bogra.

The contrast between these advanced and backward areas, with the Noakhali 'plains' villages somewhere in between, lies behind their contrasting records of agricultural growth and stagnation. This contrast can be drawn on three related axes. There is the contrast between growth and stagnation, with which we started. This contrast is what needs explanation. Behind that is the contrast between high and low levels of investment in agriculture, which may plausibly be connected to growth and stagnation. At an angle to these two axes is the question of different environmental conditions—long-established land, newly-created land, and particular vulnerabilities to flood, drought, cyclone and erosion. A third axis of contrast concerns the social conditions. In Bogra, landowners intensively cultivate small plots of land. In Noakhali, sharecroppers produce one crop per year on relatively large plots under share tenancy agreements. How do these three axes—(a) low and high levels of investment, (b) differing environmental conditions, and (c) intensively-farmed agriculture versus extensive sharecropping—relate to one another?

The history of settlement over the last two to three decades (including the 1990s) in the backward area illuminates the relation between absentee landowners and cultivating households in the *chars* of Noakhali. Erosion, followed by land reclamation, provided the settlement conditions in which a class of powerful individuals could establish an agricultural economy based on sharecropping contracts, which has some parallels with the medieval European feudal economy.

Less obviously, the much longer history of land settlement in Bogra also illuminates the agricultural conditions prevailing there. The absence of a large landowning class, the small size of holdings, and the prevalence of owner cultivation in Bogra can be traced through the evolution of landholding since the Permanent Settlement of that area. Permanent Settlement under the British Raj established a class of zamindars as owners of the land, who operated it through intermediaries, *jotedars* and others. Since independence the power of the (largely Hindu) zamindars has been undermined by legal measures (against zamindari and against moneylending) which reflect a more general decline of state support for this agrarian class. Land ownership devolved to the intermediaries, the forerunners of the contemporary rich peasant class.

The relationship between the three axes of contrast may be something like this: physical environment and social history lie behind differing patterns of land access and production relations. In turn, those patterns provide some initial understanding of why agricultural investment may be greater or lesser in each area. The analysis of surplus production in both areas provides some support for this causal chain.

ANALYSING CONTEMPORARY RURAL CLASSES

Analysis of surplus production in any society depends upon some prior categorisation of the division of labour. Within the Marxian tradition, the categories arise from the empirical identification of social classes.

There is an important and unresolved debate, focused on the countryside of South Asia, about how to identify class. The debate draws upon the theoretical and empirical work of Marx, Lenin and Mao. Social class is most commonly identified in South Asia with the ownership of land. Thus, households are placed in categories of large, medium and small farmers and the landless, using size of owned land. Two alternative methods of identification use the hiring and selling of labour power (Patnaik, 1987; Bardhan, 1984: Chapter 7) and the surplus or deficit of the household (Athreya et al., 1987; Athreya and Djurfeldt, 1990). All of these methods are subject to theoretical and empirical reservations which I begin to explore elsewhere (Crow, 1996).

In this chapter, the household is taken as the unit for the analysis of class; gender relations and class relations are assumed to be separable, and a two-dimensional measure of class is used: the classification of households by principal source of income and by hiring and selling of labour power.[1] Gender relations in the countryside of Bangladesh vary markedly with class, with rich peasant households enforcing women's seclusion to a much greater extent than poor peasant households (H. Zaman, 1996). Nonetheless, significant agricultural production decisions appear to be taken for both rich and poor households by men. Large scale market interactions are also undertaken almost entirely by men. The patriarchal farm system of Bangladesh may be 'shaky' (Kabeer, 1988) but this is still an area of 'classic patriarchy'. One careful study of gender relations in a village in Rajshahi reiterates the 'Bengali peasant world-view': 'men are considered masters of the household, the man is destined to exercise his mastery .... He has the right to make decisions about what to produce, and how to use and exploit the land' (H. Zaman, 1996: 124) The assumptions of household unity and of the separability of class and gender appear, therefore, to be reasonable working assumptions.[2]

The data used in this study come from monthly interviews over an eighteen-month period of a sample of 200 households drawn from eight villages: four in Bogra, two in the Noakhali *chars*, and two in the Noakhali plains. This is a stratified sample which seeks to select the main classes of peasant household—rich, middle, poor and landless—in approximately equal numbers. The frequency of households in each of the categories of this two-dimensional matrix is shown in Table 6.1.

Table 6.1
Numbers of Households by Class and Income Group

Social Class	Basis of Classification (Simplified)	Largest Source of Income (Cash and Product)		
		Agricultural	Non-Agricultural	Labour Power Sale
Landlord	0 family labour	2		
Rich peasant	net hiring > family labour	44	11	
Middle peasant	net hiring = family labour	29	12	7

Table 6.1 contd

Social Class	Basis of Classification (Simplified)	Largest Source of Income (Cash and Product)		
		Agricultural	Non-Agricultural	Labour Power Sale
Small peasant	net selling > family labour	16	13	16
Poor peasant	net selling >> family labour	7	4	20
Landless	0 family labour	2	1	3

Sources: The data for all tables and figures in this paper come from regular, monthly intervies in 1987–89 of the 200 households in eight villages included in the field research.

DISTRIBUTION OF OUTPUT

Table 6.2 shows the mean output of all households in our sample, and that of agricultural households (that is, households with the largest share of their income coming from cultivation). Rich peasants depending primarily on agriculture grow on an average 223 Maund per year of paddy (8,300 kg; approximately 8 tonnes) compared to the 50 Maund per year (1,865 kg) produced by poor peasant households.

Table 6.2
Gross Paddy Output, Agricultural Households by Class

Class	Gross Paddy Output (Maund/Year)	
	All Households	Cultivating Households
Landlord	180	180
Rich peasant	186	223
Middle peasant	96	138
Small peasant	49	68
Poor peasant	26	50
Landless	4	

Sales of grain

Table 6.3 shows the means of the ratio of sales to output. This table includes all households with recorded grain output. So some households

Table 6.3
Proportion of Output Sold, by Class and Region

Class	Sales as a Proportion of Output (%) by Region		
	Bogra	Noakhali Plains	Noakhali Chars
Landlord	31		
Rich peasant	44		22
Middle peasant	46	17	17
Small peasant	32	13	14
Poor peasant	30		1

fall in the non-agricultural and labour sale, as well as in the agricultural, income groups.

The table reveals that there are higher levels of sales as a proportion of output in the Bogra region than in the two Noakhali regions. In Bogra, rich and middle peasants sell respectively 44 per cent and 46 per cent of output. In the Noakhali *chars*, rich and middle peasants sell respectively 22 per cent and 17 per cent of output, and in the plains, middle peasants sell 17 per cent of their output. In Bogra, even small and poor peasants sell 32 per cent and 30 per cent respectively of their output, while in the *chars*, the corresponding figures are 14 and 1 per cent. In the plains, small peasant households average sales of 13 per cent of their output.

In Noakhali, particularly in the *char* areas, output is appropriated as product, and the grain is brought to market not by the producer but by the landlord or merchant appropriating the product. This we will explore further in the final two sections of this paper.

Payments in Kind

Figure 6.1 represents the proportion of grain output paid for land by cultivating households in each class in the three areas. The figure reveals a marked contrast between the advanced area, Bogra, and the backward area, the Noakhali *chars*. In the latter, three classes of cultivating household pay roughly the same proportion of output for the use of land. Rich, middle and small peasants in this region pay 35–40 per cent of output as a share payment. By contrast, in Bogra, there is a marked stratification in share payments. Here the poor pay more of their output for land than the rich. Nevertheless, in this advanced area, the average share payments for

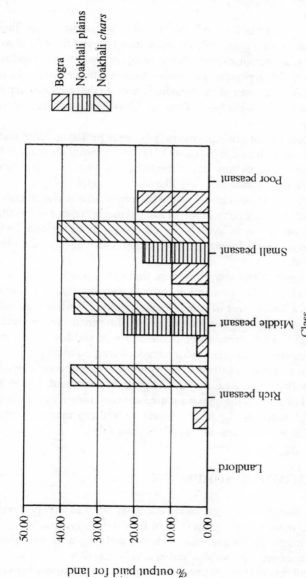

Figure 6.1: Proportion of grain output paid for land by cultivating households, by class and region

land even among the poor peasants are at a lower level, roughly 20 per cent of output.

After grain payments for land, the largest grain payments are those made for cash loans. Figure 6.2, showing loan repayment in kind as a proportion of grain output, exposes the regional concentration of this form of kind payment. Grain payments for cash loans are almost entirely concentrated in the backward area of the Noakhali *chars*. In that area loan repayments amount to 15–24 per cent of output. Elsewhere, loan repayments are negligible.

The final category of grain payments, payments for labour, commodities and services, is shown in Figure 6.3. These payments are made for a variety of goods and activities. Use of threshing machines in the advanced area is often paid for in grain, and likewise the supply of pots in the backward area. Some farm labour in the backward area is also remunerated in grain. Nevertheless, this category of payments consumes less than 5 per cent of output on an average for all classes in all regions. Grain payments are generally higher in the Noakhali *chars* than in the other regions studied.

One significant conclusion from these findings is that lower sales do not correspond to higher production for own consumption. It could be assumed that a lower level of sales from agriculture is associated with higher levels of stocks for own consumption. But this is not the picture which emerges from the comparison of the backward and advanced areas. Although sales are much higher in the advanced area, 50 to 60 per cent of grain output is retained by growers. With much lower proportions of output sold in the backward area, only 20–30 per cent is retained. The levels of stocks and of sales in the backward area are constrained by payments for land and finance. In the next section, we will examine the consequences of this for the grain surplus of households.

DISTRIBUTION OF SURPLUS

We turn now from the consideration of grain payments as a proportion of output, to the more revealing question of how these payments compare with the surplus product of each class of household. This analysis exposes large differences between surplus and deficit households.

Robert Heilbroner (1985) delineates surplus as 'the difference between the volume of production needed to maintain the work force and the volume of production the work force produces'. Figure 6.4 illustrates one measure

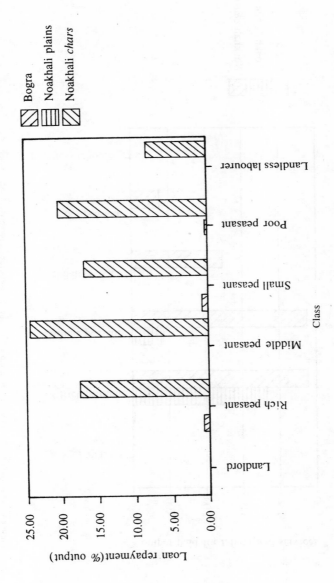

Figure 6.2: Loan repayment in grain as proportion of output, all households by class and region

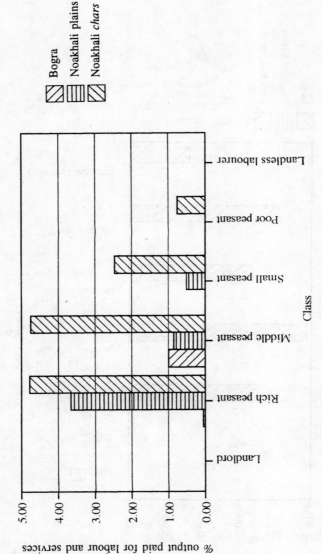

Figure 6.3: Proportion of output paid in grain for labour, commodities and services, all households

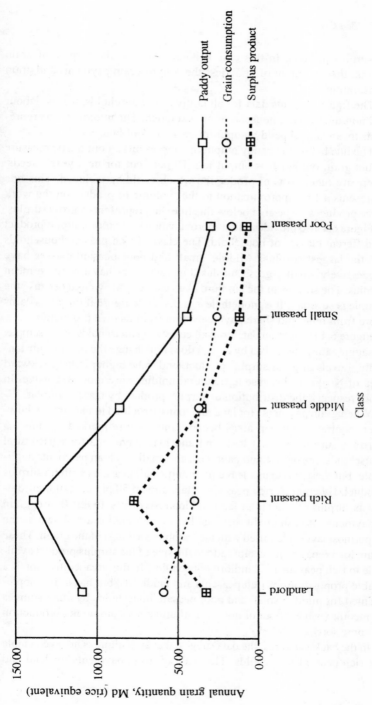

Figure 6.4: Output, grain consumption and surplus product, for all agricultural households

of surplus product. In this case it is measured as the surplus of grain output, that is, output of grain less the sum of grain payments and grain consumption.

The figure presents data for all cultivating households, that is, labour and non-agricultural households are excluded. The unbroken line represents mean annual paddy output for each class of household.

The line below the paddy output line represents an optimistic measure of net grain output. It is output less 10 per cent for next year's seeds. There are other costs of production which could be included. This line represents a first approximation to the 'volume of production the work force produces'. The area below this line then represents a gross surplus.

Figure 6.4 also provides a first approximation to the mean surplus product of different classes of household. The class of rich peasant households has the largest surplus.[3] Middle, small and poor peasant classes have successively smaller gross surplus, but each class has a positive mean surplus. The surplus of the landlord class carries little weight because the sample is so small. It is, nonetheless, a plausible figure if this class earns more from rents and share payments than from its own production.

Figure 6.4 is based on data from all cultivating households in the sample. Disaggregating these data by region does not change the picture substantially. Levels of gross surplus product tend to be higher in the backward area of Noakhali, because agricultural holdings are more extensive. In other respects, the distribution of surplus product by class is similar.

When grain payments for land and loans are added to the picture, however, contrasts between areas become much more marked. Figure 6.5 portrays surplus product less land and loan payments for agricultural households in Bogra. Land payments are small and loan payments negligible, but these payments leave poor and small peasants with no surplus product. By contrast rich peasants retain almost 50 per cent of their output as surplus product after land and loan payments. In this figure, grain repayments of cash credit received by the households are shown as the uppermost layer (identified with the 'bubble' shading) of the graph. These grain loan repayment receipts add to the size of the surplus product available to rich peasant and landlord households. In this area, cultivation is a viable proposition for rich peasants and landlords, but has to be supplemented for middle, small and poor peasant households by other sources of income (wage labour or non-agricultural petty commodity production of some sort).

In the backward area, the data suggest that agriculture is not even viable for rich peasant households. The scale of grain payments for land and

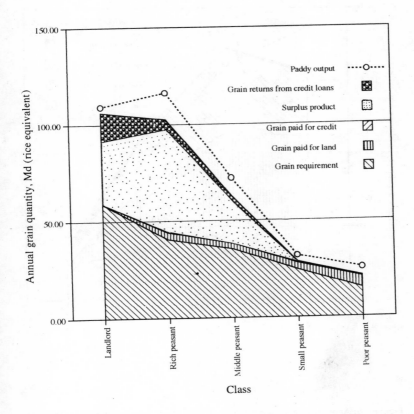

Figure 6.5: Appropriation of surplus product, agricultural households in Bogra region

loans is such that middle and small peasants are deeply in deficit, poor peasants are (possibly) just meeting basic production costs, while rich peasants generate a negligible surplus.

Figure 6.6 portrays the scale and appropriation of surplus product in the backward area. The more extensive scale of cultivation and the larger overall output from these households is overshadowed by the size of the payments all households make for land and loans. The addition of the three lowest shaded areas (grain consumption, grain paid for land and

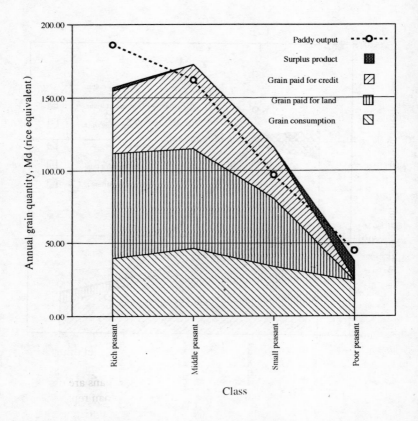

Figure 6.6: Appropriation of surplus product, agricultural households in Noakhali *chars*

grain paid for loans) exceeds the gross paddy output of the middle and small peasant households. These households are unable to meet their consumption needs from their own production if they make land and loan payments. These are, therefore, deficit households needing to purchase grain to meet their needs.

Rich peasant households in the Noakhali *chars* generate a small surplus product, but the data suggest that even those households will need to supplement their income to meet other costs of production and consumption. Poor peasant households are also shown as generating a small surplus product. However, the absence of land and loan payments generated for this class is implausible.

LOCUS OF ACCUMULATION

There is a stark contrast between the surpluses of the rich and middle peasants in the advanced area and the deficits of virtually all classes in the backward area. Only two classes of agriculturalists in the advanced area have the potential to sell a surplus and invest in agriculture. All other classes in that region and all classes in the backward region lack that potential.

In the advanced area, rich and middle peasants have a surplus product because they can hire the labour power of other classes. They can use that surplus for investment partly because they have gained a degree of freedom from high cost forms of credit. (I have shown elsewhere that peasants in this area rely firstly upon low cost loans from friends and relatives: Crow and Murshid, 1994: 1015.)

In the backward area, rich peasants hire the labour power of other classes, but the surplus product they accumulate is appropriated by others. Those who 'own' the land receive share payments. We saw in an earlier section that this category includes those who 'grabbed' the land during the settlement of this backward area. Those who lend money to these peasants receive most of the remainder of the surplus product of the rich peasants.

Here we examine how these kind payments for cash loans are distributed. Table 6.4 is a matrix describing the frequency of loan repayments in kind to each category of lender by each class in the three areas. The table shows that there are marked differences in the frequency and source of these loans among the three areas. Kind repayment of cash loans is frequent in the Noakhali *chars*, rare in the Noakhali plains, and relatively rare in Bogra. The sources also differ in the three regions. In Bogra, neighbours and relatives provide 96 per cent of these loans. In the Noakhali plains, the only loan recorded comes from a neighbour. By contrast, in the *chars*, traders of various sorts provide almost 60 per cent of these loans.

It is worth examining the distribution of lenders and creditors in the backward area of the Noakhali *chars* in more detail. Two important points can be drawn in this regard. First, traders are the most frequent lenders

Table 6.4
Kind Repayment of Loans: Frequency by Class and Region

Lender	Rich Peasant	Middle Peasant	Small Peasant	Poor Peasant	Landless Labour	%
Bogra						
Neighbour	3	1	8	2		64
Relative	1		5	1		32
Mahajan	1					5
Total	5	1	13	3		100
Noakhali plains						
Neighbour	1	1				101
Noakhali *chars*						
Aratdar	7	6	8	1		23
Grower	10	3	2	4	1	21
Samiti	9		2			12
Bepari	6	2				8
Village store	1	2	3	1		7
Mahajan	1	3	1			5
Other	11	7	2	3	0	23
Total	45	23	18	9	1	99

and thus recipients of surplus product appropriated through finance. Second, the sources of finance are distinguished by class of borrower. Traders are the major source of finance for small and middle peasants. By contrast, rich peasant households manage to get access to finance from other growers and clubs of growers. Thus, the rich peasant class may retain a share of surplus product appropriated through finance. A smaller portion of their appropriated product goes to traders.

The most frequent recipient is the *aratdar*, or commission agent, with 23 per cent of transactions. The *aratdar* is a broker of rice or paddy. Next comes the grower, that is, a cultivator of some kind, with 21 per cent. The next most frequent category of lender is the *samiti*, an informal association or club set up to lend to growers. In this area the growers' *samiti* is dominated by, and lends to, rich peasants. It should not be mistaken for a rotating loan association independent of class. The total of payments to all categories of trader (that is, excluding *samiti*, grower, relative, beggar and labour) accounts for 59 per cent of kind loan repayments. Traders big and small are the most frequent recipients of surplus product through loan repayments in kind. But the pattern is differentiated by class.

The most frequent creditors of each class can be found by examining the cell totals in each column of the table. This suggests that there is a class pattern to giving of this type of credit. Rich peasant households take credit most often from other growers (ten transactions), followed by *samiti* (nine), *aratdars* (seven), itinerant paddy traders or *beparis* (six), and a wide selection of others (eleven). So rich peasant households borrow primarily from growers and clubs of growers (42 per cent), and secondarily from traders (29 per cent). By contrast, small peasant households borrow from *aratdars* (eight transactions), then the local store, *mudi dokan*, (three), and growers (two). For this class, traders provide 56 per cent of loans and growers only 11 per cent. The distribution for middle peasants is closer to that of small peasants than to that of rich peasant households. Fifty-seven per cent of their loan repayments go to traders, and only 9 per cent to growers.

Surplus Value

The appropriation of surplus value occurs in two main forms in rural Bangladesh. Traders and financiers appropriate value through their trading and lending activities. The employers of labour appropriate the surplus labour of their employees, and this too can be measured as an appropriation of surplus value.

One way of looking at the appropriation of value in trade is simply to compare prices of sale and purchase at the two ends of market circuits. There are, broadly, two main circuits dominating the market for rice and paddy in Bangladesh. The first leads from rich peasant producers in the advanced, green revolution areas to poor peasants and cities. The second leads from rich peasants and landlords in backward areas to poor peasants and cities. From weighted mean annual prices for producers and consumers of each class (Crow, forthcoming b), it is a simple matter to derive a rough measure of the appropriation of surplus in trade by looking at prices at each end of the main circuits. Table 6.5 describes the prices at the ends of some common circuits:

1. Rich peasant to poor peasant within one village in the advanced region
2. Rich peasant in advanced region to poor peasant in backward region
3. Poor peasant to poor peasant within one advanced village
4. Poor peasant in advanced region to poor peasant in backward region

These examples do not exhaust the range of possible circuits. Nevertheless, they provide a first approximation for the sum of costs in trade and processing and surplus value appropriated in trade. The price difference in the right hand column of Table 6.5 estimates the proportion of the final consumer price used in transport, trade and processing.

Table 6.5
Appropriation and Redistribution of Surplus through Non-Tied Trade

Circuit of Exchange		Weighted Mean Prices (taka per Md)		Price Difference (%)
From	To	Producer Sale Price, Rice Equivalent	Consumer Price of Rice	
Rich peasant producer, village 1, Bogra	Poor peasant consumer, village 1, Bogra	305	348	12
Rich peasant producer, village 1, Bogra	Poor peasant consumer, Noakhali plains, village 5	305	408	25
Poor peasant producer, village 1, Bogra	Poor peasant consumer, village 1, Bogra	295	348	15
Poor peasant producer, village 1, Bogra	Poor pesant consumer, Noakhali plains, village 6	295	398	26

These prices are the average for non-tied exchanges. They do not include those prices set by prior agreement as occurs in the case of cash loans for grain payment primarily found in the Noakhali *chars*.

This analysis of paddy and rice prices at each end of two common circuits suggests that the surplus appropriated in commodity trade is substantial but probably of similar magnitude to transport costs. This may be taken as a limited corroboration of the findings of market integration studies that price differences between wholesalers in major market towns are generally similar to the prices for transport (see for example Das et al., 1997).

Taken together with earlier findings in this paper, the relatively small level of surplus value appropriated or redistributed in commodity trade suggests that the major form of surplus appropriation in exchange is an appropriation of surplus product.

CONCLUSION

Why is agrarian growth uneven? Because agrarian structure varies from one region to the next. Some articulations of rural classes encourage investment in new agricultural technologies, others discourage investment. This statement does not prophesy that this pattern of uneven growth is set in stone. However, neither does it suggest that different agrarian structures will necessarily bow down before groundwater markets and deregulation. In this concluding section, I do three things. First, I review the argument of this paper. Second, I describe the changes in the advanced and backward areas apparent from a brief visit in 1997. Finally, I note some unresearched connections between backward and advanced regimes of accumulation.

First, it is the argument of this paper that the contrasting growth rates in Bogra and Noakhali can be plausibly explained by the patterns of surplus product appropriation described here. It is only in the advanced area that an investible surplus is available to agriculturalists. The much higher levels of agricultural investment characteristic of the advanced area appear to be associated with this accumulation in the hands of agriculturalists.

In the backward area, it is the urban 'landowners' and moneylending merchants who are accumulating. These groups could invest in technical change in agriculture. At the time of this study, however, they were not doing so. The moneylending merchants were investing in agriculture: their credit made production possible. But they opposed technical and social change. Elsewhere (Crow and Murshid, 1994), I have described their effective opposition to the extension of a successful Dutch-financed land reclamation project. This project brought the water control which allowed higher productivity but also threatened to undermine, by example, the sharecropping and credit arrangements through which the urban landowning and moneylending group retained their control over peasant cultivators and their surplus product.

Differences in the forms and use of surplus in these two areas can best be understood by the particular histories of the two areas. The emergence of peasant smallholding in Bogra is the outcome of historical change primarily since 1947, not, as some new institutional accounts would suggest, of the choices of individual farmers. Similarly, the construction of an extensive, sharecropping agriculture under absentee landownership and merchant usury in the backward area can best be explained through the particular social and environmental changes of the last thirty years, not by choices with regard to risk and supervision.

Second, a brief visit to these areas in 1997 revealed some apparent changes and some continuity. In Bogra, surplus production has continued to grow and new mills have been constructed in the district we studied. The extension of micro-credit through the Grameen Bank, the Bangladesh Rural Advancement Committee (BRAC) and other agencies has sustained many poor households and fuelled the growth of new enterprises. Local milk production, village stores and human-powered transport (richshaws and rickshaw vans) are particularly prominent amongst the enterprises of the poor and landless. More substantial peasants have invested in power tillers, whose import was liberalised in recent years. These are sometimes also used in the offseason to power three-wheelers (the onomatopoeically-named '*bot-butti*') and boats.

In the Noakhali *chars*, power tillers are more slowly replacing buffalo- and ox-drawn ploughs. Better roads have made this area more accessible. But there has been no investment in irrigation and drainage and micro-credit agencies have not yet reached here. The scale of, and rates of return to, cash lending for repayment in grain appear to have declined in recent years. The agrarian structure which enables the high levels of appropriation of surplus product does appear to be on the retreat. The advance of improved communications, and the wider range of traders and information that it brings, are the most obvious causes for this withdrawal. The contrast in growth rates and poorly utilised resources, however, has not yet changed in ways which are obvious to our respondents and to brief observation.

Finally, the contrast between the advanced and backward areas, and the idea of a regime of locally-specific production relations, should not distract us from the connections between regimes. There are organic connections between backward and advanced areas which deserve much more research work.

These connections include those of migration (see Rogaly, this volume, for a discussion in relation to West Bengal), flows of surplus value and flows of grain and other commodities, as well as the more widely-studied interactions among grain price levels. The settlement and production relations of the backward area, for example, depend upon the existence of a 'reserve army of the dispossessed'. Without the geographic, demographic and social forces—riverine erosion, flood, cyclone, drought, demographic responses to vulnerability, social differentiation—which create such an army, it would have been impossible to settle the area and impose sharecropping conditions upon the settlers. Residence under the punitive economic circumstances of the area is encouraged by the continued creation of the army of the dispossessed (see also Bharadwaj, 1985, for a closely related argument). The existence of the reserve army ensures that there are no alternative livelihood opportunities which would allow exit from this area. The settlement and reproduction of the backward area is, in other words, linked to the generation of landless households, whether by geographic, demographic or social means. The rise of an urban, female proletariat providing cheap labour for garment manufacture shares its origins in the processes which generate an army of the dispossessed (see Kabeer, 1991).

There is thus an irony in the finding that some of the surplus from the backward area has enabled one of its most powerful traders to become a part-owner of garment factories in Dhaka. The forces of dispossession, which underpinned his accumulation of surplus product from backward agriculture, also provide labour for his global market factories. The most backward, apparently remote, forms of rural production turn out to be intimately connected with the most advanced forms of urban industry producing for the global market.

NOTES

1. A ratio of the hire and sale of labour power to the use of family labour power is used as follows:

$$\text{Index of labour use } (K) = \frac{\text{Days of labour hired for agricultural and non-agricultural purposes} - \text{Family labour days sold (agricultural and non-agricultural)} +/- \text{Labour days appropriated or lost through land rent}}{\text{Family labour days (agricultural + non-agricultural productive activities)}}$$

This is an extension of Patnaik's labour exploitation ratio to include non-agricultural activities.

2. There is an additional problem with the assumption that class is constructed at the level of the household. This is that multiple sources of work and income (see for example van Schendel, 1982: 38) appear to generate multiple or individualised class locations. The use of a two-dimensional measure of class, in which principal source of income can be disaggregated, reduces this problem.

3. The surplus product calculated in this paper is the net grain surplus (output less consumption and less 10 per cent for seeds and losses). This is an optimistic measure of the surplus product of the household. A household cannot live by grain alone.

REFERENCES

Ahmed, M. and A. Jenkins. 1991. 'Traditional Land Grabbing and Settlement Pattern in the South Eastern Delta' in K. Maudood Elahi, K. Saleh Ahmed and M. Mafizuddin (eds) *Riverbank Erosion, Flood and Population Displacement in Bangladesh.* Savar, Dhaka: Jahangirnagar University.

Athreya, V., G. Boklin, G. Djurfeldt, and S. Lindberg. 1987. 'Identification of Agrarian Classes: A Methodological Essay with Empirical Material from South India,' *Journal of Peasant Studies*, 17(3): 457–63.

Athreya, V.B., and G. Djurfeldt. 1990. *Barriers Broken: Production Relations and Agrarian Change.* New Delhi: Sage.

Bardhan, Pranab. 1984. *Land, Labor and Rural Poverty: Essays in Development Economics.* New York: Columbia University Press.

Bharadwaj, K. 1985. 'A View on Commercialisation in Indian Agriculture and the Development of Capitalism,' *Journal of Peasant Studies*, 12: 7–25, 4 July 1985.

Brenner, R. 1987. 'Feudalism,' John Eatwell, Milgate and Newman (eds) *The New Palgrave Dictionary of Economics: Marxian Economics.* London: Macmillan.

Chowdhury, N. 1988. 'Farmers' Participation in the Paddy Markets, their Marketed Surplus and Factors Affecting it in Bangladesh: A Comment,' *Bangladesh Development Studies*, 16: 99–111, March 1988.

Crow, Ben. 1996. 'Class in the Countryside of South Asia.' Paper presented at the 1996 American Sociological Association meetings, New York.

———. (forthcoming a) in Barbara Harriss-White (ed) *Visible Hands.*

———. (forthcoming b) 'Class and Seasonal Differences in Exchange Conditions in Rural Bangladesh.' To appear in Bernstein, H. (ed.) *Commodity Exchange and Food Systems in Developing Countries: Processes and Practices.* London: Macmillan.

Crow, Ben, and K.A.S. Murshid. 1994. 'Economic Returns to Social Power: Merchants' Finance and Interlinkage in the Grain Markets of Bangladesh,' *World Development*, 22(7): 1011–30.

Das, Jayanta, Sajjad Zohir and Bob Baluch. 1997. *The Spatial Integration of Foodgrain Markets in Bangladesh.* Dhaka: Bangladesh Institute of Development Studies, draft paper.

Giasuddin, Md and Md A. Hamid. 1986. *Foodgrain Surplus or Deficit Districts and Upazilas of Bangladesh.* Dhaka: Bangladesh Ministry of Food, Food Planning and Monitoring Unit (FPMU).

Government of Bangladesh. 1985. *Food Policy Monitoring Unit: Annual Report.* Dhaka: FPMU.

Hanchette, Suzanne, and Md Mustafa Alam. 1993. 'Riverain Sandbars (*chars*) of Bangladesh: Constraints and Potential in Socioeconomic Development.' Paper presented to the International Studies Association, Monterey, California, October 1993.

Heilbroner, R. 1985. *The Nature and Logic of Capitalism.* New York: Norton.

Islam, A., M. Hossain, N. Islam, S. Akhter, E. Harun and I.N. Efferson. 1985. '*A Benchmark*

Study of Rice Marketing in Bangladesh.' Mimeograph, Bangladesh Rice Research Institute, Bangladesh Marketing Directorate, USAID, Dhaka.

Kabeer, N. 1988. 'Subordination and Struggle: Women in Bangladesh,' *New Left Review*, No. 168: 95–121.

———. 1991. 'Cultural Dopes or Rational Fools: Women and Labor Supply in the Bangladesh Garment Industry,' *European Journal of Development Research*, 3(1): 133–60.

Patnaik, U. 1987. *Peasant Class Differentiation: A Study in Method with Reference to Haryana*. Oxford: Oxford University Press.

Quasem, M.A. 1987. 'Farmers' Participation in Paddy Markets, their Marketed Surplus and Factors Affecting it in Bangladesh,' *Bangladesh Development Studies*, 15: 83–103, March 1987.

Shaikh, A. 1987. 'Surplus Value,' John Eatwell, Milgate, and Newman (eds) *The New Palgrave Dictionary of Economics: Marxian Economics*. London: Macmillan.

van Schendel, Willem. 1982. *Peasant Mobility: The Odds of Life in Bangladesh*. New Delhi: Manohar.

Wood, Geoffrey D. 1984. Provision of Irrigation Services by the Landless: An Approach to Agrarian Reform in Bangladesh,' *Agricultural Administration*, 17: 55–80.

———. 1991. 'Agrarian Entrepreneuralism in Bangladesh,' *Indian Journal of Labour Economics*, 34(1): 13–27.

———. 1993. *Bangladesh: Whose Ideas; Whose Interests?*! Dhaka: University Press.

Zaman, H. 1996. *Women and Work in a Bangladesh Village*. Dhaka: Narigrantha Prabartana.

Zaman, M.Q. 1991. 'Social Structure and Process in Char Land Settlement in the Brahmaputra Jamuna Floodplain,' *Man*, 26(4): 673–90.

PART 2

Policies and Practices

7

Agrarian Structure and Agricultural Growth Trends in Bangladesh: The Political Economy of Technological Change and Policy Interventions

SHAPAN ADNAN

GENERAL ISSUES

The issue of whether the agrarian structure constrains agricultural growth and technological change has received renewed attention in the context of recent trends in Bangladesh.[1] Following the introduction of the new seed–fertiliser–irrigation technology in the 1960s for growing high yielding varieties (HYV) of rice, significant expansion of production has taken

---

The revised version of this paper was presented at the Workshop on Agricultural Growth and Agrarian Structure in Contemporary West Bengal and Bangladesh held in Calcutta over 9–12 January 1995. I am grateful to Ben Rogaly, Barbara Harriss-White and Sugata Bose for their help and patience in the course of revising and rewriting this paper. The work has been made possible by the support received from Shomabesh Institute, Dhaka, and the South Asian Visiting Scholars Programme of Queen Elizabeth House, International Development Centre, University of Oxford.

place. While it is evident that growth has not been constrained in terms of long term trends, there are differing interpretations of the factors promoting the overall growth, as well as of the constraints underlying periodic slowdowns (as in the mid-1980s). The dominant view attributes the primary credit for growth to a sequence of policy regimes pursued by the state in accordance with the changing dispositions of donor agencies (Hossain, 1995; Shahabuddin, this volume). However, a variant of this view claims that the initial policy regime based on public sector control was at best ineffective, and that its subsequent withdrawal provided the scope for private sector initiatives to develop the water market (based on groundwater irrigation), leading to further growth of irrigated rice production.[2]

Despite their differences, both these interpretations share in common the position that the agrarian structure has *not* been a significant factor in constraining, or otherwise influencing, expansion of technological innovation and agricultural output. When making this case, this 'structure' has been narrowly defined in terms of a number of selected quantitative variables, typically related to the unequal distribution of land (and asset) ownership, along with a restricted technical focus on one particular form of land tenure, namely sharecropping. By reducing it to a subset of 'measurable' variables, such an emasculated construct of the agrarian structure has been tailored to meet the requirements of 'multivariate analysis' concerned to evaluate its significance.[3] Not surprisingly, this method of analysis has served to exclude, by definition, the role played by *relational variables* in the process of technological change, and its impact on unequal interest groups (e.g., the unequal relationships of production, exchange and domination impinging upon different classes of the peasantry).

Furthermore, the method has also led to the drawing of questionable inferences about the causality of the processes under discussion. For example, in the interpretations just cited, public policies or private initiatives have been directly linked to quantitative results in terms of production outcomes without, however, adequately explaining the intervening steps and processes. Correspondingly, the 'growth of the market' has been viewed as a spontaneous or even 'accidental' process which has 'automatically dissolved' any structural constraints to growth (e.g., the emergence of the private sector water market as offsetting the 'malign influence of the "agrarian structure"') (Palmer-Jones, 1992: A–131).[4] In effect, these arguments have proceeded by 'knocking down a straw man' (agrarian structure), while making heroic assertions about the causality of the intervening process (how and why public policy and/or private sector initiatives have been successful in propelling technological change and agricultural growth). Consequently, little light has been shed on the role of the agrarian structure in the complex processes underlying growth (or stagnation) in agricultural production and technological change.

In the context of contemporary Bangladesh, the 'structures' mediating technological change and policy interventions extend beyond the agrarian structure, subsuming the overarching networks of the markets and bureaucracies which interlink agricultural production and exchange with the wider arenas of the state and the national and world markets. Accordingly, the problem is transformed to that of specifying the conditions under which these 'mediating structures' do, or do not, act as constraints to growth and innovation in agricultural production. For example, it would not be surprising if policy interventions which were conducive to the interests of the social groups and classes dominating these mediating structures were to prove successful, such that technological change did take place, but on *their* terms. Equally, policies which were likely to undermine their interests and dominance were susceptible to being 'distorted' by overt and covert means at the stage of implementation, such that the expansion of technology and output could become constrained.

In terms of the dynamics of the process, there could be growth, slowdown or stagnation in agricultural production at a given conjuncture, depending upon the particular alignment of interests and balance of forces obtaining at the time. Such variations over time could be quite unintended in terms of policy objectives, since the actual outcomes would be the *resultant* of the interactions between policy interventions and the responses of the social groups and classes dominating the mediating structures. As discussed further in this paper, the actual outcomes of policy interventions have diverged considerably from the expectations used to justify them. Such contradictions between policy objectives and their outcomes have served to reveal the critical influence exerted by the agrarian structure, the state and the market on the processes and consequences of technological change in agriculture. They have also raised questions about the extent to which particular policy measures have succeeded in creating an environment for *sustained* growth in agricultural production.

Furthermore, statements of intent in policy documents are not necessarily free of ideological predilections and need to be read critically to the extent that these are concerned with endowing the stated policy objectives with legitimacy and acceptability under the prevailing order. For instance, the fact that such policies had the overt objective of attaining 'self-sufficiency in foodgrain production' did not necessarily exclude the existence of a covert agenda pertaining to the actual means through which technological change to that end has been brought about. Correspondingly, there were explicit or implicit biases in policy interventions which provided scope for particular interest groups and classes to gain from the process at the expense of others. The experience of agricultural innovation in Bangladesh over recent decades exhibits the purposive promotion of 'agribusiness', involving increasing commoditisation of inputs and outputs and growing dependence on foreign aid and the world market. These outcomes had their corresponding retinues of gainers and losers, which was not unrelated to the predisposition of the concerned policy regimes and their 'unstated agenda'. Consequently, 'what has been actually intended' by the policies aimed at promoting agricultural growth and technological change is no longer self-evident, and calls for critical appraisal.

In this paper I indicate the ways in which the interlinked structures of agrarian class relationships, the market and the state in Bangladesh have influenced and refracted the policy interventions concerned with technological innovation, resulting in changing trends of agricultural production and shifts in income distribution. These general issues are addressed by focusing on specific trends in the growth rates of HYV rice production and the key technological input of irrigation. A set of exploratory hypotheses is put forward to explain these specific trends and the general issues raised by them. The interpretation put forward is situated in the context of a critical appraisal of existing views on the subject. To that end, some of the data put forward in support of the view that policy reforms have played the key role in promoting recent agricultural growth are reinterpreted to reach radically different conclusions. Relevant evidence from micro-level village studies and surveys is integrated as far as possible with macro-level data to indicate the nature of class contentions and interactive social processes shaping the overall outcomes of technological change and policy interventions. The essential objective is to put forward an interpretation which incorporates, rather than excludes, the role of the agrarian structure as well as the concomitant social relationships of production and domination, in shaping agricultural growth trends.

SPECIFIC ISSUES: EXPLAINING SHIFTS IN AGRICULTURAL TRENDS DURING THE EARLY 1990s

There has been a progressive shift in agricultural policies in Bangladesh towards privatisation, deregulation, and reduction of input subsidy which began in the mid-1970s and continued in stages up to the early 1990s. These changes have been advocated by the World Bank and many of the bilateral and multilateral donor agencies operating in the country. Along with the various political regimes in government during this period, these agencies have played a key role in formulating the successive policy packages and project interventions affecting the agricultural sector of Bangladesh.

A significant change in gear began in 1986–87, embodied in a series of policy measures which completed the privatisation of import, distribution and pricing of irrigation equipment and fertilisers by 1991–92. These measures also led to the elimination of the remaining input subsidies, while also removing virtually all the regulatory controls over utilisation of irrigation equipment exercised by public sector agencies. This round of policy interventions is henceforth referred to in brief as the 'deregulation' policy regime (also termed as 'policy reforms' in concerned official and donor circles). The achievement of relatively high growth trends in rice production during the late 1980s has been primarily attributed to this policy regime by most policy-makers, as well as by many academics, consultants and other concerned professionals:

> The Government of Bangladesh (GOB) introduced a series of policy reforms regarding the marketing of agricultural inputs and foodgrains, pricing and subsidies in the distribution of inputs and foodgrains, and in the provision of credit from financial institutions, as a conditionality of policy based lending from key donors and multilateral aid agencies. The recent success in the foodgrain sector is attributed by these agencies to the policy reforms, but the issue is still debated in the professional circles in Bangladesh (Hossain, 1995).

While agricultural growth did continue during the second half of the 1980s, the causality of the process remains to be adequately specified, particularly in terms of the way in which the deregulation policy regime interacted with the mediating structures of the market, the state and agrarian class relationships. The issues involved can be posed sharply by looking at changes in the growth trends of rice (paddy), which dominates the foodgrain production of Bangladesh, and indicators of 'modern' (mechanised) irrigation, the 'leading input' of the new technology.

Growth Trends of Rice Production and Irrigation

As described by Shahabuddin (this volume), comparison of the trends during the early 1990s (1990–91 to 1993–94) with those of the preceding decades reveals significant reversals in the growth rate of rice production. The trend growth rates of rice output during the 1970s (1972–73 to 1979–80) and the 1980s (1980–81 to 1989–90) were 2.79 and 2.32 per cent respectively. However, during the early 1990s, the trend rate of growth exhibited drastic deceleration, approaching virtual stagnation (0.37 per cent). Over the same period, rice acreage fell at the trend rate of –1.40 per cent. The causes of these trends can be ascertained by the breakdown of rice production by season and variety.

A dry winter crop for which irrigation is virtually indispensable, HYV boro has constituted the 'leading sector' of growth in rice production based on the new technology. Its trend growth rate rose dramatically from 2.12 per cent during the 1970s to 10.56 per cent in the 1980s, but then declined strikingly to 2.07 per cent by the early 1990s (Shahabuddin, 1995: Table 2.2). Even so, it accounted for 34.45 per cent of rice output during the early 1990s—the highest contribution amongst all rice crops (ibid.: Table 2.3). The deceleration in the growth rate of HYV boro was proximately mediated by the striking decline in the growth rate of its acreage from 11.26 to 0.97 per cent between the 1980s and the early 1990s.

In contrast, the acreage and output of HYV aman exhibited high trend growth rates during the early 1990s, which were sustained continuations of the corresponding rates during the 1980s.[5] This rain-fed crop usually requires irrigation only if monsoon rains are not available on time. However, the output and acreage of both local aman and local aus rice exhibited negative growth rates.[6] The areas under local aman and local aus are subject to competition from HYV aman and HYV boro respectively, both of which displayed positive growth rates in acreage during this period (Palmer-Jones, 1992: A–130). For the two local varieties, use of irrigation is a relatively minor factor (cf. ibid.), but fertiliser use can be significant. The other local and HYV rice crops were relatively insignificant in terms of output weightage during both the 1980s and the 1990s (and are not considered further here) (Shahabuddin, 1995: Table 2.3).

Purely technological considerations suggest that availability of irrigation is unlikely to have been a major factor affecting either the decline in the production of local aman and local aus, or the sustained growth rate in area and output of HYV aman during the early 1990s. However, it is likely to have had a significant relationship with the deceleration in the growth rate of acreage and output of HYV boro. It is thus pertinent to look at the evidence on the trends in growth rates of irrigation capability (stock of equipment) and its utilisation (irrigated area) during the late 1980s and early 1990s.

Data on different types of irrigation machines in operation over this period show a somewhat mixed trend. The number of shallow tubewells (STW) and 'private force-mode tubwells' rose by 51 per cent during the three-year period from 1987–88 to 1990–91.[7] Thereafter, the annual additions to the stock of STWs continued to rise up to 1992–93, but slumped in 1993–94.[8] Growth in the number of deep tubewells (DTW) and low-lift pumps (LLP) during the early 1990s was not particularly striking and eventually levelled off.[9]

The annual rate of growth in area under modern irrigation for the overlapping periods 1980–81 to 1989–90 (all of the 1980s) and 1985–86 to 1989–90 (the second half of the 1980s) were 9.67 and 9.01 per cent respectively (Shahabuddin, this volume: Table 5.2). However, by the early 1990s (1990–91 to 1993–94), it fell drastically to a 'near vanishing' 0.22 per cent. Because of claims which might imply the contrary, it is necessary to stress that the area under modern irrigation did *not* accelerate following deregulation of irrigation equipment in 1987–88.[10] If anything, the data suggest that it slowed down mildly during the second half of the 1980s and declined sharply thereafter. Lack of breakdown of irrigated area by crops does not allow the concurrent slowdowns in the areas under modern irrigation and HYV boro to be linked directly. However, there are reasonable grounds for concluding that the two trends reflect different facets of the same process.[11]

These observations over a short four-year span during the early 1990s could, of course, turn out to be a blip in retrospect (see Palmer-Jones, this volume), as the underlying trend unfolds in a more definitive manner.[12] However, they do indicate that, with a time lag of a few years following the round of policy shifts initiated in 1986–87, both the 'leading crop sector' and the 'leading input' of rice production experienced drastic slowdown. Since both growth in the late 1980s and slowdown in the early 1990s took place in the wake of the deregulation policy regime, the causal processes underlying this reversal in growth trends call for further exploration.

Alternative Explanations of the Slowdown in the Early 1990s

The slowdown in HYV boro production in the early 1990s has prompted speculation about technological, agronomic and hydrological limits to growth. Thus, it has been suggested that 'the effect of this "once for all" policy change has been plateauing with the exploitation of the potential of existing technologies' (Hossain, 1995). Another view has been that 'such [decline] may be explained in terms of declining soil fertility due to intensification of rice monoculture, and expansion of HYV *boro* into less suitable [marginal] tracts of land' (Shahabuddin, this volume). Yet another possible explanation is available in the view that the 'future spread of groundwater irrigation will soon require higher cost sources of groundwater as the most favourable groundwater areas become fully utilised' (Palmer-Jones, 1992: A–128).

These arguments are not to be dismissed lightly to the extent that they reflect accurately the technical and physical parameters circumscribing the production conditions of HYV boro and the supply of mechanised irrigation (particularly groundwater). What appears to be rather improbable is the *coincidence in time* required for these explanations to hold, i.e., that one or more of these limits became a 'binding constraint' to growth precisely at the conjuncture when the impact of complete deregulation, privatisation, and subsidy elimination had made themselves felt.[13] In fact, other observers in the late 1980s and early 1990s note the existence of considerable unexploited potential for irrigation through groundwater development.[14]

However, the fact that the (declining) additions to the stock of irrigation equipment during the early 1990s were associated with virtual stagnation in the area under modern irrigation suggests that mere availability of more machines did not necessarily translate into proportionate utilisation of the larger stock. This appears to be consistent with that part of Boyce's thesis which posits that irrigation can become the 'binding technological constraint' to agricultural growth in Bangladesh (and West Bengal) at the present level of development (Boyce, 1987: 252). However, the applicability of the remainder of Boyce's thesis to the scenario of the early 1990s—in terms of whether stagnation in irrigated area can be primarily attributed to failures in achieving 'collective action in water control' and conducive 'institutional change' (ibid.: 253, 247–8)—remains an open question. For this to have been the case, failures in collective action would have had to occur on a significantly large scale, and furthermore to coincide in time, to be able to account for the stagnation in irrigated area during the early 1990s. Moreover, the direction of the causal linkage between stagnation in irrigated area and slowdown in boro acreage cannot be determined a priori; in principle, either outcome could have led to the other. In any case, more appropriate evidence is needed to assess the nature of collective action and trends in institutional change characterising irrigation management, and this is taken up for detailed discussion in a later section.

As contrasted to these varied supply-side constraints to growth, arguments focusing on the relatively low price of rice and the reduced profitability of its production have been advanced to explain an earlier slowdown in agricultural growth during the mid-1980s.[15] These outcomes were attributed to the inadequacy of the (internal) demand for rice resulting from shifts in the distribution of income against the poorer classes, which in turn was related to the unequal distribution of the gains from the new technology in agriculture.[16] What is significant for present purposes is that these views were suggestive of a long term worsening of the income distribution against the poor following the adoption and expansion of the new technology in the preceding decades. This line of reasoning also has the merit of providing the scope for integrating the critical roles played by the state, the markets and agrarian class relationships in the determination of the distribution of income resulting from the new technology.

It is evident that the profitability of HYV boro is determined by the class distribution of income resulting from the way in which its production is socially organised. Specifically, it depends on the share of the agricultural income (value of product) retained by the grower in relation to the shares obtained by the other parties (classes) involved, as stipulated by the contracts related to all inputs (land, credit, labour, irrigation, fertiliser, etc.) as well as contracts related to the disposal of the product itself. A decline in the profitability of boro production would take place if the distribution of income amongst all the concerned parties (classes) shifted in ways which resulted in higher costs and/or declining returns for the producers. Such shifts in the distribution of agricultural income could be linked to policy interventions and their outcomes, mediated by agrarian class relationships, the various input and product markets, as well as the agencies of the state involved in the process. Accordingly, a plausible hypothesis would be that the slowdown in the growth of HYV boro production during the early 1990s was due to a sharp declining trend in its profitability which resulted from shifts in income distribution caused by the interaction between policy interventions and the mediating structures.

Furthermore, a declining trend in the profitability of boro production would imply a concomitant falling trend in the derived demand for irrigation. This suggests the hypothesis that the stagnation in irrigated area in the early 1990s was due to a declining trend in the profitability of irrigation induced by the declining profitability of boro production, rather than by any 'binding constraint' to the supply of irrigation. Such an interpretation would be consistent with the observation that annual increments to the stock of STWs, the most rapidly growing type of irrigation equipment, continued to increase up to 1992 93. While possible constraints to the supply of irrigation (as a flow) might still have contributed to this outcome, the point is that the activation of a critical demand constraint could explain the trends of stagnation in irrigated area during the early 1990s even if the supply of irrigation had been assured.

Furthermore, the very timing of the slowdown, occurring in the wake of the latest round of policy shifts, is also suggestive of the hypothesis that the deregulation policy regime had a precipitating role in bringing about the reversal in the overall growth trends between the 1980s and the early 1990s.[17] However, as already noted, the impact of this policy regime was superimposed on the longer term trends of worsening income distribution resulting from the on-going interaction between policy interventions and the mediating structures over the preceding decades. In view of this, the slowdown in the growth of HYV boro and irrigated area in the early 1990s could be regarded as the *conjunctural manifestation* of the longer term trends of declining profitability of boro production.

Stated in this fashion, these hypotheses remain rather general, lacking in precision. However, they do provide pointers towards the way in which the available evidence on policy interventions and their outcomes might be examined in terms of their impact on the class distribution of income and the profitability of HYV boro and irrigation. Furthermore, they also call for the identification of the critical features differentiating the deregulation policy regime from its predecessors, so that the particular factors which might have precipitated the slowdown in boro production and irrigated area in the early 1990s can be specified with greater precision. In order to establish this case, a critical reappraisal of the standard interpretations of the nature and impact of these successive policy regimes, as well as the claims made about them, is unavoidable.

In the line of argumentation put forward here, the general issues pertaining to the interaction between policy interventions and the mediating structures are integrally linked to the possible explanation of the specific issues concerning the slowdown in growth during the early 1990s. Accordingly, the rest of this paper is structured in such a way that the causal factors underlying these specific trends can be explored and situated in the broader context of the longer term impact of technological change and policy interventions. The next section provides a brief assessment of the objectives and predisposition of the policy regimes which initiated and expanded technological change in rice production in Bangladesh, covering the period from the 1970s to the mid-1980s (preceding the deregulation policy regime). The actual outcomes of the interaction of these policies with the structures of the state, the market and agrarian class relationships are taken up next. In the following section, the factors underlying trends in the profitability of HYV boro and irrigation during the mid-1980s are explored in the context of intensifying class contentions. These together provide an assessment of the 'initial conditions' in the context of which the subsequent section takes up the critical changes brought about by the deregulation policy regime and their specific impacts on the growth trends of HYV boro and irrigated area in the early 1990s. The perspectives emerging from the analysis of the political economy of technological change and policy interventions are summarised in conclusion.

POLICY INTERVENTIONS PRIOR TO DEREGULATION

Predisposition of the Initial Policy Regimes

The introduction and expansion of the new agricultural technology was undertaken through massive state interventions, backed by the aid resources, technical expertise and policy advice (sometimes pressure) of key donor agencies (Adnan, 1990: 147–8; Osmani, 1985). Far from being driven by demand from producers and relative factor prices generated by the market; it was largely a supply-led process. The adoption of innovation was promoted by sale of inputs at administered prices in which the 'rates of subsidy were calibrated in order to tailor demand in accordance with the structure of supply' (Osmani, 1985: 24). The overwhelming concern was that of promoting the new technology rather than ensuring that the benefits of the consequential productivity gains were equitably distributed (ibid.: 3).

Policies were also selective in terms of the markets chosen and the nature and the extent of interventions. Thus, provision of irrigation equipment and fertilisers was initiated through public sector distribution systems, paving the way for the emergence of 'brand new' markets for these commodities. However, in the case of markets which already existed, such as credit and product (rice), policy interventions were limited to the requirements of promoting and supporting the use of the new technology rather than any attempt to reform their pre-existing inequalities and serious imperfections (cf. Crow, 1989). Furthermore, the land and lease markets were virtually untouched, even though the distribution of land provided the basis for the most fundamental inequalities of wealth and power in the agrarian structure. Not least, there was little in the way of effective policy interventions in the rural labour market, such as ensuring minimum wages or maintaining the level of real wages. The explicit and implicit biases inherent in these policy interventions thus had the effect of leaving alone pre-existing inequalities in wealth and power, while providing scope for new inequalities via the input and the output markets.

During the 1960s and most of the 1970s, key inputs of the new technology such as irrigation equipment and fertilisers were distributed to producers at highly subsidised rates under the control of public sector agencies, primarily the Bangladesh Agricultural Development Corporation (BADC) (Osmani, 1985: 4–6; Hossain, 1995). This public corporation also undertook the procurement of inputs from both home and abroad which had the effect of shielding domestic agricultural producers from the direct impact of price fluctuations and other uncertainties of the world market (Hossain, 1995). While this public distribution system had its limitations (as later discussed), its failures were used to justify critical shifts in policy without adequate analysis of the rationale for such changes. Whatever existed in the way of policy documents relating to the changing role of the state in agricultural development were almost entirely the products of donor agencies.[18]

Under pressure from donor agencies, the regimes in government from the mid-1970s began to dismantle the constituent elements of the public distribution system and replace it with new ones (cf. Hossain, 1995). The key principles guiding these changes were reduction in input subsidy, deregulation of controls and the privatisation of procurement and distribution. However, these changes did not take place all in one go, nor did they take place concurrently in all the 'markets' concerned, so that the 'new' system underwent continuous modification over the subsequent period. The details of the policy changes also varied between inputs because of their technical characteristics.

Policy Changes Pertaining to Specific Inputs and the Product

In the earlier policy regime, BADC followed a system of renting out publicly owned DTWs and LLPs to co-operatives and a variety of informal farmers' groups with relatively low rental charges which incorporated a substantial subsidy (Osmani, 1985: 20; Palmer-Jones, 1992: A–130). In contrast, STWs were sold to private individuals or groups from the very outset, at prices which contained virtually no subsidy (Osmani, 1985;

Hossain, 1995). However, the private parties who were sufficiently wealthy and influential to buy STWs usually enjoyed the benefit of subsidised credit from publicly owned (nationalised) banks (Osmani, 1985: 20, fn; Palmer-Jones, 1992: A–130). The operation of all types of irrigation equipment including private STWs was subject to regulatory control by BADC (Palmer-Jones, 1992: A–130).

However, the disappointing performance of these programmes led to pressure from donor agencies 'to privatise the supply of STW and LLP, and the ownership of DTW'.[19] From 1978–79, the private sector was allowed to import and distribute STWs subject to 'standardisation' requirements.[20] The policy of increasing rental charges on DTW and LLP was pursued in parallel with that of selling them to co-operatives and private individuals assisted by access to cheap institutional credit. However, the structure of incentives (in terms of the remaining subsidy) 'was tilted heavily towards the sales programme' (Osmani, 1985: 20–1). By 1983, 43 per cent of operating DTWs and 48–56 per cent of the LLPs had been transferred to the private sector (ibid.: 21; Hossain, 1995), while STWs were almost entirely privately owned. This reflected the growing dominance of private ownership over the stock of irrigation equipment in the country.[21] None the less, there was a distinct slowdown in sales of STW and LLP from 1982–83, and loan repayment for private purchase of DTWs remained well behind schedule (Palmer-Jones, 1992: A–132, 135, Figure 8).

In the case of fertilisers, BADC controlled their distribution to producers through appointed dealers at administered prices incorporating a substantial subsidy (Osmani, 1985: 4–6, Hossain, 1995). During the first half of the 1970s, this system had been characterised by considerable shortage in supply and the existence of fairly institutionalised black markets where fertilisers from the public distribution system found their way to private traders (Adnan, 1990: 88). In 1978, a new marketing system was introduced in which the appointed dealers were removed and unrestricted private trading was ushered in. Controls on fertiliser prices were lifted and from 1983 private traders were formally allowed to charge a market clearing price (Osmani, 1985; Hossain, 1995). A series of reductions in fertiliser subsidy was initiated in the 1970s such that by 1985 there was no explicit subsidy on urea, while those on TSP (phosphate) and MP (potash) were reduced to 30 and 27 per cent respectively (Hossain, 1995). The striking consequence was a fifteen-fold increase in fertiliser prices during 1972–84, followed by another rise of about 30 per cent during 1984–85. The cumulative effect of these policy interventions was reflected in a reduction in the trend growth rate of total fertiliser consumption (see Shahabuddin, this volume, Table 5.1).

The state-initiated drive to expand the new technology was characterised by provision of liberal institutional credit for financing the increased market outlays necessary for producing HYV crops. This need became even greater with the progressive decline in the rate of input subsidy, as well as the privatisation of the ownership of irrigation equipment (Hossain, 1995). At the outset, the Bangladesh Krishi Bank (BKB) was the public sector organisation entrusted with financing agricultural credit. From 1977, nationalised commercial banks (NCB) in the public sector also began to undertake and expand comparable credit programmes.

It is worth noting that private sector commercial banks were not particularly drawn into financing agricultural credit. Consequently, these policy measures enabled primarily *public* sector funds to be used for financing the growing *private* sector involvement in agricultural production from the late 1970s onwards. The low rate of interest on institutional loans, compared to the much higher rates prevailing in non-institutional credit markets, also amounted to providing a substantial implicit subsiby (even more so if the borrowers could get away with debt default, which was not unusual) (cf. Palmer-Jones, 1992: A–136). However, the conditionalities of such institutional loans, such as provision of collateral and clearance of official formalities, incorporated a systematic bias in favour of those groups and classes which possessed greater wealth and influence as contrasted to the majority of rice producers who were in no position to match them.

Partly as a gesture to offset the reduction in input subsidy, measures aimed at providing price support to rice growers were initiated from 1975–76 (Osmani, 1985: 30). This took the form of limited purchase of rice at harvest time by the state through 'procurement' (purchasing) centres (Hossain, 1995). The intention was that producers would receive 'guaranteed prices' from the state, thus putting pressure on private traders in the rice market to maintain buying prices at a corresponding level. Purchases at procurement centres were actually made by dealers appointed by concerned public agencies. Unlike the markets for irrigation equipment and fertiliser, which were in effect created under state supervision, the 'state-as-trader' entered the pre-existing rice market, with its established controlling interests, as a relatively minor player.[22]

Overall, by the mid-1980s, the policy regime had been significantly privatised and partially deregulated as compared to the initial conditions characterised by overwhelming public sector control over the distribution and pricing of subsidise inputs. There had also been a clear shift to a composite strategy which attempted to maintain producers' incentives by providing price support while reducing input subsidy. This shift in the policy regime was part of a 'package deal' advocated by the World Bank and other donor agencies (Osmani, 1985: 29). In the process, the state handed over to the market and private trading interests many of the functions which it had initially performed in the social organisation of production centring around the new technology.

COLLECTIVE ACTION, CLASS CONTENTIONS AND POLICY OUTCOMES

While these policy interventions were intended to promote technological change in agriculture, it is also evident that they were concerned to propel it in certain directions rather than others. To the extent that the policy regimes provided unequal opportunities for gain to different interest groups and classes, these explicit and implicit biases were reflected at the level of implementation. Furthermore, those dominating the mediating structures responded in ways which were not necessarily the same as had been expected or presumed by policy-makers. In fact, intended policy objectives were often manipulated by those having the requisite influence to change the outcomes in their own interests, typically taking the form of a chain reaction of group interactions and institutional changes.

It is at this juncture that forms of 'collective action', atomistic competition and class contention aimed at appropriating the gains from the new technology assumed significance as key factors determining the actual outcomes of particular policy interventions. These manoeuvres and contestations took place in institutional arenas provided by concerned agencies of the state, specific markets and marketing systems, as well as the social institutions of the peasantry through which the agrarian class structure operated in overt and covert ways (Adnan, 1997).

The most appropriate evidence on such interactions between interest groups and classes, including forms of collective action related to irrigation management, is to be found in anthropological village studies[23] (i.e., those based on participant observation) and other kinds of micro-level studies, reports and surveys. The composite picture gleaned from the limited sources available is outlined below. In particular, observations from a cluster of ten villages in Pabna (the 'Daripalla' villages) where I conducted fieldwork during 1985–86 have been used as sources of case study evidence (Adnan, 1987), supplemented by data from comparable studies.

Evidence from villages in different parts of the country indicates that attempts were made to renegotiate the pre-existing input and output market contracts in order to increase the share of the agricultural product accruing to the social groups and classes with greater wealth and/or influence. This was because virtually all classes of peasant producers were in a position to make substantial gains in net income, even after taking account of the increased costs of shifting to the new technology of rice production. However, the very appearance of such incremental productivity provided the ground for reactive responses from all those who were concerned to raise their share of the greater agricultural product, or establish claims upon the new means of production required by the technology.

Contentions Related to Irrigation Equipment and the Water Market

In particular, getting control over irrigation equipment became a lucrative proposition, given the enhanced income flows which these new means of production made possible. This was as true of on-farm production of one's own, as it was of off-farm sale of water. Case study evidence from a Bogra village shows how a rich peasant patrilineage formed a fictitious co-operative in order to obtain allotment of an STW from an aid-financed rural development project (RD-1) (Herbon, 1985: 55-7). During 1980-81, the co-operative was found to have been disbanded, while the concerned rich peasant patrilineage had become the de facto owner of the STW, selling water to local producers. In the Pabna villages, some of the DTWs initially allotted to co-operatives were found to be under the private control of rich and influential peasants during 1985-86. Comparable instances of de facto private use of LLPs which had been formally leased by co-operatives during the 1980s were documented in rural areas of Chittagong (Adnan, 1989).

While some 'genuine' co-operatives managing their irrigation equipment continued to persist during the mid-1980s, these constituted relatively exceptional cases. In actual practice, the policy of preferential sale or leasing of irrigation equipment to co-operatives (or other user groups) had been subject to manipulation in ways which led to de facto private ownership in many cases. These typically involved covert negotiations between the aspiring 'co-operative' leadership and the public sector functionaries with control over the distribution of irrigation equipment, leading to mutually beneficial pay-offs. Consequently, the effective weightage of private ownership (including de facto control) of irrigation equipment during the 1980s was possibly even greater than that indicated in official figures. The policy of privatisation in effect gave legal (de jure) recognition to what had largely become a de facto reality.

Furthermore, the deliberate policy of selling irrigation equipment to private parties had the predictable consequence of creating and sustaining a market for water (irrigation as a service). This outcome was virtually inevitable given the 'minifundist' agricultural landscape of Bangladesh, characterised by extraordinary fragmentation and 'disarticulation' (Boyce, 1987: 44) of owned and operated holdings (see Wood, this volume). In the general case, the private owner of an irrigation machine was unlikely to be able to utilise it fully, since its command area need not entirely take up, or overlap with, his own operational holding. In an agrarian structure where there was a pre-existing market for virtually all key inputs and outputs of production, it was hardly surprising that the transactions of market exchange would also provide the means of matching the surplus capacity of irrigation machines with the unmet demand for irrigation. The emergence of a water market, along with a class of water sellers ('water-lords'), was thus no 'accident' which had been spontaneously activated by the spirit of private enterprise (as Palmer-Jones would like to believe).[24] Rather, the deliberate policy interventions leading to this outcome are consistent with an unstated policy agenda of purposive promotion of 'agribusiness' and its requisite support services. In this perspective, the 'growth of the (water) market' was a perfectly 'intended consequence' of privatisation policy given the pre-existing configuration of rights on land, rather than an 'independent causal factor' which served to overcome the possible impediments to growth posed by the agrarian structure.[25]

Thus, despite promotion of co-operative management of irrigation at the outset, subsequent changes in policy enabled private management to increasingly gain ascendancy. The potential for 'collective action' also differed between these two principal modes of irrigation management. In the case of genuine co-operatives and other user groups, there was no division comparable to that between buyers and sellers. All members (users) had in common the need to irrigate their respective plots within the command area. However, the mechanisms of allocating the water and recovering the costs from the members provided sources of potential disagreement and conflict amongst individual, or groups of, users (as well as between the 'management' and the users). In contrast, with private management aiming to make profits, the opposed interests of buyers and sellers with regard to the form and level of irrigation charges propelled the class division to the forefront.[26] To that extent, 'class action' assumed predominance over 'collective action' under private management of irrigation.

Case study evidence from the Pabna villages illustrates the nature of the problems affecting collective action under (genuine) co-operative management. Initially, the charge (fixed cash) collected from each member of a co-operative running a DTW was determined by the size of his irrigated plot, rather than the volume of water used. Consequently, there was no systematic relationship between the financial contribution made by a member and the costs of his actual use of irrigation. Disputes between members about their respective shares of the water and its costs were found to be endemic. This system of management was thrown into crisis by a coterie of rich peasants who refused to pay their charges while appropriating more water than their due share. Attempts to resolve the problem by the management of the co-operative as well as the indigenous institutions of conflict resolution amongst the peasantry (the *shalish*)[27] did not prove successful. This failure in achieving collective ac-tion was compounded by the fact that none of the other rich peasant members was willing to strain his relationship with the disruptive coterie (which also had comparable wealth and social influence). However, the bulk of ordinary members eventually forced the issue by collectively refusing to continue with this system of irrigation management, given its limitations. Confronted with pressure from the rank and file, a new system for managing irrigation was 'innovated', which established a direct link between the quantum of water used by a member and the actual costs incurred by him.

In this particular instance, the failure of collective action was eventually resolved by a combination of social pressure and institutional innovation. However, in other instances, the impasse might have remained, particularly given the nature of 'class inaction' exhibited by the rich peasantry.[28] None the less, the very scope of collective action problems to constrain the supply of ('minor') irrigation was already on the decline by the mid-1980s, given the increasing weightage of private management which required a qualitatively different institutional framework. (As discussed later, growing competition in the water market emerged as an additional factor contributing to this outcome.)

Observations from the Pabna villages also illustrate the distinctive features characterising, and occasionally constraining, private irrigation management. Under the system in operation during 1985–86, the owners of private DTWs paid for all overhead and operational costs, and received as payment a quarter of the standing crops on the users' plots. This system of irrigation payment was adopted after having experienced failure with earlier contractual arrangements, and was reported to have been 'copied' from the water market in a village ten miles away.[29] An endemic source of 'friction' under this arrangement concerned the way in which the standing crops on the users' plots were to be divided between them and the management. Since the crop on a given plot of land could grow unevenly, there were predictable disputes about 'which quarter' of the plot was to be harvested by the water sellers. In one instance, the management of a private DTW harvested the 'best quarter' of the crop on an irrigated plot without even consulting the user concerned. This provocative act led to the discontinuation of the contract by the latter from the following season. There were other instances of termination of contracts by users when water sellers failed to provide adequate and timely irrigation. However, the social-institutional dynamics of such disagreement also served to bring about mutative changes in the form of payment for irrigation, with cropshare gradually being replaced by 'on the spot' payment at fixed cash rates whenever a plot was irrigated. Comparable trends of a shift from cropshare to fixed charges by the season, and in some cases by the hour, have been noted in other areas of the country (Hossain, 1995).

In fact, available evidence from different parts of the country over the 1980s shows considerable variations in the institutional arrangements for irrigation payment under private management. In highly risk-prone parts of Bogra and Serajganj, one-third (as compared to one-fourth) cropshare was being charged for water in 1988 (Mandal, 1989: 56). In the intensively irrigated areas of Tangail, there were instances of both rise and fall in fixed cash charges collected by private deep tubewell operators, depending on their relative bargaining power compared to the respective water users (ibid.: 51). In one area, three out of four private DTW operators increased their fixed cash rates by 36 per cent between 1985 and 1988. In the fourth case, however, charges were reduced rather than increased, because of group (class) pressure from the buyers. In parts of rural Chittagong, private irrigation operators using low-lift pumps were found to be using their local level influence, inclusive of connections with the local administration and village power structure, to extract exorbitant rates of fixed cash charges from water users during 1988 (Adnan, 1989).

These variations over time and space were, to some extent, shaped by the relative (market) bargaining power between water sellers and users. In some cases, however, the (class) capability to use social and political influence, inclusive of manipulation of agencies of the state and structures of domination, was explicitly activated by the parties in conflict. In that sense, the *balance of class forces* played a critical role in determining the institutional arrangements for irrigation management, inclusive of the form and rate of water charges. The overall picture, however, is not one of private management grinding to a halt because of class contentions or failures in achieving collective action. Instead, there appears to have been an on-going process of adjustment by either or both parties, such that irrigation managed on a private basis has continued to remain operational through adaptive institutional changes. However, other factors have been at work in constraining the profitability of private irrigation, as taken up in later sections.

Lease Market Contentions

The policy regimes associated with the new technology did not incorporate any attempt to intervene in the tenurial arrangements prevailing in the lease market. However, landowners collecting ground rent did not display such indifference, and responded by renegotiating or scrapping tenancy contracts to increase their share of the incremental productivity gains. Village study evidence from different parts of the country shows that this happened in a variety of ways, involving different contractual forms and tenurial arrangments.

Sharecroppers in a Dhaka village were evicted during the late 1970s when the plots they had been cultivating fell within the command area of a newly installed DTW (Jansen, 1987: 179–81). The landlords concerned found it more remunerative to cultivate these plots with wage labour. While this was not the predominant pattern all over the country, it provides a limiting case of landlord response to the new technology.

The experience in the Pabna villages demonstrates how lease contracts were renegotiated rather than being scrapped altogether. Until the introduction of HYV boro cultivation with mechanised irrigation in the late 1960s, sharecropping had been the predominant form of tenancy in the locality. However, expansion of the new technology led to the displacement of cropshare by fixed crop rent on all irrigated land. Following its introduction in 1972, the rate of fixed rent was doubled the next year; another rise was imposed in 1984–85. The overall impact was to increase the rate of rent by 167 per cent within a twelve-year period. In effect, a significant proportion of the gains from the new technology was appropriated by landlords at the expense of the retained share of the tenants.

Both these rent hikes were initiated by an absentee landowner acting as the 'price-leader', but his bargaining power was promptly reinforced by the collaboration of locally resident rich peasants leasing out land.[30] All of the individual rent-receiving landowners closed ranks on this issue—constituting a matter of common class interest—despite their open differences and conflicts over other economic, social and political matters. On both occasions, the tenants were reported to have submitted to the demand for higher rates of rent without any significant resistance. Atomistic competition among themselves impeded the forging of any 'collective front' by the tenants to bargain with the class of colluding landowners. None the less, it is unlikely that 'market clearance' would have taken place so 'automatically', had there not been a show of force to prevent the formation of any counter-coalition by the tenants. As with irrigation, the impact of the new technology brought class contentions to the forefront of lease market contentions.

Observations from a village in Sylhet during 1984 indicated a comparable shift from sharecropping to fixed rent following the spread of the new technology (S. Islam, 1985: 49–52). However, the predominant form of tenancy here involved fixed rent in cash paid in advance, as contrasted to fixed rent in kind (paid in product at harvest time). The shift from kind to cash rents had the effect of increasing commoditisation, since tenants had to sell their product in the market to pay rent. When rent had to be paid in advance, borrowing from the informal credit market was often unavoidable.

However, discontinuation of sharecropping was not the only way in which landlords could appropriate the benefits of the new technology. Following the availability of canal irrigation from 1978–79, landlords in a Noakhali village evicted their poor sharecroppers and replaced them with new ones who had the requisite means and resources to grow HYV rice (Rahman, 1982: 32). Landowners were also found to be intervening in production decisions, e.g., stipulating that sharecroppers used the new technology. In remote *char* areas of Noakhali, where landlords exercised formidable social control, the rate of cropshare for HYV boro was found to be two-thirds of the gross output (rather than the usual half) (ibid.: 31). In parts of rural Dhaka, the tenancy contract was revised by landowners to include a fixed cash payment in advance, over and above the cropshare paid at harvest time (Jansen, 1987: 179–81).

It is evident that tenurial arrangements related to sharecropping have been adjusted by landlords with the power to impose requisite contractual stipulations to ensure that they could extract a greater share of the increased product resulting from the new technology. There does not appear to have been anything in the *phenomenal form* of sharecropping, as an element of the agrarian structure, which necessarily posed an impediment to agricultural innovation and growth.[31] In fact, other lease contracts, inclusive of the different forms of fixed rent, have also been comparably manipulated by landowners. The essential point is that landowners have not been averse to the expansion of the new technology, *so long as it has taken place on their terms*. Furthermore, they have also dispensed with tenancy altogether wherever that has been to their advantage.

However, extraction of increasing rates of rent through the modification of various forms of the tenancy contract had the effect of 'redistributing' the incremental productivity gains (incorporated in the higher levels of pre-capitalist ground rent) in favour of the class of landlords, whose role in the process of production was essentially parasitic (Marx, 1981: 926; Adnan, 1985: PE–56). This left the class of tenant producers with correspondingly lesser capability to invest in further innovation to raise productivity and undertake productive accumulation. It is in this sense that forms of tenancy, including sharecropping and fixed rent, have continued to pose constraints to the long run dynamics of agricultural growth and innovation in Bangladesh.

Contentions Related to the Institutional and the Non-institutional Credit Markets

Effective access to institutional credit has varied amongst different classes of peasant producers as well as over space and time. In Comilla villages which were part of the 'laboratory' area where the new agricultural technology was launched during the 1960s with the full backing of institutional credit, the presence of pervasive moneylending continued to be observed during the 1970s (Bertocci, 1970; Wood, 1978). In 1974–75, none of the producers in a Kushtia village were found to have received institutional credit for the purpose of using the new technology (Arens and van Beurden, 1977: 124). The formalities and social distance of institutional credit agencies made them virtually inaccessible to the majority of peasant producers in villages across the country.[32] This also provided scope for the continued operation of forms of non-institutional credit, such as usurious moneylending, which extracted effective interest rates of the order of 300–400 per cent per annum during the mid-1970s.

Case study evidence from a Jessore village indicates how the flourishing moneylending business had succeeded in linking up with sources of institutional credit in 1977, indicative of the emergence of covert 'vertical integration' between these two segments of the credit market (Siddiqui, 1982: 188–92). Local moneylenders were found to be borrowing from public sector banks at interest rates below 18 per cent and deploying such funds as capital for usurious loans, fetching interest at rates of the order of 120–240 per cent. In cases such as these, the very objective of providing support to agricultural production had been subverted, since institutional credit was being utilised for sustaining and expanding usury, rather than undermining it.

In contrast, usurious moneylending was reported to have been on the decline in some villages of Noakhali following significant expansion of institutional credit over 1974–79 (Rahman, 1982: 41–4; Arn and Mannan, 1982: 52–5, Tables 5.13–5.15). This expansion had been associated with the introduction of canal irrigation and the consequential production of HYV rice. Loans from banks and co-operatives were reported to be cheaper, and free of the contractual stipulations involving usufructuary mortgage, compared to loans from moneylenders. Comparable trends have been noted during the mid-1980s in parts of Bogra where the new agricultural technology had been extensively taken up (Crow, 1989: 223; Crow and Murshid, 1994). Producers were reported to have had access to cheap credit without disadvantageous contractual stipulations being involved. However, during the same period, loans from rice traders to poor producers involving disadvantageous contractual stipulations were found to have been significant in parts of Noakhali (Crow, 1989: 211). None the less, these contrasting patterns were indicative of the potential capability of institutional credit, when properly utilised, to undercut the grip of usury on agricultural production through the competitive cutting edge of lower interest rates and 'softer' contractual stipulations.[33]

However, the expansion of institutional credit associated with the HYV technology also generated new problems of its own. While usurious moneylending was relatively unimportant in the Pabna villages during 1985–86, the functionaries of public sector banks were found to have 'institutionalised' the practice of deducting a part of the loan amount sanctioned to peasant producers as bribes or 'appeasement charges.' The rate of appeasement was of the order of 9–15 per cent of the principal, with the rate being higher for poorer and weaker borrowers (Adnan, 1990: 113).

Despite expansion of institutional credit, moneylenders constituted the single largest source of credit in a Sylhet village in 1984, providing loans to 50 per cent of the borrowers (S. Islam, 1985: 53–4, Table 4.11). In comparison, banks and co-operatives provided loans to respectively 32 per cent and 14 per cent of borrowers during the same year. Since moneylenders had provided 100 per cent of loans only a couple of decades earlier, the situation in 1984 did reflect the considerable impact made by the growth of institutional credit. However, it was the relatively affluent sections of the peasantry which were the prime beneficiaries of this development. The majority of producers, belonging to the poor peasantry, still had little option but to meet their credit requirements from moneylenders and other non-institutional sources (cf. Hossain, 1995: Table 8).

While the policy objective of expanding institutional credit was attained to a significant extent, pre-existing moneylending and trading interests ('usurer's capital and merchant's capital') (Marx, 1976: 1022–3; 1981: 440–55, 728–35, 745) continued to operate in large segments of the agricultural credit market in the mid-1980s. The lack of adequate access to institutional credit constrained the majority of poor peasant producers to take loans from non-institutional sources on terms which allowed the lenders to establish contractual rights on their agricultural product. In particular, diverse forms of the credit–product (interlocked market) contract were utilised by the classes advancing credit (e.g., 'pure' moneylenders and rice traders) to obtain loan repayment in rice (paddy), at accounting prices which were lower than those prevailing in the open market.[34] The consequence was that indebted producers were compelled to sell rice which they might not have otherwise sold, or to sell at prices which were below what they could have otherwise obtained if there had been no such 'barriers to exit'. These 'interlocking' mechanisms based on the credit market generated and cumulatively reinforced a range of processes which have been termed as constrained or 'forced commercialisation' (Bhaduri, 1974; 1976; Bharadwaj, 1974).

These general trends applied to the producers of HYV boro as well, particularly given their need for relatively large amounts of working capital to pay for irrigation and fertilisers. The lack of effective access to institutional credit became crucial in this context for the majority of poor producers, since resorting to non-institutional sources meant not only higher interest rates but also increased chances of having to sell the product at lower prices. Either or both factors could reduce their net returns and the profitability of producing HYV boro. To the extent that this was actually the case, the lending classes (whether 'pure' moneylenders or rice traders) were able to gain from the new technology at the expense of the producing classes. The micro-level evidence cited earlier suggests that such outcomes continued to persist for poor producers, even though richer producers were able to take advantage of better access to institutional credit and its expansion over the preceding decades. As already noted, there were also 'unintended beneficiaries' of institutional credit including the rent-seeking functionaries of public sector credit agencies and the moneylenders and traders of the private sector, who operated either independently or in collusion.

Product Market Contentions

A key feature of the operative policy regime was the belief that price support programmes for rice producers would offset their cost increases resulting from the progressive reduction in input subsidy. In practice, the results of the price support programme have fallen far short of the policy objective. The quantity of rice procured by the state has usually been too small to provide effective price support to HYV boro producers at harvest time.[35] In many districts such as Bogra and Tangail, the price and amount of actual procurement were observed to be well below the needs and expectations of producers during the 1988 boro harvest (Mandal, 1989: 57-8).

The situation has been exacerbated by the existence of de facto 'barriers to entry' to the procurement centres for the majority of producers. Available evidence suggests that the concerned dealers and public functionaries have entered into collusive arrangements which gave discriminatory preference to buying rice from rich peasants and large traders, while denying effective access to producers from the ranks of the poor and middle peasantry.[36] The margin between the procurement price 'guaranteed' by the state and the lower prices prevailing in the open market served to provide the source of covert gains encouraging such collusion between public and private sector agents.[37]

The combination of barriers to entry and the low quantum of purchases by the state meant that, in practice, there was no 'guaranteed price' for most producers. This made things even worse for the majority of small and medium producers who in any case did not have much holding power at harvest time. This was precisely the conjuncture in the harvest cycle when they needed cash to meet their market obligations, including the costs of production (e.g., payment in cash of rent, irrigation charges or interest). Lacking effective access to procurement centres they were usually compelled to sell at low prices to the private traders operating in the open market. During 1985–86, rich peasants in the Pabna villages were found to be buying rice from producers among the poor peasantry for resale at higher prices.

In fact, such outcomes illustrate the way in which the policy objective of providing incentive price to rice producers has been undermined by collusive arrangements between public functionaries of the state and private trading interests of the market. They also suggest that the postulation of a binary opposition between the interests of 'the state and the market', or 'the public and the private sectors', may not be all that meaningful in actual practice (Harriss-White, this volume). Instead, it would appear that individual agents in both these hierarchies have forged common interests whenever they have been able to make private gains at the expense of public resources.

Furthermore, the very attempt by the state to regulate the market price of rice appears to have been somewhat quixotic, given its lack of will to counteract the influence of the well-established private trading interests controlling the institutions and operations of the market. Despite widely accepted views to the contrary, the fragmentary evidence available on the *substantive* structure of the rice marketing system suggests that the extent of vertical integration between traders and rice producers, as well as traders' control over price formation, were not entirely insignificant during the mid-1980s.[38] As noted before, a variety of mechanisms including interlocked (credit–product) market contracts have been deployed by trading functionaries to constrain producers to repay loans in rice, thus ensuring their own supply. In such cases, the prices received by producers are likely to have been considerably lower than the open market price, not to speak of the official procurement price.[39]

While the evidence available remains fragmentary, it is suggestive of the hypothesis that a large segment of poor producers did not have access to open market prices because of the restrictive stipulations of debt contracts with rice traders. It follows that such producers would have been unable to avail of price support *even if* such a policy had been implemented effectively. Furthermore, it would appear that the price support policy has possibly served interest groups other than those who were the intended beneficiaries. These contradictions between policy objectives and outcomes have been largely due to the 'mediating' influence of the pre-existing interest groups dominating the rice marketing system, as well as the interactions between concerned public sector functionaries and private trading interests.

Shifts in the Class Distribution of Income

By the mid-1980s, the class contentions in the input and output markets had given rise to contradictions characterising agricultural production. These arose from shifts in the distribution of income in favour of the dominant interest groups and classes which had been able to appropriate most of the incremental productivity gains resulting from technological change. This also meant a progressive reduction in the share of the agricultural product (income) accruing to the actual producers, particularly the poor majority. The latter had been faced with disadvantageous terms in both input and output markets, while being denied effective access to institutional credit and price support programmes. Progressive reduction of subsidy on fertilisers and irrigation equipment had increased costs of production, while there had not been proportionate compensatory gains in the realised value of the product because of ineffective price support and compulsive involvement in the product market. The combined and cumulative impact of these factors on the class distribution of income could be expected to have resulted in a declining trend in the profitability of production based on the new technology. These possibilities are now taken up for more detailed consideration.

TRENDS IN THE PROFITABILITY OF HYV BORO PRODUCTION AND IRRIGATION

While the impact of class contentions on policy interventions and their outcomes is suggestive of a declining trend in the profitability of HYV boro production—and, by implication, a declining trend in the profitability of irrigation—there are hardly any studies which enable further exploration of the causal linkages between these interactive social processes and their economic outcomes. However, the findings of a farm survey conducted by Mandal (1989) in the mid-1980s provides some evidence and ideas which can be built upon. In addition, a few other sample surveys from other parts of the country provide useful supplementary information.

Trends in the Profitability of HYV Boro Production and Underlying Factors

The farm survey conducted by Mandal compares the profitability of cultivating HYV boro with mechanised irrigation in a part of Tangail for the two years 1985 and 1988.[40] Data on the costs of and returns from production are provided (but, surprisingly, without working out the rate of return). The costs and returns are computed in terms of average figures for a sample of producers which includes different combinations of tenurial status, type of tubewell, and mode of irrigation payment. The producers include owner-operators and sharecroppers who purchased water from either a DTW or an STW (run by diesel or electricity), and paid irrigation charges either in the form of fixed cash per hectare or in the form of 'quarter' cropshare (one-fourth of gross output). The data do not incorporate the costs of working capital for production, whether borrowed or imputed, even though it is considered in discussion. Furthermore, in the case of owner-operators, the opportunity cost of owned land in terms of imputed rent is not considered.[41] Because of these elements of cost underestimation, the net return is overstated for both years. However, because of the systematic nature of the omissions, they are unlikely to have affected the *direction* of change in profitability between the two years.

The rate of return from HYV boro cultivation at current prices was found to have declined between 1985 and 1988.[42] This applied to all combinations of tenurial status, type of tubewell and mode of irrigation payment. For example, in the case of owner-operators using DTW irrigation (with 'quarter' cropshare as the charge), the rate of return declined from 46 per cent in 1985 to 10 per cent in 1988.[43] The corresponding figures for those using STW irrigation were 47 per cent in 1985 and 18 per cent in 1988. For owner-operators paying fixed cash charges for DTW irrigation, the rate of return declined from a phenomenal 102 per cent in 1985 to 45 per cent in 1988. In the case of sharecroppers, the rates of return were drastically lower, but showed comparable decline between the two years.[44] Some care is needed in the interpretation of these results. Apart from the systematic overestimation incorporated in the rate of return, these average figures suppress significant differences between producers of different categories having unequal access to markets and prices. The actual rates of return are likely to have been worse than the average for small and marginal producers, and better than the average for larger producers.

A declining trend in the profitability of HYV boro over the 1980s is also indicated by other sources and indicators (cited by Mandal). An official estimate (produced by the Ministry of Agriculture) indicates that the benefit–cost ratio of irrigated boro rice, at full (cash *and* kind) costs, declined from 1.2 in 1981–82 to 1.1 in 1986–87.[45] Another survey covering four different areas of the country found that the net returns to boro rice producers using irrigation from STWs declined by 1–16 per cent over the period 1981–82 to 1985–86.[46]

The Tangail data indicate that the proximate factors responsible for the decline in the rate of return were increases in the costs of specific inputs, decline in the yield of HYV boro, and less than a proportionate increase in boro price compared to input prices over 1985–88. To illustrate, in the case of owner-operators using DTW irrigation (with 'quarter' cropshare as the charge), the boro yield declined by 21 per cent and its harvest price rose by 7 per cent, while the cost of production (excluding the irrigation charge) increased by 64 per cent.[47] Amongst production costs, fertiliser prices rose by 14 per cent each for urea and Triple Super Phosphate (TSP), and by 26 per cent for Murate of Potash (MP). The daily money wage rate in this part of Tangail also rose by 50 per cent between the two years. A comparable pattern of higher rates of increase in input prices compared to that of the HYV boro price was also noted for four different regions of the country over the first half of the 1980s (between 1981–82 and 1985–86).[48]

Mandal (1989: 57) argues that amongst the numerous factors which could have contributed to the decline of boro yield in Tangail, the most proximate one was reduction in the intensity of fertiliser use, rather than in the use of irrigation (which was adequately available in the survey area). This, in turn, is partially attributed to the rise in fertiliser prices resulting from the reduction in subsidy. However, the relatively low price of boro at harvest time is regarded by Mandal as the 'the more important reason', resulting in lower investment in fertiliser use and the consequential decline in yield (ibid.; cf. Khan, 1990 and Osmani, 1990). Significantly, the overwhelming majority of HYV boro producers in Tangail failed to receive any price support from local procurement centres in 1988 (as was also observed to be the case in Bogra and other parts of the country during this period).[49]

The proximate factors underlying the declining trend in the profitability of HYV boro production for this particular sample appear to correspond closely to what might be expected from the preceding assessment of policy interventions and their outcomes. A significant exception, however, is provided by the trend in irrigation charges, which merits further consideration. In the Tangail data, irrigation payment in the form of one-fourth cropshare was clearly tied to the price and yield of boro, and hence declined over 1985–88, as did the gross value of total output. While the fixed cash mode of payment for irrigation was not so directly tied to boro price and yield, it was none the less affected by the same considerations.

Trends in the Profitability of Irrigation and the Underlying Factors

Before looking at the data, it is useful to note that the activity of selling water (delivery of irrigation as a service) is essentially a business venture, involving purchase of inputs from the market and receipt of market-mediated payments from the water users (producers) in cash or kind. Since virtually all costs of production are based on market (external) contracts, it is not enough simply to break even. An acceptable profit margin above cost recovery is needed to justify the undertaking of the activity at all (even if it is not as high as the 'normal' profit rate of the economy). In this sense, irrigation under private management is the nearest thing to *capitalist enterprise* in the agricultural sector of Bangladesh.[50] Furthermore, given the growing predominance of private management of irrigation by the mid-1980s, the irrigation sector as a whole can be regarded as becoming systematically profit-oriented in the capitalist sense of the term, and hence increasingly susceptible to factors affecting its profitability.

The farm survey data from Tangail used here also contains the average costs of and returns to a sample of private DTW and STW operators who received one-fourth cropshare as the irrigation charge. While operation and maintenance items are included, the costs are seriously under-estimated due to omission of both interest on working capital (borrowed or imputed) and annual adjustments for the capital costs of the irrigation equipment itself. Due to these considerations, the rates of return computed on the basis of the raw data turn out to be significantly inflated. However, because of the systematic nature of the biases, the direction of change in profitability is unlikely to have been affected.

The rate of return of DTW operators was found to have declined from a phenomenal 164 per cent in 1985 to 46 per cent in 1988.[51] For STW operators, the corresponding rate declined from 122 per cent to 75 per cent between 1985 and 1988. While the absolute magnitude of the rate is overestimated in all cases, the relative decline over this three-year period is striking. Survey data from other parts of Tangail and Bogra also show trends of declining profitability of irrigation during 1981–86.[52]

The proximate factors underlying the fall in profitability of tubewell irrigation in Tangail were found to be the decline in size of the command area and the rise in costs of operation and maintenance.[53] Moreover, reductions in the yield and/or price of irrigated boro rice affected directly those water sellers who received payment as one-fourth cropshare, and affected indirectly those who received irrigation charges in the form of a fixed cash rate per unit area. The average command area of DTWs and STWs during the boro season declined by 26.3 and 11.2 per cent respectively between 1985 and 1988. This was because new tubewells had been installed in the vicinity of the existing ones, reducing the size of command areas on the average. Over 1985–88, the operational and maintenance costs of DTWs increased by 25.8 per cent, while the corresponding cost increase for STWs was 3.1 per cent.[54] During the same period, the yield of HYV boro declined by 21 per cent and 15.2 per cent in DTW and STW command areas respectively. As noted earlier, the price of boro increased by only 7 per cent over the same period, resulting in a decline in the market value of the gross return (one-fourth cropshare) of tubewell management (Mandal, 1989: 61, Appendix Table 1).

Class Contentions and Factors Affecting the Profitability of Irrigation

While discussing the factors affecting the profitability of irrigation in different parts of Bangladesh, Mandal makes the following observations:

> ... when there are unexploited potentials for tubewell development, especially STWs, [as] in large areas of many north Bengal districts, the entrepreneurial tubewell managers can still make some profits from water selling. But given the high levels of natural and technical risk involved in STW operation, the pursuit of making 'normal profit' can be maintained so long as water charges can be adjusted upwards for risks and increasing costs, and kept within tolerable limits. In contrast, in intensively irrigated areas of Tangail this tolerance level of water charges had been reached and further increase of water charges from one-fourth share of crop meant inviting threats of losing irrigable plots to competing command areas (Mandal, 1989: 56–7).

These observations are suggestive of the way in which regional variations in the profitability of irrigation were related to the extent of development of the technology and the relative bargaining power of the classes of water sellers and users. Using these ideas as the point of departure, the factors affecting trends in the profitability of irrigation can be explored in terms of a set of stylised scenarios as follows.

In areas where irrigation technology has remained relatively undeveloped such that tubewells are sited at considerable distances from each other, water sellers can exercise a certain degree of 'natural monopoly' (Boyce, 1987: 242; Palmer-Jones, 1992: A–138). Their relatively greater bargaining power in relation to water users enables them to increase the rate of irrigation charge in order to pass on to the latter any rise in the costs and risks of operation. Such upward adjustments can also be made when irrigation payment received in the form of cropshare declines in value (due to reduction in yield and/or price of the crop). In relatively undeveloped areas, therefore, water sellers are in a position to maintain their profit margin (if not even more) by increasing the irrigation charge.

In contrast, the situation is significantly different in the 'advanced' areas where irrigation technology has been intensively developed. In such areas, the locational proximity of tubewells (and/or pumps) has resulted in several interlinked outcomes arising from the spatial relationships between adjoining machines and their command areas. First, the 'density' of machines on the ground has led to a decline in the average size of the command area, with significant implications for the cost of irrigation. Assuming that irrigation charges per unit area remained constant, the decline in the size of the command area had the effect of reducing the total returns from irrigation while increasing the fixed (and total) cost per unit area, also contributing to the decline in the profitability of irrigation (in terms of the return–cost ratio). Second, there has been partial overlapping between the potential command areas of neighbouring machines. As a result, the buyers (water users) have benefited from the relatively greater choice between alternative command areas within which their plots could be included. In effect, greater competition between water sellers has reduced the degree of monopoly in the water market.[55]

The interactive impact of these factors has led to a growing contradiction affecting the profitability of irrigation. Rising cost has generated a pressure to push up irrigation charges in order to maintain the profit margin of water sellers at former levels. However, growing competition meant that it was no longer possible for water sellers to shift the burden of increasing costs and/or declining returns to the users without running the risk of losing the latter's plots to other command areas. While cuts in the irrigation charge could still be accepted, a limit was posed by the capitalist nature of the business which required a minimum profit rate above cost recovery to justify the activity at all.

In order to resolve this dilemma, water sellers have attempted to devise ways of 'tying' users to their own command areas. One method has been to offer concessions and supplementary services to users.[56] Alternatively, wherever possible, they have attempted to use their social and political influence (if not overt force) to remove or prevent competition, and thus maintain a degree of monopoly in the water market (as discussed in an earlier section) (Adnan: 1989). If successful, such mechanisms have enabled them to increase the rate of irrigation charge collected from the 'captive' producers (subject to other limits discussed at a later stage).

In the general case, however, the propensity to raise irrigation charges has not necessarily been realised. The profitability of irrigation has continued only so long as it has not undermined the incentive of the cultivator to irrigate and produce. In practice, the reaching of this economic limit is likely to have been pre-empted by the social limit to what could be extracted as the rate of irrigation charge, depending upon the resistance put up by the producers.[57] The actual outcome has varied from area to area depending on the balance of class forces between water sellers and users, and the intensity of irrigation development (which determines the degree of monopoly in the water market).

These observations on the trends during the mid-1980s, including the sharpening class contentions, serve to define the 'initial conditions' which preceded the onset of the deregulation policy regime during the late 1980s and early 1990s. Its impact on the profitability of irrigation and HYV boro production, as well as the resultant outcomes, are taken up in the next section.

CONSEQUENCES OF THE DEREGULATION POLICY REGIME

The deregulation policy regime was initiated partly because of the failure of the preceding policy interventions to sustain the agricultural growth rate during the mid-1980s (Palmer-Jones, 1992: A–137). At the same time, there was a shift in the global aid climate with a growing penchant for imposing 'structural adjustment' policies. As with earlier shifts in policy, pressure from donor agencies played a key role in persuading a somewhat reluctant government to adopt these changes.[58] Most of them took place in stages over 1986–92—a period which just preceded, and partially overlapped with, the slowdown in the growth rate of HYV boro production and irrigated area during 1990–94.

The thrust of the policy shifts was concerned with handing over the remaining functions of public sector agencies to private commercial interests, while also removing virtually all the regulatory control still exercised by them. The essential objective was to 'complete the unfinished projects' of privatisation, import liberalisation, subsidy reduction and deregulation of fertiliser and irrigation equipment. The changes were highly selective, and the relative absence of effective policy intervention in markets such as credit and product also corresponded largely with the pattern observed under the preceding policy regimes.

In the case of public sector credit agencies, characterised by cumulative debt default and inadequate debt recovery, a series of financial and legal measures were undertaken from 1986–87 with the aim of improving performance. In March 1990, the system of refinancing of agricultural loans by the central bank (Bangladesh Bank) was discontinued (Hossain, 1995). These policy changes during the late 1980s had the effect of tightening up the conditions of access to institutional credit while imposing a squeeze on the volume of agricultural credit available. None the less, rent-seeking practices by the concerned functionaries continued to be observed.[59] Consequently, many agricultural producers found access to institutional credit only if they were willing to pay bribes, frequently compelling them to resort to non-institutional sources as before.

Nor was there any significant improvement in the implementation of price support policy. The average harvest prices in the market during the boro season remained below the corresponding 'procurement' price in most years since 1985–86 (Hossain, 1995). No effective measures were taken to remove the barriers to entry characterising procurement centres, or to check the collusion of public sector functionaries with rice traders of the private sector. In all likelihood, the majority of HYV boro producers failed to get 'incentive' prices for their product during the period that the deregulation policy regime served to push up the prices of specific inputs, as indicated in the following paragraphs.

Shifts in Fertiliser Policy and their Outcomes

From 1987, the upper tiers of fertiliser distribution were deregulated and privatised in sequential stages. Private traders took over the bulk distribution activities which had largely been retained by BADC.[60] 'The deregulation in marketing was completed in December 1992, when the private sector took charge of fertiliser (TSP and MP) imports following the total withdrawal of fertiliser subsidy.[61] Over 1987–92, the share of the private sector in the fertiliser market was estimated to have risen from less than 5 per cent to more than 90 per cent (Shahabuddin, this volume).

These drastic policy reversals meant that agricultural production was opened up to the fluctuations and uncertainties of the world market. The erstwhile policy of keeping the prices of imported fertilisers low and stable, particularly TSP and MP, was discarded. The private traders taking over the critical function of supplying fertilisers from home and abroad had little accountability to the producers or to anyone else. There were no checks against unwarranted profit margins being incorporated in domestic fertiliser prices by private importers and distributors.

Evidence available for the period 1987–91 indicates that the intended benefits of the policy changes were not fully passed on to producers by the private traders taking over from BADC (Hossain, 1995). Domestic fertiliser prices shot up as the full costs of importing and distribution were charged, topped up by unregulated profit margins.[62] The traders' margin on TSP increased substantially after deregulation, with a corresponding rise in its farmgate price (ibid.). Decline in the price of rice during this period also contributed to the raising of the fertiliser-rice price ratio, depressing the profitability of rice production (Shahabuddin, Palmer-Jones, this volume). Significantly, the domestic TSP–rice price ratio shot up sharply during 1990–92, even though the world price of TSP was *declining* over the same period (Hossain, 1995: Figures 2, 3). The cumulative consequence of these trends, triggered off by the deregulation policy regime, was a striking slowdown in the use of fertilisers (see Shahabuddin, this volume).[63]

Shifts in Policy on Irrigation Equipment and their Outcomes

Privatisation of the import of irrigation equipment (diesel engines) was begun in 1986, accompanied by the lifting of restrictions on their makes and models ('brands') (Hossain, 1995; Palmer-Jones, 1992: A–131). The next major steps in deregulation came in 1988, when unrestricted private import was allowed and duties were removed. Furthermore, regulations on standardisation and siting of irrigation equipment were totally withdrawn, removing the last vestiges of 'control' by public sector agencies. These changes were accompanied by complete elimination of subsidy on minor irrigation equipment (except for unsold stocks in the public sector).[64]

The deregulation policy regime is reported to have led to significant changes in the supply and pricing of irrigation equipment. The market for equipment experienced large-scale expansion and growth of competition as increasing numbers of traders took up the businesses of import, supply and distribution (Palmer-Jones, 1992: A–132; Hossain, 1995). Cheaper and smaller diesel engines from new sources flooded the market, providing buyers with a wider range of irrigation equipment to choose from (as compared to the 'standardised' brands allowed before). Elimination of duties in 1988 led to a fall in the prices of both the new and the erstwhile 'standardised' brands (Shahabuddin, this volume). By early 1989, the cost of installation of an STW was reported to have been almost 40 per cent lower than the price required by BADC before import liberalisation (Hossain, 1995). Sale of STWs in particular is reported to have picked up rapidly during the post-deregulation period.

In contrast to fertilisers, there appears to have been a set of factors conducive to a downward trend in the prices of irrigation equipment (primarily STWs) following these policy changes. This outcome is in fact not particularly surprising given that there was no subsidy to be withdrawn from STWs in the first place, which might have otherwise contributed to a rise in prices. Furthermore, to the extent that the rate of subsidy on stocks of DTWs held by the public sector was actually increased over 1991–92, their prices could be expected to have fallen.[65] In the second place, most buyers of lumpy and expensive STWs came from the ranks of the rich or middle peasantry, typically with influential urban connections. Their (class) bargaining power could thus roughly match those of the private traders selling the equipment (without any protective state intervention being required).[66]

Third, unrestricted linkages with the world market do not appear to have led to price rises similar to those of fertilisers (as yet) because of the simultaneous lifting of the standardisation requirements. The erstwhile 'standardised brands' (and the associated transnational vested interests) were exposed to competition from new and cheaper brands from different countries and suppliers, who were concerned to establish a foothold in this prospective export market. Such competition between brands of equipment and the respective suppliers in the world market, coupled with the relatively robust bargaining power of the class of domestic buyers, is likely to have resulted in the observed expansion in the supply, and reduction in prices, of irrigation equipment following deregulation.[67]

Factors Underlying the Stagnation of Irrigated Area in the Early 1990s

In the absence of adequate evidence on the processes which led to the stagnation in irrigated area during the early 1990s, a hypothetical account of what might have happened can be attempted on the basis of the stylised scenarios from the mid-1980s discussed earlier. In particular, the consequences of the availability of a burgeoning supply of cheaper irrigation machines, accompanied by the deregulation of siting, merit further attention. The availability of relatively inexpensive machines, primarily STWs, led to an initial surge of new installations, as reflected in the growth of the stock of irrigation equipment during the late 1980s and the early 1990s (prior to the eventual slump). It is likely that the buyers of these cheaper machines included those who had not been able to afford them before. The resultant effect was that of increasing 'density' or irrigation equipment on the ground, leading to a decline in the average size of the command area.

Furthermore, since there were no longer any restrictions on siting, at least some of these new machines were likely to have been installed in the vicinity of older devices. The consequence was the partial overlapping of command areas, leading to increased competition among water sellers. Under such conditions the emergent contradiction characterising the water market in the mid-1980s is likely to have sharpened at the close of the 1980s and the beginning of the 1990s. While the impact of deregulation of siting resulted in the 'overcrowding' of irrigation machines in areas where growth of HYV boro had become predominant, shortages in irrigation capacity persisted in other areas. However, even in such less developed areas, with a lower 'density' of pre-existing machines, comparable trends could be expected to develop over time as the number of new installations increased.

The essential impact of 'destandardisation' and 'unregulated' siting of irrigation equipment was therefore to increase the proportion of boro producing areas which were intensively irrigated. During the late 1980s and early 1990s more and more areas must have acquired the features which had characterised parts of Tangail during the mid-1980s. In particular, as the basis of monopoly in the water market was eroded by growing competition, the bargaining power of water sellers was reduced in relation to their buyers. Declining command areas pushed up unit costs and reduced the total returns to the water sellers, while irrigation charges could not be raised without risking the loss of customers. The simultaneous pressure of rising costs and declining returns tended to wipe out the monopolistic component in the erstwhile returns from irrigation and subject the business to the strictures of a competitive market to a greater extent.

Other things being equal, such a trend should have led to expansion in irrigated area as boro producers, including the majority of the poor peasantry, responded to the cheaper irrigation and the more 'competition-conscious' water sellers. Since, however, irrigated area virtually stagnated during the early 1990s, other forces must have been at work to offset this possibility. Consequently, other factors which could have affected the growth of irrigated area need to be considered.

The potential significance attributed to the failure to achieve collective action and conducive institutional change as a key constraint to the expansion of irrigation had been largely pre-empted by the overwhelming predominance of private irrigation management by the early 1990s. However, as indicated earlier in the discussion, class contentions in the water market assumed primacy over collective action under private irrigation management. It is thus unlikely that failures in achieving collective action would have had the scope in the first place to play a significant role in constraining the expansion of irrigated area during the early 1990s. Instead, the outcome of class contentions in the water market would have had greater significance.

In fact, as with the sample from Tangail during the mid-1980s, it was the relatively low price of HYV boro, and its slower rate of growth compared to those of input prices, which had the potential of posing the key constraint to the growth of irrigation. At a critical threshold of (reduced) profitability, boro producers could be expected to set a limit on how much they were willing and able to pay as the charge for irrigation. A declining trend in the profitability of boro production could thus have set a limit to the demand for irrigation, leading to a decline in the profitability of the latter. The intensification of such a demand constraint on water selling enterprises operating according to capitalist norms of profitability would have progressively subjected them to the market sanction of 'going out of business'.

It is possible that an interactive sequence of this kind might explain the stagnation in irrigated area in the early 1990s, as well as the consequential slowdown in capital investment at the margin despite adequate supply of cheaper irrigation equipment in the market. The hypothesis does not exclude the possible contributions of hydrological, social-institutional and/or technological constraints in affecting irrigated area. Crucially, however, it would hold even if none of the latter had assumed the proportions of a 'binding constraint' by the early 1990s, as appears to have been the case given the available evidence.

Factors Underlying the Slowdown in the Growth of HYV Boro in the Early 1990s

Assessment of the impact of technological change and policy interventions during the pre-deregulation period was suggestive of a long term trend of worsening of income distribution against the poorer classes of agricultural producers. What needs to be ascertained in this context is the specific impact of the deregulation policy regime on HYV boro production which might have 'precipitated' the slowdown in its growth rate during the early 1990s. However, given the absence of adequate evidence from this period, the argument draws upon the patterns of earlier periods as noted in the preceding discussion.

The technological indispensability of irrigation for producing HYV boro has naturally focused attention on the possible role of the former in constraining the growth of the latter. In this instance, however, it is unlikely that the slowdown in boro output and acreage can be explained by constraints to the supply of irrigation, for reasons which have already been discussed. (Rather, the causality is likely to have operated in the *reverse* direction, given the specific impact of deregulation policies related to the irrigation sector.)

However, the nature and timing of the impact of deregulation policies concerning fertiliser were qualitatively different from those observed in the case of irrigation equipment, particularly in terms of the classes affected and the implications for the profitability of HYV boro production. Unlike the better-off purchasers of the expanding supply of cheaper STWs, all classes of HYV boro producers needed to buy fertilisers, including the majority who lacked bargaining power in comparison to the traders they had to deal with. Furthermore, while unsubsidised STWs had been imported and sold by private traders from as early as 1978–79 (subject to standardisation), private import of (unsubsidised) MP and TSP began only at the end of 1992. The 'shock effect' of this policy reversal came on top of the adverse impact of the measures to privatise the bulk distribution of fertilisers which were begun in 1987. The cumulative effect of these policy shifts, given the relatively weaker bargaining power of the majority of boro producers, was to set off and sustain a rising trend in fertiliser prices, particularly of those which were mostly imported (TSP and MP). It would appear that domestic fertiliser prices increased even when world market prices were decreasing. The evidence available suggests that these trends were associated with the rising (unwarranted) profit margins of the mediating chain of importing and distributing intermediaries through whom producers had to find access to fertilisers.

Depending on the type of fertiliser, the consequence of rising prices was either a slowdown in the growth rate (urea) or an absolute decline in the amount used (TSP and MP) during the early 1990s. This aggregated pattern would suggest that insufficient quantities, and disproportionate mixes, of fertiliser were used by producers of HYV boro, the most fertiliser-intensive of the rice crops (Hossain, 1995). The rise in the cost of fertilisers relative to the price of HYV boro (the fertiliser-rice price ratio) constituted the proximate factor leading to their sub-optimal application, particularly by the majority of producers with limited access to working capital. This can be regarded as the specific impact of the deregulation of fertiliser distribution which was instrumental in precipitating the slowdown in the growth of HYV boro during the early 1990s.

However, there were other critical factors contributing to this outcome which can be attributed to 'acts of omission', rather than of commission, by the deregulation regime. Thus, it is unlikely that the majority of HYV boro producers had improved access to working capital from institutional sources, or any better price support, during the early 1990s as compared with the past. Given the 'presence' of unfavourable outcomes from fertiliser policies, the continued 'absence' of conducive policy interventions in the credit and the product markets assumed critical significance by failing to provide the means to cope with rising fertiliser prices. It was the *combination* of these 'acts of omission and commission' of the deregulation regime which provided the scope for new means of surplus appropriation (in the fertiliser market) to be added to the old ones (in the credit and product markets), to the detriment of the bulk of poorer boro producers.

Furthermore, the majority of boro producers were too poor and unorganised to have provided any effective political resistance to the detrimental outcomes of the elimination of subsidy as well as privatisation of the import and distribution of fertilisers. A comparable pattern obtained in respect of the rice market, where making price support inaccessible to the producers was in the interest of the mercantile trading classes dominating these markets and exercising varying degrees of control on its processes of price formation (sometimes in collusion with rent-seeking public sector functionaries). While the richer boro producers were also affected by the same policies, their simultaneous connections with the private trading interests in the fertiliser and/or rice markets are likely to have made their position somewhat ambivalent.

In effect, the relative balance of class forces in the structures mediating technological change and policy interventions had a crucial part to play in bringing about a severe worsening of income distribution against the bulk of poorer boro producers in the wake of the deregulation regime. In all likelihood, the consequence was a sharp downturn in the profitability of HYV boro production, leading to a slowdown in its growth rate during the early 1990s. Furthermore, to the extent that some of the underlying causal factors reflected the continuation or intensification of the state of affairs in the pre-deregulation period, this slowdown can be viewed as a *conjunctural manifestation* of longer run trends of worsening income distribution. However, it also follows that this trend could be reversed if subsequent changes in policy, as well as their structurally mediated outcomes, serve to improve the profitability of HYV boro, enabling its producers to sustain further growth.

In the composite strategy advocated by donor agencies and implemented by the state, the elimination of input subsidy on fertiliser and irrigation equipment was supposed to have been balanced by the maintenance of 'guaranteed' product prices. As pointed out earlier by Osmani, this 'policy package' was seriously flawed in terms of its economic logic and provided little incentive to the large segment of poor producers whose primary concern was to meet their subsistence needs rather than to sell on the market (particularly those who were net buyers, rather than sellers, of rice) (Osmani, 1985). Furthermore, this donor-driven 'policy package', as it was implemented in practice, deviated from its original blueprint such that the eventual outcomes were surprisingly different from, and sometimes virtually the opposite of, what had been intended. These deviations bore the stamp of 'refraction' by the unequal structures of the respective markets, by the agencies of the state charged with policy implementation, as well as by the agrarian class relationships embodying conflicts between the interests of different categories of producers. It is in this sense that the agrarian structure, interacting with the structures of the market and the state, played a crucial role in determining the actual outcomes of technological change and policy interventions.

PERSPECTIVES ON TECHNOLOGICAL CHANGE AND POLICY INTERVENTIONS

The specific issues related to the slowdown of HYV boro production and the stagnation in irrigated area during the early 1990s have also thrown into sharper relief the general issues pertaining to the possible structural constraints to agricultural growth. The long term perspectives emerging from this experience (between the 1960s and the early 1990s) indicate that the adoption and expansion of the new agricultural technology in Bangladesh have not been significantly constrained by the pre-existing structures of the market, the state and agrarian class relationships. However, the process has been able to continue and grow only to the extent that it has been conducive to the interests of the groups and classes dominating these mediating structures. Conversely, the process has slowed down whenever the nature and intensity of surplus appropriation by the dominant groups has undermined the capability of HYV growers to sustain such growth. It is unlikely that those making and changing policies were entirely unaware of the role of dominant interest groups and classes in contributing to such periodic growth and slowdown—even if no analysis to that effect is explicitly incorporated in publicly circulated policy documents. In this sense, these considerations are also symptomatic of the explicit and implicit biases of the associated policy regimes, as well as the unstated (or covert) agenda behind the promotion of the new agricultural technology, which merit brief consideration.

The thrust of policy changes over the preceding decades has been on reduction of input subsidy, removal of public sector controls and privatisation of the ownership and the distribution system of inputs. While these changes did to some extent limit the scope for rent-seeking and corruption by public sector functionaries, the benefits of such 'policy reform' did not necessarily reach most producers. Instead, the scope for extraction of undue profit margins by private traders and importers was simultaneously enlarged, as evident after the takeover of the import and distribution of fertilisers by private traders.

Furthermore, these features were not confined only to the concerned public and private agencies of Bangladesh. Vested interests in these policy regimes were to be found amongst key players in the aid business as well (cf. Jansen, 1992). For example, the regulations on standardisation of the brands of irrigation equipment provided, to some extent, 'captive' markets for particular transnational corporate interests. This was ensured in some cases through tying of bilateral aid to equipment produced in the donor country, as well as the unduly capital-intensive choice of technique promoted by multilateral donor agencies.[68] Aid financing of agricultural development projects also generated dependency on the world market for import of inputs and technical expertise, typically mediated by transnational and national corporations in the private sector.

Increasing commoditisation of inputs and outputs made agricultural production and its support services increasingly susceptible to considerations of market-based profitability characteristic of capitalist production. This was particularly evident in the case of private irrigation management, as already noted. However, the extent to which pre-existing peasant production has been transformed into capitalist 'agribusiness' is less evident. None the less, the nature of technological change has been such as to lead to the definitive outcome of making rice production in Bangladesh compulsively dependent on external inputs, including aid resources, technical knowledge and expertise. Viewed in a historical perspective, the impact of the new technology of rice production has been to shift Bangladesh's dependency on the world market from export of agricultural produce to import of production inputs.[69]

The continuing failure of policy interventions, despite numerous 'reforms', to provide an environment in which those producing with the new technology could actually retain an adequate share of the gains arising from it, and hence undertake productive accumulation in the sense of self-sustained investment in production and further innovation, poses a certain paradox. The policy responses to these failures have been restricted to dealing with the symptoms rather than addressing the root of the problem. This lies in the relative lack of assets and bargaining power of the majority of producers in relation to those controlling access to the allocation of inputs, as well as the disposal of the product in the corresponding markets and public sector agencies

In particular, it would appear that there has been a stubborn unwillingness on the part of government and donor agency policy-makers to confront the private sector interest groups (classes and their fractions) which have persistently extracted surplus from producers through control over price formation and access to resources. These include the highly unequal structures and institutions of the marketing systems of rice, non-institutional credit, (privatised) fertiliser distribution and operational land (the lease market). The relatively undisturbed operation of the interest groups dominating these institutional mechanisms has played a critical role in undermining the professed objectives of successive policy regimes. In this enterprise, they have also been successful in co-opting public sector functionaries charged with policy implementation, as evident from the existence of collusive arrangements which have blunted the effectiveness of programmes of price support and institutional credit. Furthermore, the pervasive 'institutionalisation' of such collusive arrangements between private and public sector agents, with well-defined contractual stipulations and 'rates of appeasement', has led to a certain 'intermeshing' between the hierarchies of the state and the market, such that their interests have converged against the public interest (Harriss-White, this volume).

These patterns are also indicative of the presence of certain structural continuities, given that policy shifts leading to complete privatisation and deregulation have largely been unsuccessful in actually benefiting the majority of poorer producers, while also displaying limitations comparable to those which had characterised the earlier public sector dominated policy regimes. In fact, what have actually changed to some extent are the *specific institutional mechanisms* of appropriating surplus from agricultural producers, since these have varied in accordance with the specific opportunities for gain made available by the respective policy regimes. However, neither the general process of surplus appropriation itself, nor the interest groups benefiting from the process under successive policy regimes, has changed in a substantive sense.

It is true that the roles of the large landowner and rice producer, or of the dominant moneylender, rice trader, water seller or fertiliser dealer, do not necessarily overlap or converge in the same individuals (although the extent to which this occurs is by no means insignificant). However, these roles remain interchangeable, and interrelated by ties of kinship and social affinity, as well as by common concerns in the business of extracting surplus from the majority of poorer producers and appropriating the benefits of public policy. It is these interconnected interest groups dominating the agrarian class structure and the hierarchies of the agricultural marketing systems which have largely been able to avoid being adversely affected by policy interventions while continuing to benefit from successive policy regimes. Ironically enough, the aid resources propelling the expansion of the new agricultural technology, as well as the strident policy directives of donor agencies to that end, have had the effective consequence of sustaining, if not reinforcing, this order of things.

NOTES

1. See Rogaly et al. (this volume) for an introduction to the debate covering both Bangladesh and West Bengal.
2. Palmer-Jones (1992: A–129; see also his contribution to this volume), endorsed by Harriss (1992: 218–9).
3. For example, Shahabuddin (this volume), who also cites other examples.
4. Cf. earlier debates in a different historical context, involving Dobb (1963: 38–42), Sweezy (1976: 41–6, 103), Brenner (1976: 43–5, 60; 1977: 41–7) and others in Hilton (1976). A commentary is provided in Adnan (1985).

5. Shahabuddin (1995: Table 2.2). The growth rates of area under aman in the 1980s and 1990s were 5.46 and and 5.19 per cent respectively; the corresponding output growth rates were 5.19 and 5.49 per cent respectively.
6. Shahabuddin (1995: Tables 2.2 and 2.3). The trend growth rates of local aman during the 1990s were close to the trends in the 1980s: −2.83 per cent in area and −2.94 per cent in output during the 1980s, compared to −3.02 per cent in area and −1.81 per cent in output during the 1990s. The trend growth rate of local aus declined drastically in terms of area at −10.98 per cent, and in terms of output at −11.29 per cent during the 1990s, as compared to −3.17 per cent in area and −1.25 per cent in output during the 1980s.
7. Shahabuddin (1995). Their numbers rose from 183,000 in 1987–88 to 276,000 in 1990–91.
8. Shahabuddin (1995: 12, Table 3.3). The number of operational shallow tubewells (STW) increased by 33 per cent from 270,309 in 1990–91 to 359,297 in 1993–94. However, annual increments declined from around 39,000 in the first two years to 10,422 in the last year.
9. Shahabuddin (1995: 12, Table 3.3). The number of deep tubewells (DTW) in operation rose from 21,519 in 1990–91 to 25,546 in 1991–92, but remained virtually stagnant thereafter up to 1993–94 (when it declined to 25,049). The proportion of deep tubewells (DTW) not in operation during this period was quite significant, ranging from 18.6 to 29.7 per cent. The number of low-lift pumps (LLP) remained virtually constant, with a mild overall increase of 1.7 per cent from 51,625 in 1990–91 to 52,528 in 1993–94.
10. Shahabuddin (this volume: Table 5.2). He argues that there was an acceleration in the growth rate of total irrigated area (the sum of areas under modern and traditional irrigation) 'during the late 1980s following some liberalisation policies . . . pursued by the government since 1987–88'. However, these policy changes would not have affected directly the growth rate of the area under *traditional* irrigation. Rather, it is the policy impact on the growth rate of the area under *modern* irrigation which is relevant to the point (as indicated by the data cited above).
11. The only other major cereals which use modern irrigation are wheat, which has a relatively small weightage in total foodgrain production, and HYV aman rice, which generally requires irrigation to supplement rainfall as and when necessary.
12. These were normal years (1990–94) without any serious natural calamity affecting agricultural production in general, and HYV boro in particular. However, there was a drought in the following year, 1994–95 (Shahabuddin, 1995).
13. However, it is possible that these technological, agronomic and hydrological constraints had begun to surface gradually with the expansion of boro cultivation, constituting longer term trends antedating the round of policy shifts in the mid-1980s.
14. Mandal (1989: 56). In fact, Rogers et al. (1994) have claimed that the groundwater potential of Bangladesh is far from exhausted, so that the supply of irrigation and the growth of boro rice are unlikely to be constrained by such considerations in the foreseeable future.
15. Khan (1990). Concerns about the low price and declining profitability of HYV boro affecting its growth were also expressed by Mandal (1989: 43–4).
16. Both de Vylder (1982: 101–2) and Osmani (1985) pointed out that unequal distribution of income resulting from the new technology could constrain the effective demand for foodgrain (rice), thereby limiting the potential for its longer term growth. Subsequently, Osmani (1990) and Khan (1990) argued that changes in income distribution had adversely affected the poorest classes, limiting possible increases in their effective demand for foodgrain (rice). However, these arguments were advanced

in the distinct context of assessing trends in the incidence of poverty over recent decades.
17. In the present discussion, assessment of the impact of policy changes on output and input trends has been limited by the fact that the available data on agricultural production are organised by decades (i.e., the 1970s, 1980s and early 1990s). Reorganisation of the data in terms of cut-off dates corresponding to the different policy regimes and specific policy interventions would have been more relevant and useful for addressing the issues involved. However, this has not been possible due to constraints of time.
18. Osmani (1985: 7, 21, fn) notes that virtually all that was available were documents of the World Bank.
19. Palmer-Jones (1992: A–131), citing World Bank (1981).
20. Osmani (1985: 20–1); Hossain (1995); Shahabuddin (this volume). See further discussion on the content and implications of these standardisation requirements in the section dealing with the consequence of the deregulation regime.
21. This applied to legal or de jure ownership. If de facto control over irrigation equipment is taken into account, then the dominance of private ownership during the mid-1980s was likely to have been even greater, as discussed in the following section.
22. Crow (1989). Cf. Harriss-White (this volume) on the role of the state in the agricultural markets of West Bengal.
23. It is pertinent to note here that in the debate concerning Bangladesh, Palmer-Jones (1992: A–131) appears to be explicitly dismissive about evidence from village studies without, however, adducing any argument to back up his opinion.
24. Palmer-Jones (1992: A–128, A–138) argues that certain policy documents of the government and donor agencies were apprehensive of the potentially inegalitarian impact of water markets dominated by 'water-lords', and hence their eventual emergence under the initiative of private sector water sellers constituted unintended consequences as far as policy objectives were concerned. However, the indisputable fact that privatisation of ownership of irrigation equipment continued to be promoted by the very same policy-making agencies would perhaps suggest that what is stated on paper in policy documents might deserve a critical reading rather than a literal one.
25. Cf. Dobb (1963), Sweezy (1976), Brenner (1976; 1977), and Hilton (1976) on earlier debates about the relative causal significance of the 'growth of commerce' as compared to the pre-existing class structure.
26. While individual users might have opposed needs and interests, the resolution of such conflict would have to be mediated through the management, since it was only with the latter that each of them could invoke their respective contractual rights.
27. The functions of the *shalish* are defined and discussed in Bertocci (1970) and Adnan (1990: 175–80). The factors underlying the failure of the *shalish* mechanism in this particular instance is treated in greater detail in Adnan (forthcoming).
28. Cf. survey data provided by Murshid (1985) and Sattar and Bhuiyan (1985).
29. This is suggestive of a process of diffusion of institutional innovation over a micro-region, following in the wake of the diffusion of technological innovation. This social-institutional process had the effect of spreading a particular form of management without it having to be 'independently invented' everywhere. The process of institutional change thus appears to have been operating at levels beyond the individual irrigation management unit, the village, or the locality.
30. Cf. Scott (1985) on dominant classes living in a locality needing to find ways to

disguise or tone down their aggressive economic moves in order to avoid or reduce social opprobrium.

31. This bears upon the relevance of using variables related to sharecropping in 'multivariate analysis' concerned to test the significance of the agrarian structure as a possible constraint to agricultural growth, noted at the outset.
32. Adnan et al. (1975: 118), based on observations from four villages in Barisal, Chittagong, Rangpur and Mymensingh in 1974.
33. The history of capitalist development has been characterised by such patterns (Abdel-Fadil, 1975; Adnan, 1985: PE–61).
34. Bharadwaj (1974). Case study evidence from different parts of the country is provided, amongst others, by Crow (1989: 223) and Crow and Murshid (1994).
35. Hossain (1995), citing Osmani and Quasem (1990).
36. '... less than one per cent of the 2,000 producers ... sold to government procurement centres.... Nine of the 12 centres ... purchased overwhelmingly from traders .... the procurement centres "are not acting as a support price program for the farmers. They serve mostly to supply prices to the dealers who have bought supplies from the growers at distress prices".' Crow (1989: 216), citing Islam et al. (1985: 123, 152–3).
37. Crow (1989: 216) observes that '... procurement was dominated by traders because they had preferential arrangements with government officials and could dominate queues should procurement begin in earnest'.
38. Crow (1989: 198). The case study evidence comes from different areas of the country, collected in the period between November 1985 and January 1988.
39. Crow (1989: 208). The margin between such 'producer's price' and the open market price provides the source of trading profits which are typically shared out between a chain of marketing intermediaries who are largely financed on an 'onward lending' basis by wholesalers at the apex of the rice marketing system.
40. Mandal (1989: 45–8, Tables 1 and 2). The printed tables and accompanying discussion in the text display minor inconsistencies and omission of pertinent data, and methodological problems related to the computation of rates of return. The data cited here have been recomputed to rectify these limitations as far as possible.
41. Given the coexistence of the lease market, the prevailing cropshare (rent) would define that opportunity cost.
42. Limitations of the data do not allow computation of the rates of return at constant prices. However, because of relatively higher cost increases between 1985 and 1988, the decline in the rate of return is likely to have been marginally greater at constant prices.
43. Recomputed from Mandal (1989: 47, Table 2).
44. In both years, the rate of return is negative for sharecroppers, and this is partly due to the fact that rent for land is included in the costs, unlike the computation in the case of owner-operators. Sharecroppers have positive returns only at 'cash costs', i.e., if *only* market costs are considered, excluding owned inputs such as family labour (Mandal, 1989: 46–7, Tables 1 and 2).
45. Mandal (1989: 55–6, Table 8), citing the Agro-Economic Research Section of the Ministry of Agriculture, Bangladesh (AER–MOA, 1981–82; 1986–87). The official estimate is not free of methodological problems, but it is likely to underestimate the costs rather than the returns.
46. Quasem (1987), cited by Mandal (1989: 55).
47. Mandal (1989: 46–8, 61, Tables 1 and 2, and Appendix Table 1). As discussed earlier, all the data are in current prices.

48. Quasem (1987), cited by Mandal (1989: 55).
49. Mandal (1989: 57–8). See references cited earlier for other parts of the country.
50. In contrast, co-operative or 'user group'-based irrigation management does not face the compulsion of generating a profit margin above costs, since it can survive so long as it breaks even.
51. Recomputed from Mandal (1989: 50–3, Table 6). The normalised per hectare figures have been used in the computation of the rates of return in order to control for the difference in size between DTW and STW command areas, and for the fact that there was a mix of diesel and electrically operated tubewells. Current prices were used to compute the rate of return in both the years, since data limitations did not permit computation in constant prices.
52. Mandal (1989: 55), citing Quasem (1987). In areas intensively irrigated by STWs in Bogra, net returns had declined because of relatively higher operational costs and decline in command areas, and in some cases because of decline in boro yields.
53. Mandal (1989: 52–3, Table 6). Since capital costs are not included in the data, the observed decline in the profitability of irrigation does not incorporate any possible effect of the policy of reducing subsidy on irrigation equipment. In any case, there was little subsidy on STWs.
54. Mandal (1989: 48–53, 58, Tables 4 and 6). Most of the STWs in 1988 were electricity-operated and their operational and maintenance costs did not increase significantly between 1985–88. In fact, many diesel-run tubewells in Tangail were electrified over this period. The concerned tubewell managements were reported to have had to bribe official functionaries in order to get electricity connections, while also being over-charged in the case of 'un-metred' connections. Irrigation became critically dependent on the supply of electricity which was irregular with frequent disruptions. These factors contributed to reduced crop yields as well as covert increases in costs such that the absolute magnitude of the 'electricity bills' were much higher than the corresponding fuel costs of diesel-run tubewells in 1988.
55. Growing competition amongst private operators of low-lift pumps was observed in a Chittagong village which I studied during the late 1980s and early 1990s (fieldwork notes).
56. Mandal (1989: 57) observes that STW operators in Tangail were offering supplementary irrigation for aman rice to those who would continue to use their water for boro production.
57. Kula (1976: 47) has used the evocative phrase 'coefficient of realisable coercion' to describe this social limit to surplus extraction (in a different context).
58. Palmer-Jones (1992: A–131), citing Task Force (1991, Vol. 2: 120).
59. Based on my fieldwork notes on the activities of a public sector bank in a rural marketing centre in Chittagong during the late 1980s.
60. Hossain (1995); Shahabuddin (1995). Private traders began to lift fertilisers in bulk from factories and ports with discount incentives. Their procurement prices were equalised with those of BADC in 1989. They were also allowed to take on lease 'surplus' warehouses of BADC.
61. Hossain (1995). Because of lack of domestic production capacity, two-thirds of TSP and the entire requirement of MP have to be imported from abroad, mostly with foreign aid. Some implicit economic subsidy on urea has continued because of its domestic production under controlled prices, unlike MP and TSP.
62. Substantive studies analysing the structure of this changed fertiliser marketing system, including the relationships between the market functionaries and commercial capitals from home and abroad, are not available.

63. According to Hossain (1995), 'empirical studies based on time series data show a price elasticity of demand [for fertiliser] in the range of –0.6 to –0.8, while those based on cross-section data report price elasticity at more than unity . . . . Thus withdrawal of subsidy [on fertiliser prices] would have some negative effect on fertilizer consumption'.
64. Shahabuddin (this volume). According to Palmer-Jones (1992: A–136, 139–40, fn 23), subsidy on DTW has continued in practice because of the need to sell off existing public stocks, presumably held by BADC.
65. Palmer-Jones (1992: A–136, 139–40, fn 23). The rate of subsidy on DTW in early 1991 had been 70 per cent; by March 1992, it had been increased to more than 80 per cent in the attempt to sell off remaining stocks.
66. Their position was qualitatively different from that of the vast majority of the peasantry who, for example, came to the market to buy fertilisers or sell produce with inferior bargaining power compared to the traders they had to deal with.
67. These inferences appear to be plausible, despite the lack of adequate evidence from substantive studies analysing the structure of the marketing system and the commercial interests which have taken over import, pricing and sales of irrigation equipment. The few studies available (e.g., Raha and Akbar, 1993) are not concerned with the political economy of such world market contentions. Nor do they display critical assessment of the data collected through questionnaires filled in by the suppliers of equipment themselves.
68. Boyce (1987: 238) citing illustrative evidence observes that: 'Choice of techniques in aid-financed projects has also been influenced by donor interests in export promotion . . . .'.
69. During the 19th and 20th centuries the agriculture of undivided Bengal had been linked to, and made dependent upon, the world market for the export of its products—pre-eminently jute, preceded by indigo (Bose, 1986; 1993).

REFERENCES

Abdel-Fadil, Mahmoud. 1975. *Development, Income Distribution and Social Change in Rural Egypt: A Study in the Political Economy of Agrarian Transition.* Cambridge: Cambridge University Press.

Adnan, Shapan. 1985. 'Classical and Contemporary Approaches to Agrarian Capitalism,' *Economic and Political Weekly*, 20(30): PE–53 to PE–64.

———. 1987. 'The Roots of Power: A Re-Study of Daripalla in Rural Bangladesh.' Mimeograph, Shomabesh Institute, Dhaka.

———. 1989. 'Performance, Problems and Prospects of Farmers' Cooperatives in Bangladesh,' Compendium Volume 5 of *Bangladesh Agriculture: Performance and Policies, Bangladesh Agriculture Sector Review*, UNDP/Ministry of Agriculture, pp. 1–60.

———. 1990. *Annotation of Village Studies in Bangladesh and West Bengal: A Review of Socio-Economic Trends over 1942–88.* Bangladesh Academy for Rural Development (BARD), Comilla.

———. 1997. 'Class, Caste and *Shamaj* Relation Among the Peasantry in Bangladesh: Mechanisms of Stability and Change in the Daripalla Villages, 1975–86,' Jan Breman, Peter Kloos and Ashwani Saith (eds) *The Village in Asia Revisited*. Delhi: Oxford University Press.

Adnan, Shapan, A. Kamal, M. Muqtada, and A.M. Khan. 1975. 'The Preliminary Findings of a Social and Economic Study of Four Bangladesh Villages,' *Dhaka University Studies*, Vol. 23, Part A.

Arens, J. and J. van Beurden. 1977. *Jhagrapur: Poor Peasants and Women in a Village in Bangladesh*. Amsterdam/Birmingham: Third World Publications.

Arn, A., and M.A. Mannan. 1982. *Lakshmipur Thana: A Socio-Economic Study of Two Villages*. Copenhagen: Centre for Development Research (CDR) Project Papers A.82.6.

Bertocci, Peter J. 1970. 'Elusive Villages: Social Structure and Community Organization in Rural East Pakistan.' Unpublished Ph.D. dissertation, Michigan State University.

Bhaduri, Amit. 1974. 'Towards a Theory of Pre-Capitalist Exchange,' Ashok Mitra (ed.) *Economic Theory and Planning: Essays in Honour of A.K. Dasgupta*, Calcutta: Oxford University Press.

———. 1976. 'The Evolution of Land Relations in Eastern India under British Rule,' *Indian Economic and Social History Review*, 13(1): 45–58.

Bharadwaj, Krishna. 1974. *Production Conditions in Indian Agriculture: A Study based on Farm Management Surveys*. Cambridge: Cambridge University Press.

Bose, Sugata. 1986. *Agrarian Bengal: Economy, Social Structure and Politics, 1919–1947*. Cambridge: Cambridge University Press.

———. 1993. *Peasant Labour and Colonial Capital: Rural Bengal Since 1770*. The New Cambridge History of India, Volume 3(2), Cambridge: Cambridge University Press.

Boyce, James K. 1987. *Agrarian Impasse in Bengal: Institutional Constraints to Technological Change*. Oxford: Oxford University Press.

Brenner, Robert. 1976. 'Agrarian Class Structure and Economic Development in Pre-Industrial Europe,' *Past and Present*, 104: 25–92.

———. 1977. 'The Origins of Capitalist Development: A Critique of Neo-Smithian Marxism,' *New Left Review*, 104: 25–92.

Crow, Ben. 1989. 'Plain Tales from the Rice Trade: Indications of Vertical Integration in Foodgrain Markets of Bangladesh,' *Journal of Peasant Studies*, 16(2): 198–229.

Crow, Ben, and K.A.S. Murshid. 1994. 'Economic Returns to Social Power: Merchants' Finance and Interlinkage in the Grain Markets of Bangladesh,' *World Development*, 22(7): 1011–30.

de Vylder, Stephan. 1982. *Agriculture in Chains: A Case Study in Contradictions and Constraints*. London: Zed Books.

Dobb, Maurice. 1963. *Studies in the Development of Capitalism*. London: Routledge and Kegan Paul.

Harriss, John. 1992. 'Does the "Depressor" Still Work? Agrarian Structure and Development in India: A Review of Evidence and Argument,' *Journal of Peasant Studies*, 19(2): 189–227.

Herbon, Dietmar. 1985. *The System of Exchange and Distribution in a Village in Bangladesh*. Institute for Rural Development, Geory-August-Universitat, Gottingen, and Rural Development Academy, Bogra.

Hilton, R.H. (ed.). 1976. *The Transition from Feudalism to Capitalism*. London: New Left Books.

Hossain, Mahabub. 1995. 'Agricultural Policy Reforms in Bangladesh: An Overview.' Mimeograph, revised version of paper presented at the Workshop on Agricultural Growth and Agrarian Structure in Contemporary West Bengal and Bangladesh, Centre for Studies in Social Sciences, Calcutta, 9–12 January 1995.

Islam, A., M. Hossain, N. Islam, S. Akhter, E. Harun and J.N. Efferson. 1985. '*A*

Benchmark Study of Rice Marketing in Bangladesh.' Mimeograph, Bangladesh Rice Research Institute, Bangladesh Marketing Directorate, United States Agency for International Development (USAID), Dhaka.

Islam, Sirajul. 1985. *Villages in the Haor-Basin of Bangladesh.* Studies in Socio-Cultural Change in Rural Villages in Bangladesh, No. 4, Institute for Study of Language and Cultures of Asia and Africa, Japan.

Jansen, Eirik G. 1987. *Rural Bangladesh: Competition for Scarce Resources.* Dhaka: University Press Ltd.

———. 1992. 'Interest Groups and Development Assistance: The Case of Bangladesh,' *Forum for Development Studies*, No. 2, Norwegian Institute of International Affairs (NUPI), Oslo.

Khan, A.R. 1990. 'Poverty in Bangladesh: A Consequence of and a Constraint on Growth,' *Bangladesh Development Studies*, 18(3): 19–34.

Kula, Witold. 1976. *An Economic Theory of the Feudal System: Towards a Model of the Polish Economy 1500–1800.* London: New Left Books.

Mandal, M.A.S. 1989, 'Declining Returns from Groundwater Irrigation in Bangladesh,' *Bangladesh Journal of Agricultural Economics*, 12(2): 43–61.

Marx, Karl. 1976. *Capital: A Critique of Political Economy.* Vol. 1, Harmondsworth: Penguin Books in association with New Left Review.

———. 1981.*Capital: A Critique of Political Economy.* Vol. 3, Harmondsworth: Penguin Books in association with New Left Review.

Murshid, K.A.S. 1985. 'Is there a "Structural" Constraint to Capacity Utilisation of Deep Tubewells?' *Bangladesh Development Studies*, 13(3, 4): 147–54.

Osmani, S.R. 1985. 'Pricing and Distribution Policies for Agricultural Development in Bangladesh,' *Bangladesh Development Studies*, 13(3, 4): 1–40.

———. 1990. 'Structural Change and Poverty in Bangladesh: The Case of a False Turning Point,' *Bangladesh Development Studies*, 18(3): 55–74.

Palmer-Jones, Richard. 1992. 'Sustaining Serendipity: Groundwater Irrigation, Growth of Agricultural Production and Poverty in Bangladesh,' *Economic and Political Weekly*, 27(39): A–128 to A–14.

Quasem, M.A. 1987. 'Financial Returns of Irrigation Equipment to Owners and Users: The Case of Shallow Tubewells in Bangladesh, 1981–85,' Development Research and Action Programme (DERAP) Working Paper/A373, Chr. Michelsen Institute, Bergen.

Raha, S.K., and M.A.A. Akbar. 1993. 'Marketing of Minor Irrigation Equipment in Some Selected Areas of Bangladesh,' *Bangladesh Journal of Agricultural Economics*, 16(1): 89–90.

Rahman, Hossain Zillur. 1982. 'Report from Raipur Thana.' Mimeograph, BIDS–CDR Joint Noakhali Development Study, Copenhagen: Centre for Development Research (CDR) Project Papers A.82.2.

Rogers, P., P. Lydon, D. Seckler, and K. Pitman. 1994. *Water and Development in Bangladesh: A Retrospective on the Flood Action Plan.* Arlington, Irrigation Support Project for Asia and the Near East (ISPAN) sponsored by USAID.

Sattar, M.A., and S.I. Bhuiyan. 1985. 'Constraints to Low Utilization of Deep Tubewell Water in A Selected Tubewell Project in Bangladesh,' *Bangladesh Development Studies*, 13(3, 4): 155–66.

Scott, James C. 1985. *Weapons of the Weak: Everyday Forms of Peasant Resistance.* New Haven and London: Yale University Press.

Shahabuddin, Quazi. 1995. 'Agricultural Growth Performance in Bangladesh: Some Recent Evidence.' Mimeograph (revised version of paper presented at the Workshop on Agricultural Growth and Agrarian Structure in Contemporary West Bengal and Bangladesh, Centre for Studies in Social Sciences, Calcutta, 9–12 January 1995).

Siddiqui, Kamal. 1982. *The Political Economy of Rural Poverty in Bangladesh*. Dhaka: National Institute of Local Government.

Sweezy, Paul. 1976. 'A Critique' and 'A Rejoinder', R. H. Hilton (ed.) *The Transition from Feudalism to Capitalism*, London: New Left Books.

Wood, Geoffrey D. 1978. 'Class Differentiation and Power in Bandokgram: The Minifundist Case,' M.A. Huq (ed.) *Exploitation and the Rural Poor: A Working Paper on the Rural Power Structure in Bangladesh*, Bangladesh Academy for Rural Development (BARD), Comilla.

8

Panchayati Raj and the Changing Micro-Politics of West Bengal

GLYN WILLIAMS

INTRODUCTION: EMPOWERMENT, THE STATE AND PANCHAYATI RAJ

Over the last decade, the issue of 'empowerment' has been placed firmly on the agenda of development studies. Writers such as Robert Chambers (1993) have shown that powerlessness is an integral part of poverty, with the result that empowerment, defined by Edwards as 'increasing the control which poor and powerless people (and specifically the poorest and the most powerless) are able to exert over aspects of their lives which they consider to be important to them' (Edwards, 1993: 80), is now an important aim for most development practitioners.[1] Empowerment defined in this way implies a degree of both economic and political change. The

---

[1] I would like to thank all of those who commented on the earlier draft of this paper I presented in Calcutta in January 1995, especially Shapan Adnan and Kristoffel Lieten, and also Stuart Corbridge for his guidance and encouragement throughout the course of my research.

economic changes involved have long been a topic of debate within development studies, as exemplified by the 1970s literature on 'redistribution with growth'. In the context of contemporary West Bengal, both government-led redistribution of wealth and sustained growth of agricultural output (if it were associated with increases in labour demand) could potentially contribute much to the economic empowerment of the rural poor. In this paper it is the micro-scale consequences for the poor of West Bengal's recent agricultural growth, rather than the debates as to the macro-scale patterns and causes of growth per se, that will be my principle concern.

Alongside these long-standing economic concerns, an explicitly political view of empowerment has more recently become a part of the 'alternative development' canon. John Friedmann has argued that improving the political control which the powerless have over their own lives is important both as an end in itself, and as a means by which economic empowerment can be achieved (Friedmann, 1992).[2] The wider effects of such political empowerment, whether initiated by the state or (as is more usually assumed to be the case) by grassroots organisations, can lead to a radical change in the relationship between the state and these individuals. Panchayati raj, the CPM's programme to radically restructure local government, was intended to provide such a political change through popular participation in a decentralised state developmental apparatus.[3] Here, I will examine the extent to which this programme has resulted in the empowerment of poorer households in West Bengal through a detailed investigation of the changing 'micro-politics' of three villages.

I have used the term 'micro-politics' here to define a broad range of conflicts and competitions occurring at the village scale. This would correspond to an extended definition of politics which Gillian Hart has noted as being of increased importance:

> Instead of referring simply to electoral politics and/or actions focused specifically on the state, politics has increasingly come to be used in a broader sense to refer to the processes by which struggles over resources and labour are simultaneously struggles over socially constructed meanings, definitions and identities (Hart, 1991: 95).

Formal politics—the activities of political parties and their representatives in the local area, and conflicts over these actions—are clearly an important part of micro-politics. This is particularly so in West Bengal where in theory the devolution of power to the panchayats gives many people easy access to government and the opportunity to participate directly in the formal political system. However, micro-politics can also exist in the sphere of informal, unorganised action; competition between rival factions, caste and communal tensions or labour disputes are all political in that they affect local hierarchies and power structures, but can and do exist without the participation of parties or panchayats. James Scott has stressed the importance of this informal political action in the context of rural Malaysia, arguing that a series of petty struggles over material benefits (via pilfering, foot-dragging and evasion) and ideological resources (via slander, gossip and public shows of deference) make up the poor's 'everyday forms of resistance' to their social superiors (Scott, 1985). In practice, the division between these formal and informal aspects of micro-politics is far from straightforward. Party-political and government actions are subject to reinterpretations and reworking by local people to suit their own ends, and local non-party conflicts can be deliberately politicised by the use of party-political terms.

As the later work of Scott (and others sympathetic to his ideas of 'everyday forms of resistance') has largely focused on peasant resistance to states which are actively disempowering individuals,[4] the study of micro-political change in rural West Bengal could potentially provide a useful contrast. According to party documents, the CPM has three objectives for its rural development policy: 'to involve the entire people in the process of development by democratic decentralisation of the power structure', 'to bring about a change in the correlation of class forces in favour of the poor and working people' (which would include material improvements for the poor deemed essential for their participation in political struggles) and to raise class-consciousness through struggles over development (Mishra, 1991: 9). The CPM claimed that its panchayati raj programme was central to achieving these development aims: by instituting a system of democratic local government, the party hoped there would be mass participation in the panchayats and increased class-consciousness as a result.

The CPM's publicly stated aims for its rural development programme in West Bengal therefore appear to be broadly supportive of the empowerment of the rural poor (see Bhattacharyya, this volume, for a discussion of their covert aims). However, if this programme is to be successful, it will require the alteration of existing relationships between society and the state, and the structure of micro-politics, in the Bengali countryside.

The implementation of a 'democratic' local state apparatus is not in itself enough to ensure this empowerment if the discourse of micro-politics, and the value system upon which it is based, is left unaltered.[5] Some idea of the nature of the changes required is provided by Davis' study of a Medinipur village in the early 1970s (Davis, 1983), which identified a conflict between 'traditional' and 'modern' value systems in determining the structure of micro-politics. The traditional value system, *gramer kaj* ('village politics'), constantly looked back to a golden age where members of all castes supposedly acted in accordance with their dharma. This therefore idealised a system of unequal, but harmonious, relationships in which the rule and authority of those in power is naturalised and legitimated as being the fulfilment of the 'proper' role of high-status individuals. By contrast the modern *sorkari kaj* ('government politics') involved appeals to authorities beyond the village, which were in turn based on the ideal of the state as enforcer of the individual's right to fair and equal treatment. In this idealised form, sorkari kaj aims to empower many of those marginalised by gramer kaj through its attempts 'to level, if not completely eliminate [the latter's] system of inequalities and to substitute in its stead a social and political order grounded on the premise of equality' (ibid.: 215).

Such distinctions are not only of theoretical interest, but also have important practical implications. Arild Ruud has used elements of Davis' micro-politics framework to explain the growth of CPM support in Bardhaman during the second United Front period (Ruud, 1994). Growth in support, he argues, was not merely a result of economic gains for the peasantry through the 'land-grabs' of the period, but instead:

> The Marxists' mass support seems rather to have been a result of the manner in which they sought to form their political authority. It seems to have been a matter of the Marxists' ability not only to penetrate the rural areas as organisations, but their ability to establish themselves as major actors in what may be seen as *traditional village politics* (Ruud, 1994: 369, emphasis added).

By taking on dominant (and Congress-supporting) factions in the villages in all disputes, regardless of their overtly 'political' content, and by acting as patrons to groups previously excluded from much of village politics, Ruud argues that CPM party workers were operating within the 'rules' of gramer kaj, and winning support as a result. This shows the importance of the CPM's sensitivity to micro-politics in the party's growing popularity in the 1960s, but questions remain as to the extent to which the CPM has been able to transform the nature of micro-politics to the more broadly empowering sorkari kaj since the Left Front achieved power in 1977.

This paper attempts to evaluate the degree to which the CPM has been able to restructure the micro-politics of rural West Bengal through its panchayati raj programme. In the analysis that follows, I investigate three main themes: people's perceptions of the panchayat programme, the degree and nature of grassroots support for political parties, and the nature of 'everyday' social conflict in the Bengali countryside, focusing on conflicts over agricultural labour and tenancy. By looking at these three themes and the interactions between them, a detailed picture of the panchayats' contribution to the empowerment of the poorest in contemporary West Bengal will hopefully emerge.

To investigate these micro-political issues in detail it is insufficient to look at the actions of a panchayat, or local voting patterns, in isolation; instead, a sensitive investigation of the full breadth of micro-political actions and motives is required to place this information in its proper context. As a result, my study is based around a set of detailed interviews with people living in three villages in Birbhum district.[6] Birbhum was chosen as it is predominantly rural, and the CPM has only had a strong political presence there in the last twenty years. The three villages of the study were selected on the basis of their different party-political histories (Table 8.1), and their mixed social composition: each village had a mixture

Table 8.1
Party Affiliation of Village Gram Panchayat Members

Village	1978	1983	1988	1993
Ramnagar	Congress	Independent	CPI	BJP
Durgagram	Congress	Congress	Congress	CPM
(2 members)	Congress	Congress	Congress	CPM
Kalipara	CPM	CPM	CPM	Congress

of scheduled and general caste households, and Durgagram had a sizeable Muslim community. All three villages are relatively easily accessible (Durgagram and Kalipara were within a few kilometres of Siuri, the district capital, and Ramnagar lay on the bus route from Sainthia to Rampurhat) but agricultural conditions, which were in turn highly dependent upon irrigation, varied greatly between them (Table 8.2).

Table 8.2
Statistical Comparison of Agriculture in the Survey Villages

Village	Source of Irrigation	Percentage of Cropped Area under HYV	Cropping Intensity[1]
Durgagram	tank	7	1.11
Kalipara	tank, canal	69	1.45
Ramnagar	tank, canal	28	1.40

[1]Calculated here as (gross cropped area operated)/(total area operated) for all households in village.

GOVERNMENT AND PARTICIPATION: THE PANCHAYATS IN (IN-)ACTION

Under the Left Front's panchayat legislation, the lowest tier of local government is the gram panchayat, a council of representatives of around ten villages, covering a population of around 12,000. In theory, this tier of local government has a high degree of autonomy: it is responsible for drawing up annual development plans for its own area, and distributing state and national funds to realise their goals. In fulfilling this role at the village level, individual members of the panchayat are intended to be in close consultation with villagers to ensure that 'grassroots' opinions form the fundamental basis of all development work.

There are important distinctions between this official role of the gram panchayat members, and the expectations placed upon them by their constituents. While the proper distribution of development funds and consultation over development plans were seen as an important part of a panchayat member's role, all interviewees also expected a good member to promote the interests of the village at gram panchayat meetings, and have the authority and standing to sort out any disputes that arose in the village, preferably without resorting to outside agencies. Significantly, this idealised role requires the member to be actively involved in the full breadth of 'village politics', beyond his/her more limited official duties.

In practice, panchayat members in the study villages had a varied degree of success in meeting this idealised role. In terms of being able to meet the development needs of villagers, panchayat members are heavily reliant upon the resources provided by the Jawahar Rozgar Yojana (JRY) and the Integrated Rural Development Programme (IRDP), both of which are national programmes aimed at poverty relief.[7] These provided the bulk of the direct development support distributed at the village level, and although the sums of money involved are relatively small, this remains one of the most visible (and hence contested) parts of a panchayat member's work.[8] Within these financial constraints, the most recent panchayat members from Durgagram had been relatively successful, and were complimented by many for paying attention to the needs of scheduled caste and Muslim families to a greater degree than any previous members. In Kalipara and Ramnagar, the villagers were less content with their representatives, complaining that they had not done sufficient work to secure development funds, and had misspent those which had been distributed.

If we ignore for the moment the individual details of the villages, a few more general points arise. First, the Left Front's model of the panchayat as a medium for popular participation was not being achieved in practice in any of the survey villages. For most people in rural West Bengal the village's gram panchayat member is their first point of contact with panchayati raj, and many interviewees stated that the panchayat member was the only state representative they had ever consulted. None of the members had ever held a gram sabha meeting, where all villages should be invited to discuss their views of panchayat work with their member and the *pradhan* (gram panchayat chairperson) in public, although many said they would welcome the chance of participating in this way. Consultation over development work was universally restricted to the close friends of the member; for most people their only involvement with the panchayat was voting every five years, and requesting help from their representatives. There were instances of villagers going directly to political party offices or the panchayat office to contest the decision of a member, but such actions remained the exception rather than the rule. Moreover, it was usually the more confident and better educated villagers who complained, so that those most in need of extra assistance felt least able to command it.

Second, due to this lack of participation, panchayat members came to play a role in the village similar to a traditional headman. Members were expected to sort out a whole variety of 'non-political' disputes and, as Davis notes, acting as arbiter in these arguments is a way in which the 'big men'[9] of a village indicate and contest their rank. The control of development funds for JRY and IRDP schemes gave members a degree of economic influence beyond that of most landlords, and thus supported them in fulfilling this leadership role. Also, any development work a member undertook became highly personalised: rather than the fulfilment of an objective set of criteria, it was seen by potential beneficiaries as 'help' (*sahar*), especially by those amongst the labouring classes.[10]

This is significant in that by requesting 'help' from their member, villagers were using a 'language of claims' equivalent to that used by, for example, a tenant requesting a loan from his landlord. Such requests are indicative of the way in which the whole panchayat system is viewed by many: rather than being an institution in which they actively participate, it is seen as a distributor of personalised benefits.

Third, with representatives' power being highly personalised, the integrity of the individual member becomes all-important and the potential for malpractice grows. The issue of defining corruption is a difficult one; in some cases, deliberately self-serving behaviour and outright theft can be identified. In all three villages, there was gossip concerning instances where panchayat members had allegedly used funds illegally for personal gain, and the scope for such action is certainly widespread.[11] In Kalipara, tales of their last member's greed were told openly and reached epic proportions: 'All panchayat members today are corrupt—but ours has stolen so much money from the panchayat fund that he could afford to marry all three of his daughters at once!'(agricultural labourer, Kalipara).

More subtle than such outright theft, and more commonplace, was the use of panchayat funds to secure political support: a general complaint of all villagers not belonging to the faction of the panchayat member was that all development money was being directed exclusively to the members' friends and supporters. In the majority of these cases, funds were still going to the economically deserving (and were therefore, in a sense, not corrupt), but within this target group, it was only the politically well-connected that benefited. Such partisan behaviour is at first glance no different from the selective distribution of resources by any powerful villager. However, unlike the 'help' given by a landowner to a sharecropper or labourer, this faction-building by panchayat members was much more open to public criticism and challenge from opponents wishing to see a 'fairer' allocation of benefits. When pressed on the issue, many admitted that such partisan behaviour had always been present in panchayat work, but complained that it was becoming more widespread as a direct result of the increasing party-politicisation of panchayat members. This increase in 'party feeling' is widely condemned as promoting factional disputes and disrupting village harmony.

It should be noted, however, that the word used by interviewees to describe these party factions was *dol*, a term which is not specific to party politics but has more general connotations of 'side' or 'team'. To gain an insight into what belonging to a political dol means, and whether this has any more significance than more traditional factional identities, it is necessary to investigate the issue of party politics in greater detail.

POLITICAL PARTIES: SUPPORT AND CONFLICT IN THE 1993 PANCHAYAT ELECTIONS

As already mentioned, the interactions between formal and informal micro-politics are often far from straightforward: support for a political party can come from a variety of different motives. In what follows, therefore, I will use the interview responses to build up a picture of the different party dol as they actually existed, and also to investigate participants' reasons for supporting a given party. As the research period coincided with the 1993 panchayat elections, political divisions in the village were brought into the open, and people's own interest in politics was high, making this investigation easier.

Campaigning for the 1993 elections was intense throughout the study area. Local and state newspapers provided detailed coverage of events, and a series of political rallies and marches were staged in Siuri and other parts of the district, all of which raised tensions during the election period.[12] In the villages themselves, political graffiti appeared on the houses of all representatives and their core supporters, and awareness of the different political factions was high through their involvement in village disputes; as one Ramnagar resident put it, 'at election time, every argument becomes a political matter.' Coupled with this activity, local party representatives and candidates had a detailed understanding of the balance of electoral support, being able to identify the percentages of voters supporting each party on a neighbourhood-by-neighbourhood basis.

The elections themselves in the three villages were also free from ballot-rigging or coercion of voters, and had a remarkably high reported turnout, with only one elderly widow not voting from the households interviewed. The elections saw a change of party in all three villages (Table 8.1). In Durgagram and Ramnagar, the elections were a four-way contest between the Left Front, Congress, independent candidates and, for the first time, the BJP, whereas in Kalipara there was a straight competition between the CPM and Congress. The general picture of panchayat elections in the study villages is therefore one of a securely established, actively contested but largely peaceful electoral system.[13]

This sophistication in the mechanics of the electoral process does not however imply that the logic of gramer kaj had been altogether displaced (see also Bhattacharyya, this volume). For many people, voting at the gram panchayat level was not on the basis of party affiliation at all, but rather on the candidates involved, whether or not they would be good leaders in the terms outlined above, or would be likely to offer them help.[14] Men generally decided who their wives and daughters should vote for, and their decisions were in turn often made on the basis of discussions within a dol comprising of their extended family and neighbours. Caste and kinship affiliations were important in defining the membership of such informal voting groups, and traditional bases of status often determined who had influence within them: several younger men stated that they were prepared to follow their elders who were 'experts in such matters'. Many of these dol were undecided at the start of the election campaign as to whom they would vote for, and had no open party affiliation. There were also a few committed party supporters in each village who played an active role in the selection of candidates: these individuals and their close supporters were much more open about their party loyalties which appeared to be common knowledge in the villages.

With the importance of individual candidates and culturally-based group loyalties within the elections, it is perhaps no surprise that understandings of the different parties were much less well developed. When directly questioned on their perceptions of different parties, interviewees' responses were also somewhat surprising. The BJP was supported by some on religious grounds: in the area around Ramnagar, Hindu–Muslim tensions ran high, and several local Left Front members were portrayed by these BJP supporters as having links with a 'criminal element' in Muslim society. This 'association' was gaining the BJP some sympathy amongst the richer general castes:

> It is good that the BJP have a movement to end favouritism towards minorities.... Muslim Indians support Pakistan at cricket, and are choosing to be separate from other Indians—if this is how they act they don't deserve special attention.... Hinduism is a part of India's past and culture—to be a good Indian you must recognise this (landlord, Ramnagar).

It is however important not to stress this 'communal' element of BJP support too much, as many people supported the BJP due to frustration with the other two parties, a frustration felt most acutely by marginal farmers. In particular, the Congress was seen as having failed them as it had been unable to control demands from the labouring classes: 'Congress was best for farmers before—now they are not looking after farmers' interests so I will support the other rightist party [i.e., the BJP]' (cultivator, Durgagram); and, 'The existing parties only help the tribals and the labouring castes—no-one looks after the interests of small farmers' (cultivator, Durgagram).

Such sentiments, and the fact that no labouring families gave the BJP their support, suggest that the growth of the BJP was not exclusively the result of increased communal tensions in West Bengal's politics, as some in the CPM feared (personal interview with Benoy Konar, West Bengal Kisan Sabha Secretary, 20 October 1992). It might indicate instead that the BJP's rapid growth was in part due to the absence of an effective political movement that would openly support farmers' interests when they came into conflict with those of agricultural labourers.[15]

Support for the Congress was again based more on pragmatic grounds than on political ideology. In Durgagram, two attached labourers admitted that they only voted for Congress because their bosses told them to; local CPM members said that such 'passive' voting had been more common in the past, and today it was loyalty to their bosses rather than physical force or other threats that influenced their decision. Within Kalipara, the Congress vote was largely a vote against the corruption of their previous member rather than anything more positive. More generally, many Congress voters mentioned that the party acted to promote 'calm' in village life in general, an association which won the party support from many agricultural labourers as well as the cultivating and landlord classes: 'This party helps all people—rich and poor—and doesn't cause chaos and agitation, so it is a good party' (labourer, Durgagram). There were also suggestions that the Congress was now more actively involved in supporting the labouring classes: 'Now both the Congress and the CPM are involving themselves with the needs of poor people, so there is not much difference between them. This is better for us' (labourer, Kalipara).

Turning to the CPM, the identification of the CPM as the 'party of the poor' was far from universal among the labouring classes. In Kalipara, where the last CPM member was accused of corruption, any such association had been replaced by a robust cynicism towards party politics in general: 'Everyone who gets elected is good for two or three years then they become corrupt. It doesn't matter which party they're from, they're all the same' (labourer, Kalipara). By contrast, the CPM was perceived as having a unique identity by richer families, but this was couched in negative terms. They claimed that the CPM had 'politicised' village life through stirring up trouble and the competition of local elections; a situation that had not existed when the Congress was in power at the state level. This issue of stability versus disruption is clearly bound up with people's material interests: the protection of the status quo is to the advantage of the currently privileged, as will be discussed further in the analysis of agrarian relations. However, the continued support for calls for 'calm' amongst some labouring families suggests that the conflictual 'class' politics called for in the CPM's literature does not enjoy universal legitimacy even amongst those who would stand to benefit most from it.[16]

The CPM's aim of 'raising class-consciousness' through the panchayats therefore appears to have been limited in narrowly party terms: few people had a detailed understanding of the different political parties' programmes, and even fewer voted on this basis. Even some core followers of panchayat members were unable to explain their political behaviour, as one labourer in Kalipara explained: 'I don't know what is good or bad about the party [the CPM], I just stamp on the symbol.' In the absence of a more formal or abstracted view of politics, most people's political support was based primarily on their perceptions of a given panchayat member, and the potential for achieving 'help' that this support may give, regardless of the party. As a result, the politics of panchayat elections remained a politics of dol, of building and maintaining a dominant faction. This process involves many elements of 'traditional' politics: kinship linkages were still important in all three villages and, to a limited extent, so were traditional patron–client relationships, as the voting of attached labourers in Durgagram showed. Panchayats and political parties were actively involved in this process, not least through supplying the resources for building support, but they participated largely within the framework of competing dol, rather than the party- or issue-based politics the CPM's leadership would have hoped for. However, it is important not to under-estimate the degree of change in social relations represented by the elections: although Davis' gramer kaj may still dominate panchayat elections, the fact that there was almost universal voting in the villages studied, and that the scheduled castes played an active role within this 'village politics' is a considerable achievement in itself. This is in sharp contrast to the period before 1978, when the participation of these households in village affairs was much more limited. These significant but restricted changes in the sphere of formal politics will now be compared with an aspect of informal politics on which the CPM claims to have had an even greater impact: agrarian relations.

MICRO-POLITICS AND AGRARIAN CONFLICT

Tenancy and landownership were both targeted by the CPM at the beginning of the Left Front regime as areas in need of direct action to protect the interests of the poorest. In spite of the academic debate surrounding Operation Barga and the government's land redistribution record, these programmes were of little current interest to the villagers in 1993. The impact of the land redistribution on the villages was limited due to the small size of landholdings in general (Tables 8.3 and 8.4). Panchayat

Table 8.3
Landownership Structure in the Survey Villages[1]

Village	Proportion of	Landless	Marginal (< 3 acres)	Poor (3–5 acres)	Middle (5–7.5 acres)	Rich (> 7.5 acres)
Durgagram	Households	63% (158)	25% (63)	7% (18)	3% (8)	1%(2)
	Land	0%	31%	33%	26%	11%
Kalipara	Households	59% (55)	34% (32)	4% (4)	2% (2)	0
	Land	0%	46%	31%	25%	0%
Ramnagar	Households	24% (31)	64% (84)	10%(13)	2% (2)	1% (1)
	Land	0%	53%	30%	8%	10%

[1] Figures in parentheses show the total number of households in each class. Figures for land show the proportion of all villager-owned land in each class.

Table 8.4
Impact of Land Redistribution on the Survey Villages

Village	Redistributed Land (Acres)		Number of Families		
	Total	Cultivable	Still Landless	Receiving Land	Farming that Land
Durgagram	11.6	8.4	136	26	18
Kalipara	3.3	1.2	48	16	7
Ramnagar	1.8	0.7	27	3	2

pradhans stated that there was little scope for future redistribution of land, a fact confirmed by my socio-economic survey of the villages which found no families owning more than the legal ceiling. Tenancy reform had brought some material benefits: around half of the tenants had a greater share of the crop than they did fifteen years ago, but only one in three tenants received the full government share of 75 per cent.[17] The most dramatic change in relationships over land had come in Kalipara: when the CPM gained power locally several *mahindars* ('attached' agricultural labourers)[18] used this to their own advantage, illegally registering themselves as sharecroppers during Operation Barga. By getting themselves registered with the support of the local CPM, the mahindars achieved a major political coup: they became tenants with a hereditary right to rent, enabling them to exercise a degree of control over the land previously unheard of. Unsurprisingly, this had caused a great degree of ill-feeling at the time, but ten years later most general caste farmers were largely resigned to the loss of control of land, and had sought out service employment instead. Even here, where landlords' power had been effectively challenged, only two families received the full government share.

By and large, these changes in relationships over land were no longer a major source of conflict in the villages at the time of my interviews. There had been some tension in the early 1980s when the mass registration programme was in operation, but now the legal procedures for registration were being carried out peacefully with the help of the Kisan Sabha. Sporadic disagreements between landlords and tenants occurred, but this did not result in mass action to redress grievances, which were usually felt and acted upon on an individual basis.

Despite the lack of dramatic changes in the land market, there did appear to be a general improvement in the material condition of the poor, closely bound up with changes in the labour market. Labouring families in all three villages stated that work availability had increased over the last fifteen years: in Kalipara and Ramnagar this was due largely to the spread of modern agricultural methods, in particular HYV rice (which is marginally more labour-intensive) and multiple cropping. An increase in off-farm employment in the area was also important, with many unskilled workers taking jobs in *gul* (coke) factories, oil mills or as unskilled construction workers. As a result of these changes, many mahindars in the villages had willingly transferred to casual labour. The wages that could be earned from casual agricultural and off-farm employment were greater than those of a mahindar, and most of those remaining as attached labourers were older men who were prepared to forego these benefits in return for the security of work.[19]

In all three villages, the conditions of work for casual labourers had also changed. In particular, there were now annual increases in the wage rate: in Ramnagar and Durgagram, labourers dated this back to a specific *andolan* (1990 in Durgagram, and 1984 in Ramnagar),[20] and although in both cases Kisan Sabha members had helped to organise the strikes, this was not described by labourers as being a 'political' matter. Employers had tried to resist these andolans, but subsequently negotiated annual wage increases without an all-out strike. These increases have been based on the local non-agricultural wage rate, rather than on the government minimum wage which the vast majority of labourers were unaware of.[21] There had been no overt party-political involvement in the annual wage increases since the initial action; significantly, the role of local Left Front members in disputes appeared to be one of negotiation rather than aggressive politicisation.

Assessing the magnitude of these economic changes is virtually impossible from the interviews, which were more concerned with qualitative changes in working conditions than their absolute value. Certainly, it was not the case that poverty amongst labouring households had ended, but there were various indications for relative improvements in labourers' positions. One of these was the change in their credit arrangements: earlier studies of West Bengal suggest that the usury of *mahajans* (professional moneylenders) had historically been an important means of capital accumulation in the countryside (for example, see Bose, 1993: 122–39). Today this no longer appears to be the case: the most common forms of credit mentioned in interviews were wage advances from regular employers or small loans from relatives, both of which were generally given interest-free. The financial status of labouring families appears to have improved over the period, such that today many households invest in small savings schemes.[22] Also, the fact that 'attached' forms of labour appear to be both less common and no longer linked to debt-bondage contrasts sharply with labourers' situation in the 1970s (Rudra and Bardhan, 1983). At the village scale, the CPM has no doubt encouraged this change in relations of production, as the party's facilitating role in wage negotiations showed, but it did not appear to be sufficiently active to be creating the change.

One micro-political effect of these economic changes has been to change the nature of dependency of labouring classes on richer villagers: today the control of the richer families is openly challenged through demands for better wages and shorter hours. This is particularly important as the division between agricultural labourers and the rest of society corresponds to the existing cultural division in the villages between the scheduled caste *chhotolok* (literally 'little people') and the rest of the villagers.[23] All villagers agreed that the scheduled castes were not subject to the same levels of social discrimination as they were previously, and this is reflected in their attitudes; as one labourer explained, 'before we were not able to speak our minds [to general caste families], but now we can do so without fear.' Nevertheless, the fact that this social divide exists and is reinforced through labourer/employer divisions is significant in the way in which changes in labour relations are interpreted politically by villagers.

In general, labourers and employers alike complained that the atmosphere (*abhaoa*) between the two was unpleasant. Labourers claimed that relationships between workers and employers had worsened, one labourer summarising the situation: 'Before there was friendship [between landowner and labourer]; now there is just work and pay.' As noted earlier this has led to higher overall wages for casual labourers, but also to the withdrawal of the financial security a 'good relationship' with an employer could provide, through help with medical costs or loans. This change reflects the experiences of struggles labourers have had in off-farm employment, where partial unionisation has led to some improvements in working conditions.[24] In general, labourers were at pains to stress the need to have wage increases to match price-rises in food and other commodities, and that their demands were limited to the 'fair' wage rate given for off-farm employment. Even so, the need for active, collective campaigning to achieve these increases was widely recognised: 'This year we haven't been given the full rate of Rs 22, instead we only got Rs 20. We need an andolan, but it is not possible to keep all the labourers together and the landlords are very strict about keeping wages down' (labourer, Durgagram). This campaigning for a 'fair' wage was, unsurprisingly, seen in a far different light by the upper and middle class households. For the poorer general caste families, many of whom would be marginal landowners according to the CPM's own classification, even modest increases in wages reduce their incomes significantly, and there was a degree of bad feeling as a result:

> Casual labour is much worse today: there are wage increases, but the work they do is less and they need constant supervision. This is partly a political matter, but the availability of non-agricultural work in [the neighbouring village] and Siuri has pushed up the rate (cultivator, Durgagram).

Perversely, it was the richer general caste households who complained loudest about what they saw as the decline of agrarian relations, and also stressed the importance of politics in this, as the following three families stated:

> Before, the relationship between an owner and his tenants and labourers was not a political relationship at all—they would be trusted to guard caste houses and other things. Now this has been spoiled (*nostho*). The growth

of all political parties and political idealism has been very rapid over the last fifteen years. The growth of chaos and quarrels has come from this politicisation (landlord, Durgagram).

I use one mahindar and he is a good worker—much better than the casual labourers. The relationship with the casual labourers is not good—they are always being cunning, and try to get away without working for their money. You need to supervise them all the time. Now I can't grow vegetable crops: if I did, women of the labouring families would steal them. If you complain the CPM will stick up for them: it's not fair (landlord, Ramnagar).

Things have changed: before people were very poor and would work for anyone. Today they are earning more and are politically motivated thanks to the party system. I think that this is not good—not everyone should be equal. There is no respect these days (cultivator, Ramnagar).

Such complaints form part of the 'war of words' that Scott suggests is a universal feature of micro-politics, and as such are important in terms of their intent as much as their objective validity. Despite the stress on the party-political element in labour conflicts there is little evidence of active Marxist attempts to disrupt landowners' interests. In Ramnagar, one landowner went as far as to suggest that if Left party activists stirred up trouble with labourers in the village he would hire in workers from Santhal Parganas, about fifteen miles away over the border with Bihar.[25] This threat has not been acted on, but it shows that the potential for a more aggressive village politics of labour than today's 'war of words' lies just below the surface. Significantly, in this war it is the richer families that have been most active, and in many of their complaints over labour relations the party-politicisation of village life, scheduled caste assertiveness and Left Front activism are recurring and linked themes. 'Leftists' are seen by many richer families as disrupting peaceful relationships between workers and owners, despite the limited evidence of such involvement from the villages studied. The content of these complaints and the often repeated desire to 'turn the clock back' to the 1970s, or in some cases even the British raj, suggest that fifteen years of CPM rule have hurt the material interests of these families somewhat.[26]

It is ironic that many villagers should associate disruption of labour relations with the CPM, when so little of the party's own literature has addressed this issue, pressing instead for 'all-peasant unity' amongst small and middle farmers, tenants and labourers (see Bhattacharyya, this volume). Indeed, the agrarian issue given most attention by the CPM over the last fifteen years, land reform, appears to be of much less political importance in the villages today. With over 30 per cent of the villagers employing casual agricultural labourers, the CPM's hesitancy in actively politicising this issue is perhaps understandable for reasons of its electoral survival. Whatever the political calculations involved, the CPM's developmental aim of 'changing the balance of class forces in the countryside' appears to be progressing in the villages, but not in the manner originally anticipated by the party's leadership. It is labour, rather than land, that has become the main political issue in the villages today, and here it is local, informal action rather than state legislation that has been significant in empowering the poor.

CONCLUSION

The preceding discussion has shown some of the important micro-political issues within three villages in contemporary West Bengal. Panchayati raj itself, as experienced by most villagers, is not a framework within which development strategies are planned and actively debated, but has rather become an important part of a more traditional village politics based on faction-building. Similarly, an investigation of political support in the villages suggests that, at the grassroots, the CPM has not been exclusively or universally identified by the poor as 'their' party, nor has it initiated a system of local government with a 'party democracy' in the western sense. As a result, political consciousness is largely confined to awareness of the various competing dol, rather than the raised class-consciousness the CPM claims is essential to furthering socialist development. However, this somewhat depressing picture was balanced by the real changes observed in 'class relations', especially those between workers and employers, that suggest that the former are gaining in confidence and independence. The micro-politics of village life therefore seems to present a paradox: the implementation of panchayati raj has not led to the replacement of Davis' traditional village politics with the form of modern class-based politics advocated by the CPM, and yet the statements of many richer villagers suggest that they fear the potential or actual loss of control over village life.

This paradox is resolved when we look at the changing nature of village micro-politics: despite failing to break the mould of Davis' 'gramer kaj', panchayati raj has undoubtedly broadened its scope. Davis observed factional disputes in the early 1970s that were dominated by 'big men'; these were conflicts between individuals at the top of caste and class hierarchies, in which the vast bulk of rural Bengal's population would have had little involvement. Panchayati raj has changed this dramatically: even if politics has little meaning beyond patronage networks and building factions, elections and universal suffrage mean that the 'big men' must actively court the votes of those previously marginal to the political process. In addition, the resources routed through panchayats and the involvement of political parties have enabled a far wider number of villagers to become active in village politics. The control of panchayat resources makes any elected member a powerful figure regardless of his or her personal wealth, and political parties, even if their programmes are poorly understood by the majority, have further widened the avenues for villagers to become 'big men' through access to powerful outsiders. In contrast to the gramer kaj of the past, it is therefore possible for economically powerful households with high ritual status to find themselves marginalised in micro-political contests. This was most clearly seen in Kalipara where general caste households had been powerless to stop the illegal registrations of sharecroppers, and had subsequently largely withdrawn from the arena of village politics.

This broadened village politics has undoubtedly been supported by economic change, especially in the labour market. Although labourers obviously remain dependent on employers for their well-being, and are usually placed in an inferior position to them both economically and socially, the nature of this dependency appears to be changing. The increased availability of work through the spread of 'green revolution' agriculture and off-farm employment, coupled with the decline of 'attached' forms of labour, has strengthened the bargaining position of labourers. These changes may be reversed in future if there is widespread mechanisation of agriculture, but for the time being labourers have some degree of influence over their choice of employer and conditions of employment. As a result, the ties of loyalty and dependency (often reinforced through debt) which may have bound labourers to a single employer in the past have been loosened. The political implication of this is that labourers are more able to change their support between party camps in order to corner some of the benefits of panchayat development programmes. It is thus a mutual interaction of economic change and political institutions that has been important in changing the micro-politics of the villages studied: without economic change the freedom of labourers to switch support groups would have been constrained, and the power of their vote reduced accordingly. Without panchayati raj, it is unlikely that changes in the rural economy would have been experienced in as positive a fashion by labourers: it is through the power of their votes that their increased influence on village affairs has been maintained.

This examination of the micro-politics of the three villages has thus identified the panchayats as playing an important role in the empowerment of the rural poor, but not necessarily in the way in which the CPM would have intended. With this being so, some important question-marks remain as to the permanence of this improvement. The observed changes in the labour market that have supported labourers' interests may be both localised and ephemeral. Also, as there is limited evidence of support for parties on the basis of programmes and policies, a well-organised political opposition could potentially erode much of the CPM's rural support base, especially if it were to mobilise support on one side of the overlapping divisions of caste identity and labour relations. Such difficulties will certainly remain important over the next decade of panchayati raj, and it will need an active and sensitive government to ensure that the challenges they raise can be met. Nevertheless, there can be no doubt that panchayati raj has helped to produce positive changes in the livelihoods of labouring families in the villages studied. It is these changes that the CPM needs to build upon over the coming years if its rhetorical claims of empowering the rural poor are to be realised in practice.

NOTES

1. Within this broad consensus, there are widely divergent views of how empowerment can best be achieved, and which aspects of empowerment should be emphasised. On the political right there is more emphasis on 'negative freedom' (especially individuals' freedom from the state), whereas on the left 'positive freedoms' (such as access to resources) are stressed (Dasgupta, 1993).
2. Within Friedmann's work, 'inclusive democracy' and 'appropriate economic growth' form part of a mutually supportive set of normative (re)orientations for alternative development. Political change plays a central role in his view of empowerment, but is seen as contributing both directly and indirectly to economic growth once poverty and the economy have been reconceptualised in terms broader than those used by neoclassical economists.
3. Panchayats (local councils) have a long history in the Indian countryside. The Left Front government enacted a series of legal changes to form a three-tier (district, block, and village-cluster) structure of councils whose members are elected on party tickets every five years. The first panchayat elections were held in 1978, and from 1985 the councils have been given a major role to play through the decentralisation of West Bengal's development planning (see Webster, 1992: 33–5, Appendix 1).

4. See Colburn (1989); Scott (1990). Scott's work has subsequently been criticised on a number of grounds (for example, see Brass, 1991 and Hart, 1991); there is not space here for a detailed evaluation of his work, but the interest in informal political activity it has encouraged is in itself a useful contribution to the study of political change.
5. The enormity of this task should not be underestimated. Kaviraj (1991) shows how at the all-India level the Congress failed to achieve such restructuring at the time of independence, and how the resulting exclusion of the bulk of the population from the (western) discourse of the governing elite has opened the political space for a challenge to the principles of a secular, democratic state.
6. This formed part of my doctoral research on the political and economic impacts of the Left Front government on the rural poor in West Bengal. Intensive fieldwork was conducted in two main stages, the first being a detailed socio-economic survey of all 473 households in the villages. This was followed by a series of semi-structured interviews, where a number of issues, including political and panchayat matters, were discussed in greater depth with a 25 per cent random sample of households from each village. It is primarily these interview materials, supplemented by discussions with gram panchayat members and local political representatives, that I will draw on in this paper. All of the interviews were conducted in Bengali with the help of my long-suffering field assistants, Tapan Bhattacharya and Surajit Adhikary. Pseudonyms have been used for various individuals and the three survey villages ('Durgagram', 'Ramnagar' and 'Kalipara') but all other names are unchanged.
7. Since 1989, JRY has been India's central programme for the provision of slack-season employment in rural areas. The IRDP aims to enhance the earnings of the poorest households through providing credit for the purchase of income-earning assets.
8. Elsewhere in my research I have attempted to evaluate the impact on households' 'entitlement sets' (Sen, 1981) made by the IRDP and JRY. IRDP projects reached less than 15 per cent of households, and at their most successful could provide incomes reaching Rs 8,000 per year, not counting loan repayments and other income-earning opportunities foregone. JRY support by comparison was wider-spread, reaching 45 per cent of scheduled caste labouring households, but contributed under Rs 150 per year to most of these.
9. In Davis' study, only men are mentioned as participating in these contests. In all three villages studied here, party and panchayat politics remained an all-male affair: there was one female panchayat samiti candidate, but she took no part in her own election campaign, which was conducted entirely by her father (a CPM member). When the issue of women's role in politics was raised, the majority of men argued that the work involved in being a panchayat member (such as going from house to house unaccompanied) was incompatible with 'decent' female behaviour, especially for a general caste or Muslim woman. Ironically, there were many amongst these men who hung portraits of Indira Gandhi on their walls, and claimed that Mamata Banerjee was West Bengal's best political leader. Since 1993, one-third of panchayat seats in all three tiers have been reserved for women.
10. I have used a rough three-fold division of society here: an upper class of 'landlords' (landowners, often only operating small areas of land, who do not directly participate in cultivation), a middle class of 'cultivators' (those directly cultivating owned or rented land, but not working for others) and a lower class of 'labourers' (those with or without access to land who work as paid labourers for others). This system has useful similarities with local interpretations of class (households describing themselves as *mallick, kisan* and *munish* would correspond closely to my upper, middle and lower

classes), but the range of agricultural and non-agricultural incomes within each 'class' means that a household's economic condition is not simply a result of its position in land and labour markets. For a fuller discussion of class and wealth in the villages, see Williams (1996: Chapter 4).

11. JRY drainage and road repairing work is a prime example: the real amount of work done is often a fraction of that recorded in the panchayat records, and the difference in wages can be pocketed by the gram panchayat member, who stands little chance of detection (interview with panchayat samiti member, Kalipara). As an example of the scale of corruption, zilla parishad (district council) officials alleged that in the north of the district where corruption was particularly acute, panchayat members paid between 50,000 and 100,000 rupees to secure the position of pradhan—on the assumption that they would make far more than this in kickbacks over five years in office!

12. Rallies were used, as were *bandhs* (strikes), as shows of strength by the parties; although these were ostensibly peaceful protests, the threat of violent disorder was always present.

13. This was certainly the case in the villages themselves, but elsewhere there were sporadic incidents of violence. In Ramnagar, it was alleged that a BJP candidate from an outlying village had been kidnapped and wounded by CPM-supported thugs, and there were newspaper reports of a number of campaign- and polling-related murders in the state as a whole.

14. When voting for members of the block and district councils, the vast majority of villagers would not know their candidates. Here, voting on a party basis was more common, but stories about the reputation of candidates were also important in influencing decisions.

15. Ironically, the CPM appears to have no qualms about supporting medium or even rich peasants in its development policy, one notable example being the party's failure to levy agricultural income tax (Mallick, 1990; Basu, 1992: 40–1). Such areas of CPM 'non-performance' (Basu, 1992: 50) and their implicit regressive transfers of wealth do much to discredit the party's claims to be doing as much as possible for the poor within the constraints of the parliamentary system. My argument here is not that labouring households were the sole beneficiaries of CPM rule, but rather that the refusal of local CPM (and Congress) activists to take an aggressive *and explicit* anti-labourer stance in their political campaigning had allowed the BJP to win some support amongst famers.

16. The particular form of rural 'class conflict' which the CPM advocates in contemporary West Bengal is that between the 'feudal elements' of rural society and peasants/workers (see the Programme of the CPM, Paragraph 110). Both theoretical and practical problems arise from their claims to support workers' interests whilst simultaneously calling for 'all-peasant unity' (see Bhattacharyya, this volume, and Williams, 1996, for a fuller discussion).

17. This was often through mutual agreement. The government share is based on the tenant providing all inputs, but tenants were often willing to take less than 75 per cent of the crop in return for a more even division of inputs. If these 'goodwill' arrangements broke down, registered tenants would invoke the government terms to protect their interests in the last resort.

18. Mahindars were hired on an annual basis to do agricultural and other work at the beck and call of their employer; frequently, the same worker would be employed for many years, but no mahindars mentioned that they were forced into such arrangements through debt-bondage. See Rogaly (1996) for a more detailed discussion of labour arrangements.

19. Casual labour rates varied from Rs 20 to Rs 22 for men and from Rs 15 to Rs 20 for

women. Total annual income for a male casual labourer (assuming six months of agricultural work and four months' off-farm employment) was around 10 per cent above that of a mahindar.
20. Andolan covers a range of English meanings, including 'agitation', 'strike' and 'movement'. Here, the protests had been non-violent strikes by all labourers and domestic servants: they had lasted for a few days, after which the landowners had agreed to wage increases.
21. Usually, the wage rate for unskilled building labourers (work which many labourers in all three villages undertook) was seen as 'fair', and was equivalent to Rs 22 per day. This rate was in turn set as a fraction of the skilled builders' wage, which was negotiated through union action. The government minimum wage of Rs 27 per day was much higher than this 'fair' rate, but was only mentioned by one interviewee (a CPM supporter) who thought it was too high to be the basis of realistic agricultural wage negotiations.
22. Details of these schemes—run by private companies—were not investigated, but in interviews a number of scheduled caste labourers mentioned that they had been saving over the last ten years.
23. The terms 'scheduled castes', 'chhotolok' and 'labourers' were not perfectly synonymous in the villages, although all three had associations of low status. One group of scheduled caste families in Ramnagar was *not* classed as chhotolok by villagers, who said that their inclusion amongst the scheduled castes was a mistake by government officials. If this group is disregarded, it is significant that only scheduled caste households (and the lowest status Muslims) would work as agricultural labourers, and employment of agricultural labourers *by* scheduled caste families was virtually nil. See also Ruud (this volume), and Rogaly (this volume) on the variations in labour hiring practices between different scheduled caste groups in Bardhaman and Puruliya districts.
24. Some gul factories and oil mills had unions, mainly affiliated to either the CPM or Congress. Here reductions in the working day from 12 to 8 hours had been achieved, often at the cost of employers withdrawing their 'goodwill' and its associated informal benefits, such as the advance payment of wages.
25. See Rogaly (this volume) on the hiring of seasonal migrant wage workers in rural West Bengal.
26. The 'class' composition of these richer households was varied: many were petty landlords, some were richer cultivators of their own land, but the vast majority were not the rural 'gentry' of ex-zamindars/*jotedars* for whom the CPM reserves the fiercest of its rhetorical attacks. Any 'hurt' which these families had suffered was no doubt relative rather than absolute—by their own admission they had not seen their profits from agriculture decline under the Left Front—but the loosening of their control over labourers was seen by them as significant in itself.

REFERENCES

Basu, Amrita. 1992. *Two Faces of Protest: Contrasting Modes of Women's Activism in India*. Berkeley: University of California Press.

Bose, Sugata. 1993. *Peasant Labour and Colonial Capital: Rural Bengal since 1770*. The New Cambridge History of India, Volume 3(2), Cambridge: Cambridge University Press.

Brass, Tom. 1991. 'Moral Economists, Subalterns, New Social Movements and the

(Re)Emergence of a (Post-)Modernised (Middle-)Peasantry,' *Journal of Peasant Studies*, 18(2): 173–205.

Chambers, Robert. 1993. *Rural Development: Putting the Last First*. London: Intermediate Technology Publications.

Colburn, Forrest D. (ed.). 1989. *Everyday Forms of Peasant Resistance*. Armonk NY: M.E. Sharpe.

Dasgupta, Partha. 1993. *An Inquiry into Well-Being and Destitution*. Oxford: Clarendon.

Davis, Marvin. 1983. *Rank and Rivalry: The Politics of Inequality in Rural West Bengal*. Cambridge: Cambridge University Press.

Edwards, Michael. 1993. 'How Relevant is Development Research?' Frans J. Schuurman (ed.) *Beyond the Impasse: New Directions in Development Theory*. London: Zed Books.

Friedmann John. 1992. *Empowerment: The Politics of Alternative Development*. Oxford: Blackwell.

Hart, Gillian. 1991. 'Engendering Everyday Resistance: Gender, Patronage and Production Politics in Rural Malaysia,' *Journal of Peasant Studies*, 19(1): 93–121.

Kaviraj, Sudipta. 1991. 'On State, Society and Discourse in India,' James Manor (ed.) *Rethinking Third World Politics*. Harlow: Longman.

Mallick, Ross. 1990. 'Limits to Radical Intervention: Agricultural Taxation in West Bengal,' *Development and Change*, 21(1): 147–64.

Mishra, Surjya Kantra. 1991. *An Alternative Approach to Development: Land Reforms and Panchayats*. Calcutta: Government of West Bengal (Information and Cultural Affairs Department).

Rogaly, Ben. 1996. 'Agricultural Growth and the Structure of "Casual" Labour Hiring in Rural West Bengal,' *Journal of Peasant Studies*, 23(4): 141–65.

Rudra, Ashok, and **Pranab Bardhan.** 1983. *Agrarian Relations in West Bengal: Results of Two Surveys*. Bombay: Somaiya Publications.

Ruud, Arild Engelsen. 1994. 'Land and Power: The Marxist Conquest of Rural Bengal,' *Modern Asian Studies* 28(2): 357–80.

Scott, James C. 1985. *Weapons of the Weak: Everyday Forms of Peasant Resistance*. New Haven: Yale University Press.

———. 1990. *Domination and the Arts of Resistance: Hidden Transcripts*. New Haven and London: Yale University Press.

Sen, Amartya. 1981. *Poverty and Famines: An Essay on Entitlements and Deprivation*. Oxford: Clarendon Press.

Webster, Neil. 1992. *Panchayati Raj and the Decentralisation of Development Planning in West Bengal*. Calcutta: K.P. Bagchi.

Williams, Glyn. 1996. 'Socialist Development? Economic and Political Change in Rural West Bengal under the Left Front Government.' Unpublished Ph.D. thesis, University of Cambridge.

9

From Untouchable to Communist: Wealth, Power and Status among Supporters of the Communist Party (Marxist) in Rural West Bengal

ARILD ENGELSEN RUUD

INTRODUCTION

Social aspirations cannot be ignored as a constitutive element in rural and village politics, not even in the entrenched position enjoyed by the Communist Party of India (Marxist) (CPM) in rural West Bengal. This essay argues that the CPM represents a path for upward social mobility for groups of poor people, and also that it is very sensitive to local status considerations in its on-the-ground establishment of long-term relationships with crucial support groups. Other factors have also contributed towards

---

The primary material for this paper was collected in West Bengal while researching for my Ph.D. dissertation, 'Socio-Cultural Changes in Rural West Bengal.' This research has been funded by the Norwegian Research Council. I am grateful to John Harriss and Biplab Dasgupta and in particular Ben Rogaly for valuable comments given to earlier drafts of this paper.

the party's position, such as economic reform and land redistribution, political mobilisation and organisation. These are important factors—not least symbolically—and have in general benefited poor rural people. However, actual redistribution has fallen far short of being radical and has often enough been criticised for being rather mild-mannered. Moreover, an interesting feature of the redistribution has been that not all poor people have benefited to the same extent from these policies. The divisions that distinguish pro-CPM groups from other poor people follow approximately *jati* divisions, a point particularly evident in political representation.[1] The favoured groups of poor are more active in the CPM than other groups, and are more closely identified with the party.[2] Their support is reciprocated by the local party leadership which considers these groups' interests—both their economic interests and their social or cultural status.

The literature on West Bengal's now two-decade-old CPM regime has largely ignored the issue of how the party also represents a cultural programme, that is, a certain set of moral standards constructed within a socially hierarchical society. A partial but very important exception is Amrita Basu's *Two Faces of Protest* (1992). Her study is partly devoted to modes of women's activism in three Medinipur villages, but also more generally to relationships between various socio-economic groups in the villages and the local party leadership. According to Basu, major factors in the varied response of villagers to the CPM's calls for mobilisation are their social status, their identity vis-à-vis and closeness to the dominant sections of society, and the socio-cultural identity of the local party leadership. To a considerable extent this echoes the findings of the present study, although Basu's also includes many other factors that cannot be coped with here due to limited space. However, my contribution will be to emphasise again the importance of cultural and social factors in both the mobilisation, and not least in the maintenance, of durable support for the CPM. Particular attention will be paid to the mobilisation and maintenance of support from the broad social groups formerly known as 'untouchables' and now as Scheduled Castes (or *dalits* in Basu's terminology)—groups that constitute almost one quarter of West Bengal's population and that are mostly solid CPM supporters—and to a lesser extent to the much smaller social groups formerly known as 'tribals' and now more commonly as Scheduled Tribes (or *adivasis* in Basu's study), whom she correctly identifies as being much more ambivalent towards the party.

Social and cultural factors become crucial when we consider how the party has drawn its local leaders and its local representation—that is, most of its local ethos—from among lesser landlord and middle-to-rich peasant households, i.e., from a certain socio-cultural stratum of society. The culture of this stratum was and is very different from that of the low-caste landless the CPM set out to mobilise. True, there are also other sources of influence on the party's ideology in its local manifestations. By historical association and ideological conviction the party has come to be associated with causes such as anti-casteism, social equality for the poor, teetotalism, and literacy—objectives that do not derive entirely from within a class society. The party has also sought to implement these objectives, successfully to a significant degree. But the context of these objectives should be noted carefully. Even the party did not adopt its objectives out of thin air. These are objectives which were and are associated with certain conceptions of good living, moral standards and respectability developed within class society. These conceptions gave rise to desires for cultural and social reform, desires that the party and its local leadership inherited and that were shared by their prospective supporters. Because of the limitations in the party's economic reform programme and the limited control groups of poor have on the party, it is here, in the social implications of the party's policy, that we need to look for the reason why so many poor people and certain jatis in particular identify themselves with the party.

It is important to remember that although Bengal may once have been 'golden', it was so only for some in the population (at least during most of the 20th century). Sections such as the tribals, the 'untouchables' and other groups placed far down on the ritual and social ladder were excluded from the riches of the soil and the socially very significant status of landownership. The present study will show how formerly poor and ritually low groups in two villages in Bardhaman have gained land and political positions and, more importantly, why they are rapidly losing their stigmatising identity as a result. The reign of the party does represent a significant period of agrarian change: the practical abolition of untouchability, enhanced social, economic and political status for the former 'untouchables', improvements in the position of women, and increased literacy. Though far from all positive changes are attributable to the party, these have accelerated under the mature CPM raj, and to many the attainment of higher social status is associated with the reign of the CPM (facilitated as the process has no doubt been by the party's policy of economic redistribution and its pro-poor orientation). In short, the net effect has been that many, but not all, low-caste land-poor outsiders have turned into standard if poor Hindu cultivators with a communist leaning.

From a study of the history of two villages, Udaynala and Gopinathpur (fictitious names), in the Raina *thana* of Bardhaman district I will outline the social, political and economic gains made by the poor over a thirty-year period, and how these show marked differences between specific groups of poor people. I will focus primarily on the Bagdi and Muchi jatis, both of which are classified as Scheduled Castes (SC), but between which the difference in attention paid by local politicians, in particular CPM politicians, comes out quite starkly. But first we turn to the historical construction of the clean-caste cultivator and his other, the low-caste labourer.

HISTORICAL CONTEXT

From a study of economic relations in western Bengal at the turn of the century, Sugata Bose found a tripartite system which he terms 'the peasant smallholding–demesne labour complex' (Bose, 1986: 18ff.). This complex consisted of a few major landowners, fairly broad sections of owner-cultivators, and large numbers (but not a majority) of landless labourers or sharecroppers. These divisions corresponded roughly to the social divisions. The landlords were often of high or at least clean caste, the owner-cultivators were mainly '. . . chashis [cultivators] of the agricultural castes, such as the Mahishyas, Sadgops and Aguris . . .' (ibid.: 19), while the landless were mostly low castes such as 'the Bagdis, Bauris, and tribal people' (there were of course overlaps) (ibid.: 29). The salient point for what follows is the social difference among that part of the population ranked below the landlord classes.[3]

The importance of the distinction between the *chashis* (cultivators, peasants, the ritually clean) and the lower castes and tribals (whom I will term *majurs*, labourers) becomes particularly clear in Rajat Ray's readings of Tarashankar Banerjee's novels, set in Birbhum in the early decades of the 20th century (Ray, 1987; 1992: 275–314). Ray too found three social groups, the *bhadralok* (gentlefolk of high caste), the chashis and the lower castes. One of the most striking aspects of Birbhum's social world, Ray points out, is that these groups were so different as to have distinct emotional patterns. The romanticism of the bhadralok contrasted with the frugal restraint of the chashis, which again contrasted with the abandon of the majurs.

It is the chashi–majur contrast which interests us here. In general the lower castes did not follow the rules and norms of the clean-caste owner-cultivator society to which they were subjected. At one point Ray states that 'Param and his wife Kaloshashi are a fairly typical couple—Param is a dacoit, Kaloshashi a prostitute' (Ray, 1987: 714). The Kahar Bauris of Banerjee's novels had few eating inhibitions, practised extensive commensality, drank heavily, and had a defiling caste occupation. Economically they were poor and dependent and entangled in endless webs of debt and patronage. They were illiterate and their religious beliefs were of a world of spirits of the dead and of animals, of ghosts, and of the erratic ways of wrathful and vengeful gods.

It is in sexual mores that we find the ultimate contrast to chashi society, a contrast which expressed social position. The sexual norms of the Kahars were lax compared to those of the chashis. They largely accepted pre- and extramarital sex, divorce and widow remarriage (see particularly Ray, 1987: 738 ff.) At the same time, the selling of Kahar women's sexual favours to clean-caste men or their exploitation by the latter formed an integral aspect of the Kahars' relationship with the local higher castes. Sexual favours could not be sold freely to anyone with money, only to the *babus* and *mandals* of the locality. 'Sexual servility and sexual freedom sustain[-ed] each other,' writes Ray (ibid.: 739). Moreover, 'the sexual factor in social formation has not been adequately comprehended' (ibid.: 745), and Ray argues correctly that sexual mores—especially those concerning women—are central to perceptions of caste ranking and to the construction of subjectiveness within the larger construct of a ritual hierarchy. Sexual mores were both instruments of social formation and tangible expressions of status.

Liberal sexual mores were paralleled by low ritual status, landlessness and dependency, and vice versa. All these elements marked the difference between the majurs and the chashis who were known for their sexual restraint (on the part of women) and preoccupation with ritual cleanliness, their economic independence and frugal morality, dominance of village affairs, and their brahminical rituals and Puranic gods. The lower castes were not only lower, they were different, following mores that marked their origins as the tribal other and also implied their lowliness and their dependency.

CULTURAL REFORM PROCESSES AND BENGALI MODELS

Srinivas' (1965) concept of sanskritisation denotes efforts made by jatis to enhance their social and ritual status. It is this type of effort or ambition which most low-caste Bengalis have harboured throughout the 20th century.[4] However, the term sanskritisation, as Chris Fuller (1992: 24–8) points out, confuses rhetoric with actual practice as there is no agreed Sanskritic Hinduism. Moreover, ritual-religious reform constitutes but one such process; others would include 'modernisation' processes.[5] It is also generally agreed that the models aspired to are not universal, or all-Indian. The specific content is locally formed, a replication of the lifestyle of a dominant caste or some other locally well-known group. These limitations notwithstanding, the concept has gained wide currency in the literature and refers in general to efforts towards social status enhancement engaged in by whole groups, more often than not jatis. We should perhaps term these cultural reform processes exactly that, 'cultural reform processes', with the proviso that they are locally and historically formed. These processes aim at locally well-known models (that may change over time), at the lifestyle and/or religious practices represented by those models.

Of the three distinct socio-cultural groups noted above, bhadralok and chashi functioned as the models for cultural reform processes. The chashi was stereotypically a sturdy cultivator of his own lands, frugal, morally upright and conscientious in his religious observances. The bhadralok, on the other hand, enjoyed a service-position or enough land to keep him comfortable in non-manual engagement. He was not only literate but had a lively interest in literary pursuits, poetry, drama and song.

A result of the vast increase in village primary schools after independence[6] was that many elements of the bhadralok model, including literary pursuits and political leftism,[7] blended with a basically chashi lifestyle. (It should be noted that the archetypical landlord bhadralok with his leisured life and urban orientation has vanished from the rural scene.) Literacy, in the extended sense of knowledge of poetry, drama (*jatra*) and Tagore songs, has now become a fundamental ingredient in an ideal chashi lifestyle. Shame on the village that does not stage at least one jatra performance during its main festival, that does not have a function on Rabindranath's anniversary, that is still without a library.[8] Both Gopinathpur and Udaynala have a number of trained singers, many would-be and some published poets, and a few dramatists, some of whom have had their plays staged. Most of these belong to cultivator castes, not the archetypical bhadralok castes. Few males of the peasant castes have not at one point acted in a jatra performance.

Such cultured activities were at one time not available to all castes. Some discrimination did take place, though more as a result of difficulties in bridging a long-standing social gap than from the conscious exclusion of certain castes. This caste barrier has gradually been removed, but illiteracy naturally constituted an enduring obstacle. Disability on this front not only precluded participation in activities such as jatra or poetry recitals but hampered the inclusion in the status group that was associated with such activities. In other words, aspirations towards higher status became increasingly difficult because for majurs it meant chashi status—chashis were becoming increasingly literate and 'literature-conscious'. Literacy became a sine qua non of an even marginally improved social status, the hallmark of the upwardly mobile.

BAGDI AND MUCHI SOCIAL CHANGES: 1960-90

Efforts to change their cultural mores were evident among the lower castes in Udaynala and Gopinathpur some thirty years back. In Gopinathpur two Bagdi families were 'sanskritising', in Udaynala, one. But these families were the exceptions. Most Bagdis and all Muchis were very poor, politically deprived and socially stigmatised as untouchable. Few owned land, and even fewer owned enough to be self-reliant. Their poverty was reflected in their appearance, housing and lifestyle. Hardly any were literate and most drank heavily during festivals and on other occasions. All low-caste labourers were served on banana leaves, sitting in the middle of the employer's courtyard; payment was dropped into their hands to avoid touch, and they were not allowed near the temple for fear of polluting the deity.

While all low castes fitted the stereotype, the Bagdis had a particularly bad reputation. They were known for their lax sexual norms, fierceness, heavy drinking, and thieving.[9] Because of this their festivals were shunned by other villagers, and landowning employers complained about insubordinate behaviour. When theft occurred one immediately suspected the Bagdis. They were included in the guarding of ripe paddy fields only in the 1980s. Bagdis also constituted an important element in dacoity, a widespread activity in the area that vanished only in the late 1970s.

In local political history, the Bagdis had a role as *lathiyals* (fighters) for village leaders who depended on this support in struggles over leadership. In particular, their militancy (or threat thereof) was crucial for intimidating opposition, rivals and insubordinate subjects. For this willingness to fight, Bagdis received particular patronage: employment (in spite of insubordinate manners), and loans and festival contributions (when village leadership was contested).

In spite of their poverty, the Bagdis of these two villages do not have a history of continuous relations with the CPM. In Gopinathpur they were generally aloof from the entire history of mobilisation in the late 1960s, and no land occupation took place there. Rather, CPM agitators were resisted at the behest of Bagdi-supported Aguri village leaders. In Udaynala on the other hand, the Bagdis were very active, and constituted perhaps the most important jati in the unrest (under the leadership of middle-class Muslims). However, by the early 1970s, the Udaynala Bagdis were active in the repression of communists and they disassociated themselves from local politics in the mid-1970s. Only by 1977 were the Bagdis of these two villages coaxed into active support of the party.

The present-day CPM leadership depends on them for much the same reasons as the village leaders of old: for their numbers, but particularly for their usefulness as activists. Bagdis were especially active in demonstrations, strikes and instances of intimidation of recalcitrant opponents and landowners during the early years of CPM rule.[10] Their fighting capacities and willingness to participate actively when called upon have earned them particular attention from the local party leaders, just as they had from village leaders in the past.

As mentioned earlier, there were among the Bagdis of these two villages three 'sanskritising' families. Since the coming of the CPM raj, all Bagdis of these two villages (and adjacent villages) have become standard Hindus. Their promiscuous sexual norms have been curbed, widows are no longer permitted to remarry, dowry is paid instead of the symbolic brideprice of thirty years ago, and girls marry young, rather than in their late teens and early twenties as their mothers did. Bagdi festivals are no longer occasions for excessive drinking; rather, their festivals are enjoyed nowadays by other villagers, even women and children, particularly the expensive jatras and the frequent video shows.

The Muchi jati constitutes a different case from the Bagdis and have a different approach to village leaders. They were probably more stigmatised than the Bagdis because of their low ritual occupations (carrying away carrion, making shoes from hides, and drumming at Hindu rituals); but at the same time Muchis were also integral to the ritual well-being of Hindu society. The performance of these polluting tasks made it possible for the rest of Hindu society to stay ritually clean. In short, theirs was a ritually ambiguous position (Fuller, 1992: 138). While Bagdis have come to be regarded as standard Hindus, Muchis have not. They attempted cultural reform fifteen or twenty years ago (they ceased eating carrion, curbed their drinking and took new names), but were met with scorn and ridicule and ultimately gave up. Their improved position has come from outside. Any Hindu is now allowed close to the temple, any labourer irrespective of his jati is seated on the porch and served on plates. But the position of Muchis is still ambiguous; they are still explicitly polluting and this is reflected in everyday behaviour. As Muchis informed me, while Bagdi labourers nowadays leave their plates unwashed, Muchi labourers wash their plates themselves before leaving these for the women of the employer's household to pick up. Bagdis socialise with anyone anywhere, restrained only by an inferior economic status, while Muchis are restrained by both an inferior economic status and an implicit, never-expressed ritual distance. The touch of a low-caste no longer causes high-caste Hindus to rush off for a ritual bath, but Muchis will themselves take care to avoid touch, thus sparing high-castes an embarrassing situation. The same reluctance is evident in approaching temples. Muchis approach one by one, give their offerings and return to a distance while other Hindus, including Bagdis, linger around in front of it.

'Bagdis are not low caste or poor any more,' as one village leader put it, whereas Muchis are still stigmatised as 'untouchables'. The Bagdi reform process' aim of a higher social status has been recognised, while the Muchi's aspirations are ridiculed and scorned. There are several reasons for this. One is that the Muchis had and still have a ritually polluting caste occupation. Another is that they have received relatively less of the attention of local party activists that had helped the Bagdis reduce their stigmatisation.

REDISTRIBUTION OF WEALTH: WAGES, LAND AND IRDP LOANS

The general pattern of land redistribution under the Left Front government (LFG) since 1977 has consisted of distributing land in minuscule plots ranging from 8–10 *katha* to a *bigha*-and-a-half.[11] Since the mid-1980s, land redistribution has been replaced by subsidised loans through the Integrated Rural Development Programme (IRDP) as the main tool of redistribution of wealth. The Udaynala figures for the (re-)distribution of land and IRDP loans are given in Tables 9.1 and 9.2 (unfortunately I do not possess comparably reliable figures for Gopinathpur).[12]

Table 9.1
Recipients of Redistributed Khas by Jati Group, Udaynala, 1993

Jati Group	Population N	Khas Recipients		
		N	% of All Recipients	% of Jati Population
Clean caste[1]	94	0	0.0	0.0
Sekh Muslim[2]	935	40	41.2	4.3
Namasudra	299	12	12.4	4.0
Mallik Muslim[2]	190	11	11.3	5.8
Bagdi	304	23	23.7	7.6
Muchi	91	5	5.2	5.5
Santal	93	5	5.2	5.4
Total	2006	96	99	4.8

[1]'Clean caste' comprises Bamun, Kayastha, Bene and Kalu.
[2]Sekh and Mallik are separate jatis of Muslims.

Table 9.2
Recipients of IRDP Loans by Jati Group, Udaynala, 1993

Jati Group	Population N	IRDP Recipients		
		N	% of All Recipients	% of Jati Population
Clean caste	94	0	0.0	0.0
Sekh Muslim	935	51	40.2	5.5
Namasudra	299	22	17.3	7.4
Mallik Muslim	190	12	9.4	6.3
Bagdi	304	27	21.3	8.9
Muchi	91	6	4.7	6.6
Santal	93	9	7.1	9.7
Total	2006	127	100.0	6.3

The figures in Tables 9.1 and 9.2 show that in these villages the distribution of land and IRDP loans has overwhelmingly favoured jatis that comprise poor households.[13] Clean-caste households have received neither land nor IRDP benefits,[14] and most low-caste recipients were poor. There are, however, interesting variations. Bagdis stand out in the tables as having received more land grants per population than others, and more IRDP loans (with the exception of the Santals who, classified as Scheduled Tribe [ST], come under special provisions).

Local and district party activists and leaders acknowledge that favouring party affiliates in the distribution of land and IRDP loans amounts almost to an unofficial party line.[15] This should not be exaggerated; the tendency is rather to disfavour anyone affiliated to the opposition. And anyway, contrary to the attention the issue has received, it does not appear from these villages that land redistribution by itself accounted for much. In Gopinathpur only 17 bighas (or 1.3 per cent of total village lands) and in Udaynala 53 bighas (2.8 per cent) were confiscated and redistributed. In this respect, the distribution of IRDP loans seems more important. However, there are qualifications. An average IRDP loan stood at about Rs 9,000 in 1993. This sum was not insignificant even considering the deductions (fees, 'commission' to some local politicians), but still amounted to only a portion of the total costs of the non-productive purposes for which it was often used (dowry, house-building).

Nonetheless, there has been a significant levelling of economic inequalities over the last thirty-odd years in these two villages. In Tables 9.3 and

Table 9.3
Landownership by Person and Jati, Gopinathpur, 1960 and 1993

	High Caste[1]	Aguri	Napit	Bagdi	Dule	Muchi	
Population in numbers							Total
1960	71	137	79	203	47	69	606
1993	111	214	165	371	85	165	1111
Land per person[2] (in bigha)							Average
1960	2.53	3.18	1.66	0.71	0.38	0.30	1.53
1993	1.30	1.83	1.53	0.85	0.36	0.19	0.99
Percentage of village land owned							Total
1960	19.3	46.8	14.1	15.7	2.0	2.1	100.0
1993	13.2	29.1	23.1	28.9	2.8	2.9	100.0

[1]'High caste' includes Bamun and Kayastha.
[2]Refers to total population of individual jatis.

9.4, the jatis are given in order of ritual purity (approximate in the case of Muslims). The 1960 figures show marked economic differences corresponding to this hierarchy while by 1993 the differences are much less marked, and certain low-caste jatis have fared quite well.

However, there are interesting variations among the lower castes. In Gopinathpur the Bagdis almost doubled their ownership of village land over the thirty-three-year period, and now control nearly the same amount

Table 9.4
Landownership by Person and Jati, Udaynala, 1957 and 1993

	Clean[1] caste[1]	Muslim	Nama-sudra	Bagdi	Muchi	Santal	
Population in numbers							Total
1957	97	538	217	165	48	35	1100
1993	94	1125	299	304	91	93	2006
Land per person[2] (in bigha)							Average
1957	1.64	2.14	1.39	0.99	0.35	0.40	1.15
1993	1.40	1.20	0.55	0.93	0.60	0.15	0.79
Percentage of village land owned							Total
1957	8.6	64.3	16.4	8.9	0.9	0.8	99.9
1993	6.6	67.5	8.3	14.2	2.8	0.7	100.1

[1]"Clean caste' includes Bamun, Kayastha, Kalu, and Bene.
[2]Refers to total population of individual jatis.

as the former 'dominant caste', the Aguris. Contrary to the general trend Bagdis have seen an increase in land per person over the same period. This is quite a feat in view of a near doubling of their population. The Muchis of Gopinathpur have seen a small increase in their proportion of village land, but because of population growth their 'land per person' figure has decreased at the average rate.

In Udaynala also Bagdis have gained control over more village land. Their land per person ratio has decreased slightly (from 0.99 to 0.93), a negligible decline compared to the all-village decrease (from 1.15 to 0.79). Only the Muchis have fared better, with their land per person ratio having nearly doubled. However, they still fall short of the all-village average, whereas the Bagdis by 1993 were better off than average.

Population increases over the three decades have caused a substantial decrease in landholding per person. The formerly dominant jatis have been severely affected and have lost their dominant position, while the lower castes have in the main gained.[16] Among the various jatis, Udaynala Muchis have the highest rate of increase in landownership, but their absolute ownership levels still fall short of the all-village per person average. Their counterparts in Gopinathpur have become relatively poorer over the period in question. Bagdis, on the other hand, have done very well in both villages. While in Udaynala Muchis have done well, Bagdis have reached an above average land per person ratio; while in Gopinathpur Muchis have fared poorly, Bagdis have increased their land per person ratio. Unfortunately, I do not have figures that can confirm whether or not these were developments significantly under way before 1977. However, local history does suggest that all major absentee landowners sold their lands after the rural unrest and the CPM-led mobilisation of the late 1960s, leaving little to be redistributed. This may explain some of the changes.[17]

Other elements in the LFG/CPM programmes have been much more effective for the redistribution of wealth, in particular increases in daily wages for agricultural labourers. In 1977 local wages normally consisted of Rs 2, 1 kg of paddy, ten country cigarettes and some body oil. By 1993 standard wages stood at Rs 12 and 2 kg paddy, with cigarettes and oil dropped (this was still a bit under the official minimum of Rs 14 and 2 kg paddy). If calculated in paddy purchasing power, wages have been raised from about 1.75–2 kg of paddy (plus the cigarettes and oil) in 1977 to about 4 kg of paddy in 1993; a doubling made possibly by CPM pressure on four occasions between 1978 and 1993, mainly against landowner opposition. More significantly, perhaps, the number of working days (and thus the level of income) has vastly increased. The traditional *aman* crop requires about 20–25 man-days per bigha, while the more intensive *boro* crop needs 35–40 man-days. Until the late 1970s the aman crop was all-important, and there was no boro crop. The recent expansion of irrigation facilities and the availability of HYVs have made possible intensive boro cultivation. Whereas previously labourers were unemployed—'we were sitting around'—for six to seven months a year, most are now fully employed for at least nine to ten months in a year.

These developments have been crucial in raising the living standards among the poorer sections, irrespective of jati. The raising of wages concerns non-divisible boons and has benefited all poor sections, while in the distribution of land and IRDP we find a favouring of certain groups. There is no doubt that here the Bagdis have benefited more from the CPM raj than comparable groups. A similar difference is also found in the distribution of political representation.

PANCHAYAT REPRESENTATION

West Bengal's broader political representation as represented in the *panchayat* system has received much scholarly attention.[18] In a recent article G.K. Lieten (1994: 1835) found that in terms of caste composition the panchayats in Memari II (adjacent to Raina) saw a 'remarkable' increase in SC/ST representation. The percentage of SC/ST representatives rose to 51.1 per cent—in Lieten's words, 'well above their proportion in the population' (ibid.). This over-representation is interesting. Similar trends of increased SC representation are evident in the gram panchayat to which Udaynala and Gopinathpur belong (see Table 9.5).

Table 9.5
Representation by Jati in the Gram Panchayat to which Udaynala and Gopinathpur Belong

	1988–93	1993–98
Namasudra (SC)	1	2
Bagdi (SC)	6	8
Aguri	6	4
Muslim	3	2
Kayastha	–	1
Bamun	2	2
Total	18	19

However, a breakdown of the SC/ST group into individual jatis shows that their well-above-average representation pertains to certain jatis rather than the entire group.

Table 9.6 shows that as a bloc the ST/SC are represented in proportion to their share in the total population. However, apart from one representative elected in 1993, all SC members were Bagdis. The Udaynala Bagdis held one of the village's two seats for three periods, although constituting only 15 per cent of the population. In Gopinathpur only Bagdis among the SCs have been represented. For these two villages, seven out of eight SC members have come from a jati that constitutes less than 50 per cent of the SC population. Moreover, as representation in the panchayats constitutes only one form of political influence, it is worth noting, although space does not permit a detailed outline, that Bagdis are also well represented in other fora, such as the boards of village cooperative societies and the recently created gram committees. But even when considering panchayat representation alone, we must conclude that Bagdis have enjoyed a privileged representation under CPM raj.

The question then is what this over-representation amounts to. In general the higher castes still enjoy the real power and hold the more important positions; and in terms of actual political influence at an all-village or extra-village level the picture is qualified. Gopinathpur's Bagdi gram panchayat member of 1988, who in 1993 was elected to the jela parishad, is an

Table 9.6
Gram Panchayat Representation by Jati, Udaynala and Gopinathpur, 1978–93+ (in periods)

	1978	1983	1988	1993[1]	Total Seats	Percentage of Village Population 1993
Gopinathpur						
Aguri	1	1	–	1	3	19.3
Bagdi	–	–	1	2	3	33.4
Unrepresented groups						
High castes						10.0
Intermediary castes						14.9
Other SC						22.5
Total SC population						55.9
Udaynala						
Muslim Sekh	1	1	1	1	4	56.1
Bagdi	1	1	1	–	3	15.2
Namasudra	–	–	–	1	1	14.9
Unrepresented groups						
Clean castes						4.7
Other SC/ST						9.2
Total SC/ST population						39.3

[1] From originally three gram panchayat seats in 1978, an extra seat was created in Gopinathpur in 1993. In addition, one former gram panchayat member was elected member of the jela parishad. Two seats were reserved for women in 1993.

articulate and prominent party member yielding much influence with the party in the area. In addition, his brother is full-time Secretary of the local cooperative society that covers four villages. On the other hand, Udaynala's two Bagdi gram panchayat members yielded little influence outside their own neighbourhood. The first member, who sat for one period only, was elderly and semi-illiterate, and although powerful among Bagdis he had limited interest and capacity for the affairs of the larger polity. The second member was only 17 when first elected, and his youth effectively prevented much clout with senior party workers of administrators even after ten years as member.

In her study of low-caste and tribal representation versus control over panchayats, Kirsten Westergaard (1987: 109–10) argues that in spite of fair representation of the poor, '... by and large this representation has not resulted in any significant increase in their [the poor's] control over these institutions.' This seems accurate for one village here, although not for the other. But lack of control does not seem to have engendered dissatisfaction. On the contrary, Bagdis of both Udaynala and Gopinathpur appear to be staunch supporters of the party. What we need to investigate is the social meaning of gram panchayat representation, including the context in which such representation is important even without increased control.

Panchayat members are not representatives of their individual communities as much as they are the party's people. It is well known that in West Bengal's panchayat system (as in most of the democratic world), where the party holds the majority, decisions are taken in internal party meetings and only presented as fait accompli at the official panchayat meetings. This policy becomes particularly interesting when considering that although ultimately relying on popular support, individual panchayat members are not *elected* by the people qua individuals but as representatives of a party that *selected* them as its candidates.[19] The CPM normally selects candidates with long records of party-affiliated work (mostly in one of the auxiliary organisations such as the Kisan Sabha or the Mahila Samiti). In short, panchayat members are primarily the party's representatives, and the voters are represented by the party. The tendency is for people to vote for the party, not for the individual candidate. But this analysis seems to be contradicted by the Bagdis' over-representation.

Representation entails status. Bagdis are tied to the party by being given representation in, but not necessarily control over, public positions. Considerations of jati for the selection of a candidate are important and give rise to both pride and hurt feelings. To take one example from Udaynala: the party decided to relieve the Bagdi panchayat member of his position for the 1993 election because of corruption and unpopularity among other social groups. A number of Bagdis launched an independent candidate and were coaxed back only by being promised the gram committee chairmanship as compensation. At the gram committee election meeting, Sekh Muslims from one neighbourhood advocated (from behind the scenes) the 'election' of one of their own people (these committees are officially elected, but in practice appointed by important party activists). They argued that they (the Sekh Muslims) had a longer record of party work than Bagdis (which was true). At the same time Bagdis also advocated the candidacy of one of their people.

The advocacy was entirely along jati lines, 'us' and 'them', and was never a question of differences over policy. What was important was not 'control' but 'representation'. The gram committee chairman does have some powers but in reality these are null and void unless backed by other sources of power, such as the party or the panchayat chairman. Nonetheless, the bestowal of the gram committee chairmanship on the Sekh Muslims of this neighbourhood would—according to them—have reflected their importance in the party's local history, their allegiance to and support of the party. Representation would have been an acknowledgement by the party leadership of this importance. Only one individual would have gained such a position, but the implicit acknowledgement would have reflected on all. As indeed it did in the case of the Bagdi 'winner', just as the gram panchayat membership had reflected on this jati since 1978.

We should remember that previously all political positions, formal or informal, were dominated by the higher castes (or high-status Muslims in Udaynala). Nowadays the lower castes and the poor are given formal political positions and participate in public meetings on a par with everyone else. Political representation gains its social relevance from these historical circumstances. It is not merely a question of gaining influence or of control over institutions. Representation without control does not lead to dissatisfaction because for a group previously not included in the fora of power, public representation is in itself a step forward, an enhancement of status, and an acknowledgement of importance.

From this outline the relationship may appear as an ordinary patron–client relationship with exchange of votes and support for both material and political benefits. However, the case may not be that simple. As stated earlier, most Bagdis and a number of other poor low-caste people are staunch defenders of the party and identify with it. But, as we shall see in the next section, Bagdis in particular have adopted broader aspects of the party's ideology. An investigation into these aspects reveals that Bagdis have continued to pursue their cultural reform ambitions under the communists. This has been possible because the basic tenets in the cultural world from which the party emerged in the first place include very strong assumptions about status and strong moral categories that have important bearings on perceptions of society. By becoming communists, the Bagdis (and to an extent other supporters) also became chashis, partly because the communist ideal of a good party affiliate was groomed in a cultural world where austerity, cleanliness and restraint already had a high moral premium. The two—the good chashi and the good communist—had so much in common that they were complementary. By becoming communist one also became chashi.

THE CPM IN A CULTURAL SETTING

The local CPM leadership was instrumental in curbing the excessive drinking among both Muchis and Bagdis after 1977. The two communities exploited a local political lacuna between 1974 and 1977 to engage in the liquor business, but in 1977 local CPM activists campaigned for its abolition. The Muchis yielded without protest, but in Udaynala the Bagdis put up resistance, waved lathis and threw pieces of dried mud at the delegation of anti-liquor notables, which approached their neighbourhood in the autumn of 1977. A week later the police were brought in and the production utensils broken. A month or so after this incident, a rapprochement took place. The Bagdi leader who had collaborated with the Congress until three years earlier, ended all liquor production among Bagdis and advocated support of the party instead. In exchange, he was elected as the party's candidate in the 1978 panchayat elections. The Bagdis' representation in the panchayats has already been outlined.

This is not an 'exchange' in any common sense of the word, but represents a complex relationship of mutuality and shared values. Teetotalism runs through the CPM's on-the-ground programme and is strongly advocated by its local representatives. The local representatives are mostly middle-class peasants of clean caste—chashis. For them, as communists, alcohol consumption contributes to the further impoverishment of the poor and to a continuation of their dejected position. For them, as chashis, inspired as they are by the bhadralok model, alcohol consumption is abhorrent, the typical sign of low status, lack of restraint, and wasteful immorality. Progressivist ideas in the CPM's ideology have found strength and support in complementary values integral to the now dominant sociocultural model. This model, the chashi model, is again profoundly influenced by the example, ways and mores of those powerful in contemporary society—the communists. On the whole the relationship between the CPM's local manifestation and the lower castes, in particular the Bagdis in this study, is governed by exactly such subtle understandings. Over the years the Bagdis have not only become staunch supporters of the CPM, but they have also reformed their lifestyles towards the chashi model—a reform made possible through their connection with the CPM.

There is not in every field a coincidence between the CPM and the chashi model, for instance, in the field of religious observances. In contemporary Udaynala and Gopinathpur, money is diverted by the low-caste CPM supporters into the building of new brick temples which are becoming increasingly larger, and they hold more expensive celebrations with an immense growth in the number of sacrificed goats. The Udaynala Bagdis fifteen to twenty years back commonly sacrificed five to ten goats with the goddess placed in a temporarily constructed shed. In 1993 they sacrificed thirty-two goats with the goddess placed in a brick temple. All-Hindu Gopinathpur's largest temple was built by Bagdis.

Perhaps the most important field where CPM policy and the tenets of the chashi model coincide is literacy. As already noted, literacy has become the hallmark of a good chashi lifestyle in present-day West Bengal. When I did my first round of informal interviews in the poor neighbourhoods of Udaynala, I was told by Bagdis that thirty years ago, 'No one here knew how to read or write. When we received a letter or a notice from the authorities, we had to go to the central neighbourhoods to have it read to us.' This was true for most people, the exception being the sanskritising families. Nowadays most young Bagdis and a majority of young Muchis attend classes and literacy is increasing rapidly.

Table 9.7 shows the level of literacy or grade passed by Bagdis of

Table 9.7
Education and Age Groups (in Percentages) of Men and Women of Bagdi Jati Above 16 Years of Age, Udaynala and Gopinathpur, 1993

Level of Literacy/Grade	Age Groups					
	16–25	26–35	36–45	46–55	56–90	Total
None	19.4	33.9	49.2	39.1	60.9	35.0
Some[1]	11.5	22.9	18.5	41.3	26.1	20.8
1–5	26.6	16.9	15.4	15.2	10.9	19.1
6–10	26.6	22.0	13.8	2.2	2.2	17.9
11+	2.2	4.2	3.1	2.2	–	2.7
Under education	13.7					4.6
N	139	118	65	46	46	414

[1] 'Some' indicates literacy achieved either at a *pathshala* (older form of informal village school in existence until the 1950s) or at home (through a relative), or at the present-day literacy campaign centres.

Udaynala and Gopinathpur. Included in the table are both men and women of 16 years of age or more.[20] The correlation between age and education is striking. The Table shows that until forty years ago, less than 20 per cent of Bagdis in these villages had had any formal education. Sixty per cent of those now 56 years old or more are illiterate, while one-fourth know 'some'—a category ranging from the ability to write one's name to being fairly fluent. Local knowledge suggests however that few Bagdis attended *pathshalas* (an old form of informal village school); instead a number of the older Bagdis have learnt to read and write at the literacy campaign centres.[21] Among the younger generations only one in three receives neither formal education nor has been taught at the literacy campaign centres, and an increasing proportion reaches higher levels, with a little more than one in four having reached Grade 6 or higher (the 'under education' category consists mainly of people who have failed lower level exams). Eleven individuals (4.6 per cent) have reached Grade 11 or higher, of whom eight are 35 or below.

Table 9.8 gives comparable figures for Muchis. It reflects the same

Table 9.8
Education and Age Groups (in Percentages) of Men and Women of Muchi Jati Above 16 Years of Age, Udaynala and Gopinathpur, 1993

Level of Literacy/Grade	Age Groups					
	16–25	26–35	36–45	46–55	56–90	Total
None	37.5	71.4	73.3	84.0	86.7	65.0
Some	6.3	9.5	10.0	16.0	13.3	10.0
1–5	27.1	16.7	16.7	–	–	15.6
6–10	12.5	2.4	–	–	–	4.4
11+	–	–	–	–	–	–
Under education	16.7	–	–	–	–	5.0
N	48	42	30	25	15	160

trend of increased literacy and education, although at a much slower rate. Until 1960 no Muchis had had any formal training, and four out of five were illiterate. This situation has slowly changed, with remarkable developments confined to the last batch, those now aged 25 or under. This batch entered school-age at around the time of the installation of the LFG. It is noteworthy, however, that even this batch contains a substantial number of illiterates, more than one in three. Moreover, only three individuals have attended the literacy campaign classes. The proportion of illiterate Muchis in this age group is twice that of Bagdis. And till today no Muchi has gone for higher education (which would take them out of the villages).

These developments do not compare favourably with the corresponding figures for the 'dominant castes'. Here I will briefly mention some of the trends. Among the Sekh Muslims, illiteracy is down to 7.3 per cent, and 8.8 per cent have been trained in literacy campaign centres. The rest have had formal education. More than one in ten aged 35 or less have reached Grade 11 or higher. Among the three dominant castes of Gopinathpur, the situation is even more striking. Illiteracy has been eradicated among the youngest batch, and only very few individuals of these jatis have attended the literacy campaign classes. Some 45 per cent of the youngest batch have passed Grade 6 or more. More than one in ten of those aged 55 or less have passed Grade 11 or higher.

Literacy is not merely about the ability to read and write, but it is also about knowing and asserting one's rights, making one's daily life less strenuous or preparing one for unusual events. As long as literacy is not taken for granted, not an ability automatically available to every member of society, it also has a social meaning, and makes a statement about difference and status. In rural West Bengal, as in most agrarian societies, it was the well-off who were literate. This difference was underlined by a parallel division in lifestyle between the literate bhadralok, the increasingly literate chashi and the archetypically illiterate majur.

Furthermore, as Poromesh Acharya (1981, 1985 and 1986) argues, due to historical circumstances, schooling and the mode of education in (West) Bengal were attuned to a particular style of learning and to the literary interests of the bhadralok, not to the practical interests of other social groups. The spread of literacy under the aegis of a mass party has not changed this markedly.[23] The increase in the number of village schools (since 1958 in Udaynala and since 1964 in Gopinathpur) has done much to universalise literacy and thus to reduce its social significance. However, it still retains quite a lot of potential as a social marker. This is evident from the way teenagers who have been to school engage themselves in clubs, in jatra performances and in bringing libraries and books to their villages. Again and again I witnessed how eager youth, irrespective of caste but dependent on education, engaged in cultural functions where they played music, read poetry (some self-composed) and made speeches about the need for cultural upliftment. Literacy is part of a 'culture', it is about social belonging and status.

It is therefore significant that the Bagdis have achieved a higher level of literacy than Muchis, and that they have invested relatively more in schooling. We find in this investment a closer identification with the objectives of the party they support. A higher level of literacy and other marks of cultural reform together with an improved economic position have contributed to the creation for Bagdis of the outward signs of chashis, a much improved social status compared to the majur status they previously held.

CONCLUSION

It has been the aim of this paper to suggest a linkage between political allegiance, cultural reform and social status. By themselves, the redistributive programmes (land and IRDP) and the political representation of low-caste groups do not account for the changes. They were too small, often of greater symbolic than material importance. Taken together their effect has been to raise somewhat the living standards of previously excluded sections of society. However, the more significant effect was that of raising social status, and in enabling groups with a particular political role to change their social stereotype in accordance with well-known socio-cultural models. Enduring political allegiances of the type found in rural West Bengal between the CPM and poor low-caste groups cannot be properly understood unless the social significance of representation, cultural reform and economic improvements is considered.

The literature on the politics of West Bengal has largely ignored how the CPM also represents a cultural programme. By historical association and ideological conviction the CPM has come to represent and implement causes—anti-casteism, social equality for the poor, teetotalism, and literacy—which, one should be careful to note, follow close upon earlier efforts toward cultural and social reform. Indeed, many of those policies grow out of the same body of cultural perceptions. Because of the limitations in the party's economic reform programme and the limited 'control' jatis of poor people have on the party, it is here, in the social implications of the party's policies and in its conscious exploitation of these ambitions on the ground, that we need to look for why so many poor people, and in particular in this study the Bagdi jati, identify themselves with the party.

This essay does not purport to give the ultimate answer to the CPM's entrenchment in rural West Bengal or Bardhaman district. Fundamentally, all poor people irrespective of jati have experienced enhancements in their social position, economic well-being and political representation. However, beyond this, a close link—a 'special relationship'—was created in the studied locality between the local CPM leadership and one particular jati. Undoubtedly in other localities the relationship is with other groups, not necessarily jatis, but the process and linkage are likely to be the same. Favoured groups stem from and constitute the party's local vigilantes. In return these activists receive particular attention from the party's local leadership. The attention may be of an economic nature, except that economic means available to the local leadership for such distribution is normally quite limited. Political representation will be a more likely tool, a tool with strong implications for social status. More to the core of my argument: what I have described here is a case in which one jati has been favoured over other comparable jatis within the same locality. This case may not be representative of all cases, or even a majority of them. But in spite of idiosyncracies, the case has functioned to bring out the following point: that in a society which has experienced and which to some extent still experiences strong and ranked social differences, enduring political support for what is at best a party devoted to the path of slow reform cannot be understood solely in terms of limited economic redistribution. A political movement of such size and entrenchment must also be understood in terms of wider social and cultural implications, in terms of a 'fit' between the values represented by the party (or by its local apparitions) and the values and aspirations of its supporters, and in terms of the local political adaptions to this context and to the strictly speaking non-political aims and desires of its followers. West Bengal's CPM has raised the social standing of formerly low-ranking groups—in some localities all such groups, in others only some. The importance of political representation even without real influence, or the importance of distribution of minuscule plots of land together with the increased propensity towards extravagant and 'cultured' festivals among CPM supporters formerly in the fringes of Hindu society must be seen against the backdrop of a society that has historically been highly unequal and highly ranked.

NOTES

1. It may need underlining that when I talk about jati I do not intend an essentialist interpretation. On the contrary, my own experience in Bengali villages and the material presented here suggest that jati is above all a social construct that can be used—manipulated—to obtain advantages (see Ruud, 1999). I have concerned myself with jati identities for the same reason that the local CPM considers them: they are important social markers. The status of one's jati reflects upon oneself and constitutes part of one's identity in society. Jati is not the whole story of one's identity or social standing, but it is an important part of the story.
2. In the following, 'the party' refers to the CPM.
3. The typical urban-centric view that the rural world consists of only two social groups, the *bhadralok* and the *chhotolok*, is vastly oversimplified.
4. See for instance Bandyopadhyay (1985); for earlier efforts of the same type in Bengal see Sanyal (1971; 1981).
5. Or 'westernisation'. See Cohn (1990).
6. Acharya (1985: 1785) records the growth in primary schools in West Bengal, from 14,700 in 1950–51 to 'about 50,000' in 1980–81.
7. For the leftist inclination of the bhadralok, see Franda (1971), Kohli (1990), and Ruud (1995).
8. This at least is the case in the Dakshin Damodar region of Bardhaman district that I know.

9. I was allowed to read diaries written by one Udaynala schoolteacher between 1956 and 1983. Several entries mention suspicion of Bagdi theft, and occasionally some Bagdis were arrested.
10. For a study of phases in the CPM's policies on agrarian relations, see Bhattacharya (1993; this volume).
11. There are 12 katha to 1 bigha. In general, the party line has been to distribute marginal plots of land instead of 'viable' plots, so as to reach as many people as possible (see Baruah, 1990, and Lieten, 1992: 140–1). Sengupta (1981: A–69) calculates that one in three landless households has benefited from land redistribution. However, the plots, which range in size from 0.33 to 2 acres, are 'hardly more than homesteads' (Lieten, 1990: 2268).
12. All figures given in the tables in this paper are field data.
13. The comparison in the tables is made on the basis of individuals, not households. It may be noted that (as Jayoti Gupta pointed out for West Bengal as a whole in her paper at the Calcutta Workshop) no women were given land titles in these villages until 1993.
14. After my survey, two clean-caste households obtained the panchayat member's signatures on their IRDP applications (that signature being the sine qua non of the application); one of the recipients was a moderately well-off moneylender.
15. CPM MP Somnath Chatterjee argued that 'Of course people become CPM supporters when there is a question of redistributing land.' Interview in Guskara, Birbhum, 1989. Mallick (1990 and 1992) argues that well-off families get IRDPs, but Swaminathan (1990) finds such practices not to be prominent. See also Westergaard (1987).
16. The Udaynala Namasudras constitute an exception, but that was due to the rather unusual decline of a once wealthy family.
17. Most of the major absentee landowners were originally locals and their lands have thus been included in the 1960 figures of Tables 9.3 and 9.4.
18. For some of the debates see Lieten (1990, 1992 and 1994); Sengupta (1981); B. Bose (1981); Dasgupta (1984a and b); Mallick (1992); Webster (1992); Bhattacharyya (1993); Acharya (1994).
19. But see Dwaipayan Bhattacharyya (this volume), whose case study implies that prominent individuals are chosen to attract votes.
20. No breakdown by sex is attempted here as this section aims to illustrate the social importance of literacy in terms of jati ranking. However, data presented by Gazdar and Sengupta (this volume) suggest clear differences between men and women across castes with significantly lower literacy rates for women than men.
21. The state government and the CPM combined to launch a literacy campaign in Bardhaman district in 1990. In most villages teaching centres were organised by the gram committees; kerosene was supplied through the panchayats, and inexpensive reading material was provided by the government. Many youths contributed as volunteer teachers.
22. Only the material for the current literacy campaign has much improved in this respect compared to primary school material. However, even here the interests derived from the bhadralok's literary traditions are evident. For instance, about one-third of the campaign material's 'Part Two' was dedicated to poems by famous Bengali poets and to one of the lengthier stories by Saratchandra Chatterjee, *Saksharatar dvitiya path*, State Research Centre (West Bengal), Calcutta 1990.

REFERENCES

Acharya, Poromesh. 1981. 'Politics of Primary Education in West Bengal,' *Economic and Political Weekly*, 16(24): 1069–75.
———. 1985. 'Education: Politics and Social Structure,' *Economic and Political Weekly*, 20(42): 1785–9.
———. 1986. 'Development of Modern Language Text-Books and the Social Context in 19th Century Bengal,' *Economic and Political Weekly*, 21(17): 745–51.
———. 1994. 'Elusive New Horizons: Panchayats in West Bengal,' *Economic and Political Weekly*, 24(5): 231–4.
Bandyopadhyay, Sekhar. 1985. 'Caste and Society in Colonial Bengal: Change and Continuity,' *Journal of Social Studies*, 28: 64–101, April 1985.
Baruah, Sanjib. 1990. 'The End of the Road in Land Reform? Limits to Redistribution in West Bengal,' *Development and Change*, 21: 119–46.
Basu, Amrita. 1992. *Two Faces of Protest: Contrasting Modes of Women's Activism in India.* Berkeley: University of California Press.
Bhattacharyya, Dwaipayan. 1993. 'Agrarian Reforms and Politics of the Left in West Bengal.' Ph.D. thesis, University of Cambridge.
Bose, Buddhadeb. 1981. 'Agrarian Programme of Left Front Government in West Bengal,' *Economic and Political Weekly*, 16(50): 2053–60.
Bose, Sugata. 1986. *Agrarian Bengal: Economy, Social Structure and Politics, 1919–1947.* Cambridge: Cambridge University Press.
Cohn, Bernard S. 1990. *An Anthropologist among the Historians and Other Essays.* Delhi: Oxford University Press. First published in 1987.
Dasgupta, Biplab. 1984a. 'Sharecropping in West Bengal during the Colonial Period,' *Economic and Political Weekly*, 19(13): A–2 to A–8.
———. 1984b. 'Sharecropping in West Bengal: From Independence to Operation Barga,' *Economic and Political Weekly*, 19(26): A–85 to A–96.
Franda, Marcus F. 1971. *Radical Politics in West Bengal.* Cambridge, MA: MIT Press.
Fuller, Chris J. 1992. *The Camphor Flame: Popular Hinduism and Society in India.* Princeton: Princeton University Press.
Kohli, Atul. 1990. 'From Elite Activism to Democratic Consolidation: The Rise of Reform Communism in West Bengal,' Francine R. Frankel and M.S.A. Rao (eds) *Dominance and State Power in Modern India: Decline of a Social Order* (volume 2). Delhi: Oxford University Press.
Lieten, G.K. 1990. 'Depeasantisation Discontinued: Land Reforms in West Bengal,' *Economic and Political Weekly*, 25(40): 2265–71.
———. 1992. *Continuity and Change in Rural West Bengal.* New Delhi: Sage.
———. 1994. 'For a New Debate on West Bengal,' *Economic and Political Weekly*, 29(29): 1835–8.
Mallick, Ross. 1990. 'Limits to Radical Intervention: Agricultural Taxation in West Bengal,' *Development and Change*, 21(1): 147–64.
———. 1992. 'Agrarian Reform in West Bengal: The End of an Illusion,' *World Development*, 20(5): 735–49.
Ray, Rajat K. 1992. 'The Rural World of Tarashankar Banerjee: Social Divisions and Psychological Cross-Currents,' Peter Robb (ed) *Rural India: Land, Power and Society under British Rule.* Delhi: Oxford University Press. First published in 1983.
———. 1987. 'The Kahar Chronicle,' *Modern Asian Studies*, 21(4): 711–49.

Ruud, Arild Engelsen. 1995. 'Socio-Cultural Changes in Rural West Bengal.' Ph.D dissertation, London School of Economics.
———. 1999. 'The Indian Hierarchy: Culture, Ideology and Consciousness in Bengali Village Politics.' Modern Asian Studies (forthcoming).
Sanyal, Hitesranjan. 1971. 'Continuities of Social Mobility in Traditional and Modern Society in India: Two Case Studies of Caste Mobility in Bengal,' *Journal of Asian Studies*, 30(2): 315–39.
———. 1981. *Social Mobility in Bengal*. Calcutta: Papyrus.
Sengupta, Sunil. 1981. 'West Bengal Land Reforms and the Agrarian Scene,' *Economic and Political Weekly*, 16(25–26): A–69 to A–75.
Srinivas, M.N. 1965. *Religion and Society among the Coorgs of South India*. Bombay: Asia. First published in 1952.
Swaminathan, Madhura. 1990. 'Village Level Implementation of IRDP: A Comparison of West Bengal and Tamil Nadu,' *Economic and Political Weekly*, 25(13): A–17 to A–27.
Webster, Neil. 1992. *Panchayati Raj and the Decentralisation of Development Planning in West Bengal*. Calcutta: K.P. Bagchi.
Westergaard, Kirsten. 1987. 'Marxist Government and People's Participation: The Case of West Bengal,' *Journal of Social Studies*, 38: 95–113, October 1987.

10

Politics of Middleness: The Changing Character of the Communist Party of India (Marxist) in Rural West Bengal (1977-90)

DWAIPAYAN BHATTACHARYYA

> ... political discourse ... must persuade. In other words, it must get the listener to agree to the speaker's point of view, even if other options remain available. It is, therefore, a form of rhetorical discourse .... However, the political discourse that replaces persuasive speech with incantatory (or, worse still, with magic formulae containing secret messages passed from witch to witch) represents a linguistic and civic reality that every democratic community must attack with the weapon of clear-sighted analyses and demystification.
>
> Umberto Eco, 1995

1

In August 1977, defending the first budget of the newly-elected Left Front government (LFG) in the legislative assembly, Benoy Konar, a peasant-wing leader of the Communist Party of India (Marxist) [(CPI[M])], said:

> Some people could not find any ray of hope in this budget. Some people even tried to search a revolution in it. Let me tell you, please don't take trouble to look for revolution in the pages of this budget.... You will get revolution ... in the struggle of the millions of Bengal's workers and peasants who will use this budget as a pretext to take to the peak their fight against the billionaires and the owners and the landlords, they will intensify class struggle and dig the graveyards for the looters, the billionaires, the *jotdars* and the landlords of our country; this government will arouse them, will tread by the common man with all its limited opportunities and constrained resources ... (WBLA, 1977: 206).

This emphatic declaration for accelerating class struggle stands in contrast to what Benoy Krishna Chowdhury, the octogenarian and vastly experienced minister, said in 1989 after twelve years of spearheading the Left Front's agrarian reforms:

> I can tell you even as I stand in the premises of this House that we are bound to act as an agent of a class *some day*. That can't be done merely by making laws. On the other hand one has to realise the impossibility of taking struggle too lightly .... We must look into the conditions at the field level. It is necessary therefore to develop a united movement rather than talking too loud ... it will not help us to foster fresh enemies .... (WBLA, 1989b: 7–8; emphasis added).

> There are other ways to live and grow in a community. That can't be done by confrontation and conflict, rather that has to be parallel and consistent. The essence of co-ordination is the fulfillment and beauty of life (WBLA, 1989a: 92: 208).

It would be naive to treat this difference in emphasis as an isolated instance of the varying stylistic preferences of two individuals who otherwise happen to represent the same political space in the legislative assembly. Equally gullible would be an assessment which takes both pronouncements at face value, without going beneath the rhetoric, without critically

---

This paper partly draws from Bhattacharyya (1993). Thanks are due to Geoffrey Hawthorn for supervising my dissertation with extreme care. I would like to thank Partha Chatterjee for discussing several themes in the paper, Damayanti Datta for her criticisms, Barbara Harriss-White for her detailed notes on an earlier version of this paper and Ben Rogaly for being very supportive with his comments. I alone, needless to say, am responsible for all errors and interpretations.

looking at the contexts in which they were made. These statements, separated by twelve intervening years, are indicative of the changing thrust in the CPI(M)'s mode of political mobilisation which has accompanied the continuing popularity of its LFG over the 1980s and 1990s.[1]

This paper seeks to explore how the CPI(M) used a set of agrarian reforms to build legitimacy and consensus upon which it could expand its base among almost all peasant classes. There is an obvious paradox here. Redistributive reforms generally compensate the poor at the expense of the prosperous classes and thereby drive a wedge between class positions. The CPI(M)'s reforms benefited the dispossessed but simultaneously created durable conditions for electoral support for the LFG from a broad alliance of classes. Further complications surfaced as electoral support showed no sign of a slump even ten years after the reforms ran out of steam. This calls for a study of some aspects of the CPI(M)'s agrarian politics, and policies as a part of such politics, including its complex manoeuvres, calculated shifts and poignant rhetoric, which combined to form a distinct art of governance. Such art is manifested in what I will characterise as 'the politics of middleness' in rural West Bengal with the panchayat as its progenitor.[2] This paper will take that up (section 5) after discussing the tension between 'class struggle' and 'peasant unity' within the CPI(M) and its resolution (section 2), alterations in the CPI(M)'s attitude to landlord, sharecropper and agricultural labourer (section 3) and field observations of the CPI(M)'s pragmatic moves regarding the selection of candidates for panchayat elections and the use of politically emasculated ritual strikes in wage negotiations (section 4). The reforms themselves have been summarised by Gazdar and Sengupta (this volume) and need no further elaboration here.

2

In 1964, when it was born, the CPI(M) was projected in its programme as a revolutionary Marxist party aiming to build 'socialism and Communism . . . under proletarian statehood' (CPI[M], 1989: 33). It donned a 'democratic–centralist' organisation with a leadership avowedly committed to Marxism–Leninism and the Stalinist version of the 'dictatorship of the proletariat'. After more than thirty years since its inception, the party commanded only a localised influence in India and failed to expand in the manner it had anticipated. But in the locales under its sway the CPI(M) was deeply entrenched: the party's presence in these areas was fairly ubiquitous. Here (in West Bengal, Kerala and Tripura) the CPI(M) formed state-level governments of varying tenure in alliance with other 'like minded' combines. Yet the party had to operate under the constant surveillance of the union government in New Delhi which had (until the early 1990s) always been more or less hostile to the CPI(M) and its governance. This imposed several limitations on the actions of the Marxists. Their political choices, for instance, were constrained (strategies of 'class insurrection' having to be jettisoned to reduce the threat of dismissal of the state government by New Delhi) and several innovative policy options, modes of popular engagement and mass action had to be devised strictly within the limits of existing legality. In addition, the Left had to shed from its vocabulary all euphoric display of 'revolutionary zeal' or 'radical temper', phrases associated with its more familiar rhetoric.

To track shifts in the CPI(M)'s public pronouncements, a brief look at the original bearings of the party may prove instructive. This is not to condemn the party for its alleged 'dilution of ideology' nor to condone its 'pro-democracy shifts' nor to venerate its 'ideological steadfastness'; the aim is rather to trace certain changes in the CIP(M)'s political practices in West Bengal and ask how far they mattered in its ability to lead the ruling coalition.

Taking advantage of changing political contexts, the CPI(M) periodically introduced new 'strategies' and 'tactics'. Take the celebrated polemics which coloured the CPI(M)'s self-definition as a 'revolutionary' as opposed to a 'revisionist' party. In the late 1960s the party was troubled by the debate over whether, and for what purpose, to join a 'bourgeois' state government. The bone of contention in the debate was section 112 of the CPI(M) programme which stated:

> Even while keeping before the people the task of dislodging the present ruling classes and establishing a new democratic state and government based on firm alliance of the working class and peasantry, the Party will utilise all the opportunities that present themselves of bringing into existence governments pledged to carry out a modest programme of giving immediate relief to the people. The formation of such governments will give great fillip to the revolutionary movement of the working people and thus help the process of building the democratic front. It, however, would not solve the economic and political problems of the nation in any fundamental manner. The Party, therefore, will continue to educate the mass of the people on the need for replacing the present bourgeois–landlord state and government headed by big bourgeoisie even while utilising all opportunities for forming such governments of a transitional character which

gives immediate relief to the people and thus strengthen the mass movement (CPI[M], 1989: 47–8).

According to the Left critics of the party, section 112 suffered from an unhealthy intent to embrace 'parliamentary cretinism', a reformist vice which allegedly spelled death for a revolutionary organisation. The protagonists of the official line in the CPI(M), on the contrary, claimed that the parliamentary governments proposed in the section were to be provincial and transient in order to give immediate relief to the people and strengthen popular movements. They argued that a 'revisionist' move would have endeavoured to institute such government at the 'centre' while the party nurtured no such ambition (Dasgupta, 1989: 37–8). A close reading of section 112, however, reveals two different possibilities. First, in tune with the above claim, it may be recognised as pressing the need for installing 'modest' governments to give 'immediate relief to the people' at the provincial or state level, while at the 'national' level (taking its cue from such governments), a 'democratic front' of a revolutionary character will be forged in order to replace the 'bourgeois–landlord state and government'. Second, and from a contrary position, one may read into the section an urge to form transitional governments simultaneously with organising revolutionary struggle on a different plane, but both at the level of the 'nation'. Otherwise, the reference to the impossibility, of 'solv[ing] the economic and political problems of the nation in any fundamental manner' by installing 'transient' governments sounds awkward. By receiving ample revolutionary 'fillip' from such relief-oriented governments which can only address (but not solve) national problems, the 'state' of the bourgeoisie will eventually be replaced. Hence, the interpretation is open, and not decisive as the party claimed.

At the time of this debate, section 112 came in handy to the CPI(M) in defending the party's participation in the United Front in West Bengal.[3] There was no possibility of instituting any such alternative to the Congress at the national level and hence there was no problem in castigating such action as 'revisionist'. When the political climate changed considerably post-1977, an alternative alliance at the national level under the tutelage of the Left parties was no longer perceived as obstinate impossibility. If it were to happen, the CPI(M) had to find ways to defend its participation in a parliamentary government at the centre. In its 10th Congress in Jallandhar in 1978 the CPI(M) agreed in principle to participate in any possible all-India parliamentary 'Left and Democratic Front'.[4] What was untouchable as a euphemism for 'revisionism' in the late 1960s now found acceptance within the CPI(M) in a somewhat changed context.

More recently, in 1996, the CPI(M) central committee prevented the party from joining the United Front government at the centre (by thirty-five votes to twenty) on grounds of its being 'bourgeois'. Jyoti Basu, the West Bengal chief minister and a member of the CPI(M) Politburo, who opposed the central committee, publicly pronounced this as a 'historic blunder'. Basu held: 'Unfortunately in our party programme we talk about states: in section 112. Earlier we thought they would never allow us to function even in the states, but things changed. But we never discussed about the Centre, we thought it was absolutely a dream now, it would come later. But things have happened since then: in the Centre also we have to play a part. That is not in our party document, so we said we have to update that programme' (*Asian Age*, 1 January 1997: 12).

There is a surfeit of comparable instances whereby the CPI(M) clipped its programmatic angularities to gain political acceptance in a competitive democracy. Of particular importance is the issue of 'peasant unity' which played a key role in the party's rural mobilisations. For the CPI(M) and the Krishak Sabha, 'unity' was far from an aggregation of demands, whereby all sections of the peasants (barring the non-cultivating landlords) could be given equal space and a uniform stake so as to strike a neat balance. On the contrary, such unity was presented as a necessary item in the agenda for 'peasant struggle'. 'Peasant struggle', an oft-repeated phrase in the Sabha literature, presupposed the poor peasant and the agricultural labourer as its axis. While unity was desirable,[5] it was not to be purchased at the cost of the poor peasants' and agricultural labourers' interests.

Although this clearly was the official position of the Krishak Sabha, in practice it dithered. It always strove to combine 'struggle' and 'unity', and on most occasions ended up making an unhappy amalgamation. In the Medinipur session of the All India Krishak Sabha (AIKS) in 1982, for instance, a leading delegate attacked any attempt to woo the rich and the middle peasants (AIKS, 1982a: 28). According to him, the 'West Bengal experience' illustrated that mobilisation around the issues of the poor peasants and agricultural workers had always strengthened the Krishak Sabha while 'very few rich and middle peasants stayed back with us' once their demands voiced by the Sabha were met. This observation, however, came under fire from H.K. Surjeet. 'What the comrade says,' argued Surjeet, 'reflects neither the West Bengal experience nor the experience of the peasant movement in the rest of the country.'

Surjeet instead talked about giving the peasant movement a 'new orientation' which called for the awareness 'that without raising the demands of the peasantry as a whole, including the rich and the middle peasants, and without merging the different currents into one, we can neither advance towards the agrarian revolution nor will we be able to raise the movement to the level of land occupation' (AIKS, 1982a: 28–9). The 'new orientation' was perceived as less divisive, more inclusive and, indeed, electorally more promising. 'It is true,' the Sabha declared after the CPI(M)'s second consecutive victory in West Bengal, 'but for the support of an influential section of the middle peasants, this victory would have never been possible' (WBPKS, 1982: 122). It was the electoral exigency to unite various peasant classes which consequently prevailed over most other considerations.

Such needs were felt most acutely to give strategic shapes to the so-called 'new orientation' in the successive Sabha sessions (WBPKS, 1986: 56, 76).[6] The AIKS in one of its top-level meetings stated in no uncertain terms: 'On wage demand . . . [it] may be necessary to settle for a wage which is lower than the basic demand to win over the rich and the middle peasants' (AIKS, 1986: 70). Needless to say, the effort to combine 'struggle' with 'unity' became a serious problem for the Sabha in a political ambience which called for a privileging of the latter. It may be interesting to note that there were cases of local dissent to such decree from the top. The Bolepur Thana Committee of the Pradeshik Krishak Sabha, for instance, resolved to campaign exclusively around the issues of 'wage, bargadar and patta' even if they ran counter to the principles of so-called peasant unity (WBPKS, 1984: 13).

Similar shifts were visible in the party's stance on land reforms. Initially, the CPI(M) demanded 'fundamental' (*maulik*) land reforms through abolition of landlordism without any compensation whatsoever (CPI[M], 1973: 3) and treated the Congress-made land ceiling laws and legislation on minimum wages, fair prices, taxation, etc. as 'hollow' and 'pitiful'(ibid.: 4). The 'ruling class' reforms were considered to be a sham in the absence of a relentless 'mass movement' from below on the lines of 'class struggle'. This was at a time when the CPI(M) was mainly in the opposition. Once placed in the state government such expressions seemed too menacing. It was now the CPI(M)'s turn to take an active interest in augmenting and embellishing such laws, plugging their loopholes by cleverly and effectively making amendments, sparing no pains to come to terms with them and seldom challenging their underlying principles by vigorous class action.

3

In re-positioning its class priorities, it was necessary for the CPI(M) to simultaneously amend its 'revolutionary' demands for 'land to the tiller', abolition of landlordism, and a strong organisation for agricultural labourers. Before the Left Front came to power, the party regarded all landlords as enemies. A central committee resolution of the CPI(M) declared: 'The breaking of the monopoly of the land enjoyed by the landlord class—both of the old feudal as well as new capitalist type—is in the common interest of the entire peasantry including the agricultural labourers (CPI[M], 1973: 6).[7] Another such resolution distinguished the capitalist landlords (who did not take part in major agricultural operations) from the rich peasants (who or whose family members did).[8] The central committee, while issuing 'alternative amendments' to the Congress-sponsored ceiling legislations of the early 1970s, lobbied for a total abolition of landlordism to 'ensure land gratis to the landless agricultural labour and poor peasant' (CPI[M], 1973: 3–5).

When in 1979 the Pradeshik Krishak Sabha met for the first time after the LFG was installed, it continued its call for an abolition of landlordism 'immediately and completely' but with a significant qualification added to its 'Proposals on Land Reforms': 'This conference believes that the Pradeshik Krishak Sabha can in no way undertake this alternative policy for implementation within the present state structure. This can be successful . . . only through the establishment of a people's democratic state' (WBPKS, 1979: 92–3). The 'people's democracy' or any struggle for its realisation was no longer on the immediate agenda of the party, nor was abolition of landlordism any more a concern of the Front government.[9] Instead, the Krishak Sabha sought to devote its energy to identifying the 'new class' of capitalist landlords and merely to *campaign* in a planned and organised manner among the peasants to emphasise the need to *marginalise* this class (WBPKS, 1979: 46–8). 'Abolition of landlordism', like 'struggle for people's democracy', was subsequently to be regarded as a 'slogan for propaganda' and not necessarily for immediate action.

While distancing itself from the mode of its pre–Left Front politics, the CPI(M) since 1977 has also been busy explaining why even a United Front–like agitational politics on class lines was not feasible any more. As a progeny of the short-lived United Front, the Left Front was expected to know that governmental principles can ill afford to include unrestrained mass mobilisation. The repetitions of similar 'mistakes', which were allegedly committed during the United Front phase, were by no means desirable. The CPI(M) leaders were now at pains to convince the Krishak Sabha workers that the United Front-like spontaneous movement was no more possible with the Left Front in power (WBPKS, 1979: 46), first, because identification of the landlords, who were very large and isolated from other peasant classes, was much easier during the United Front period than it was now (WBPKS, 1989b: 92). Second, the 'land grabbing' movement of the late 1960s was possible because popular opinion had already formed against the big landlords who concealed large chunks of land to escape the reach of the Estate Acquisition Act. By contrast, during the present Left Front regime, the 'land of the big *raiyats* vested in the government is small in amount' and therefore precluded any impressive response to a movement (WBPKS, 1979: 46–7). Finally, encouraging 'the popular passion' to 'forcibly take away zamindars' *khas*[10] land for cultivation' was necessary during the United Front regime because despite its land-lessness, 'the poor lot' found it difficult to believe that land belonging to the 'oppressive zamindars' could truly be its own (WBPKS, 1989b: 91).

The class of sharecroppers also came under review during the Left Front period. Sharecroppers in West Bengal did not constitute a 'pure class'. Certain exceptional cases of reverse tenancy apart, the peasants who leased in land did so either in addition to their smallholding or to their income from wages as land labourers. They were traditionally subject to various forms of exploitation: eviction, non-payment of dues, denial of hereditary tenancy rights, etc. A majority of the lessors were in no way big, or even middle, landowners. Many were poor peasants who, for various reasons, let out land for sharecropping. This absence of a clear-cut divide between the classes of sharecroppers and the landowners might have made any drastic legislation, such as the abolition of tenancy as an extension of the 'land to the tiller' policy, difficult.

'Abolition of all intermediaries on land' and 'land to the tiller' were the slogans of the Congress Agrarian Reforms Committee in the wake of independence. The Communist Party of India also adopted a 'land to the tiller policy' in its document *On the Agrarian Question in India* in 1949 (Herring, 1983: 163). On assuming office in Kerala in 1957, the first communist government of the state proclaimed the Stay of Eviction Proceedings Ordinance to protect the tenants and the labourers with huts on landlords' land as a prelude to a land-to-the-tiller legislation. The objective was abortive in Kerala and the state government toppled. Such attempts were made again in 1969, this time with greater success. In West Bengal prior to the 1969 mid-term elections, the CPI(M) also reportedly adopted a 'land to the tiller' policy to gain the support of sharecroppers (Frankel, 1971: 183). The resolution on 'Certain Agrarian Issues' adopted by the CPI(M) in March 1973 clearly held that the 'right of the tenant to the ownership of the land he is cultivating is to be guaranteed, except to those who are lease-holders from small owners'. For the latter group, particularly those who intended to cultivate by their manual labour, 'such land under tenancy [will] be equally divided between the tenants and the smallholder' and the 'tenant will continue to pay a fair rent on the portion of the land he cultivates' (CPI[M], 1973: 7).

Both the United Front and the Left Front, however, clearly expressed their disapproval of the 'abolition' policy although they differed on the choice of means to protect the sharecroppers. Harekrishna Konar (1979: 189), the CPI(M) and Krishak Sabha leader as well as a minister in the United Front government, observed that the 'very complex issue' cannot 'be ended by a stroke of pen' so long as the present social structure existed. 'If banned by legislation,' he argued, 'it will go on covertly and also more crudely.' The important thing according to Konar was 'to think how this system could be better regulated in the interest of the peasantry.' Benoy Chowdhury, the land reforms minister of the LFG, also held a similar view. He maintained that even an informal suggestion of tenancy abolition by the state government without first properly recording all the tenants would prompt the panic-stricken landowners to evict the cultivating tenants. Occupancy rights could be granted to the sharecroppers only after such recording was completed (Chowdhury, 1985: 29).

While both Harekrishna Konar and Benoy Chowdhury converged in their insistence that the abolition of tenancy should wait, they differed in what they expected from the sharecroppers. Konar believed that the eviction of tenants could be stopped only by an organised resistance of the tenants themselves. In a legislative assembly debate in 1967 he argued: 'You cannot prevent eviction simply by making laws if the sharecroppers do not fight against their own eviction.' To this end the United Front resolved to stop the police from arresting a bargadar in her or his fight against the landlord. 'We will not allow the police to get there,' he declared, 'we shall tell the sharecropper you resist your eviction, you take a stand, if the landowner complains about injustice, I will simply say: we are not prepared to send the police' (WBLA, 1967: 531). The Left Front, on the contrary, was in no mood to allow a section to take the law into its own hands. It made new provisions for the existing laws and started recording the sharecroppers on an unprecedented scale to make eviction more difficult and questionable in court. This marked a significant departure for the CPI(M) from its pre-Left Front mode of politics.

The bottom line is clear. The CPI(M) in its central committee resolution (CPI[M], 1973) demanded a radical abolition of sharecropping. The CPI(M) in the United Front maintained that share tenancy could not be abolished without changing the 'present social structure'. The CPI(M) in the Left Front promised serious consideration for the land to the tiller scheme once all the sharecroppers were recorded by the state officials. It regarded recording as a vital security for the sharecroppers against the constant threat of eviction by the landlord, while the United Front had placed more emphasis on the initiative of the sharecroppers themselves. This, in a nutshell, describes the shifting position of the CPI(M) during 1960s, 1970s and 1980s vis-à-vis the sharecroppers in West Bengal.

It was in its attitude to the class of agricultural labourer that the CPI(M)'s change of emphasis (or metamorphosis) becomes most glaring. The party maintained that its peasant organisation should aspire to become a representative organisation for the agricultural labourer and the poor peasant. Yet it acknowledged in the early 1980s its failure to make these sections the 'driving force for the entire peasant movement' (CPI[M], 1981–82: 16). Here, a better performance was necessary not only for ideological reasons but also from a pragmatic point of view. The Krishak Sabha, as we mentioned before, was convinced that a larger and stronger organisation could be obtained only if it cared to 'spread struggle and arouse awareness of the agricultural labourer and the poor peasant' (WBPKS, 1982: 120).

As a vindication of this principled emphasis on the poorer section of the rural society the Pradeshik Krishak Sabha claimed that about 75 per cent of its primary members were agricultural labourers (WBPKS, 1986: 62). The reality, however, was quite different. Between 1982 and 1989 the Sabha held three state conferences in which, of all those attending, less than 9 per cent were agricultural labourers and less than 24 per cent, poor peasants.[11] A similar pattern was discerned (at the all-India level) as well in the 26th Conference (1989) of the AIKS (1989: 105).[12] An overwhelming majority of participants in these meets were the middle peasants and the non-peasant middle classes. It is not surprising, therefore, that the middle classes had the dominant voice in the deliberations on rural issues. In 1981, a separate organisation exclusively for agricultural labourers was formed within the AIKS, whose membership touched 1.5 million by 1989 (AIKS, 1989: 65). In West Bengal, however, the CPI(M) forged no separate organisation for the agricultural labourer. Nevertheless, talk of such an organisation 'to protect and safeguard the interests of agricultural workers' (AIKS, 1986: 86) and to draw them 'into the movement and wean them away from the influence of the bourgeois–landlord parties' (AIKS, 1982a: 30) filled the air. Paradoxically, and this is important, the real (though seldom articulated) objective was to protect the interests of the rural middle classes, and not of the agricultural labourer. In one of the conferences of the Pradeshik Krishak Sabha it was held that the lack of a suitable organisation for agricultural labourers made it difficult for the middle class to express its opinion without inhibition (WBPKS, 1986: 31). A similar concern to preserve the domain of middle-class influence was expressed in an AIKS conference which discussed the issue at length (AIKS, 1982b: 22). Middle-class domination in the Krishak Sabha was, therefore, programmatically indefensible but pragmatically fortified.

4

So far I have confined myself to the CPI(M)'s rural politics at the macro level, taking West Bengal as a whole, to show that the party's continuing success in electoral politics is partly a result of its ability to modify its policies to suit the demands of parliamentary democracy. I have also pointed out that the resultant shift of the CPI(M), from being a party of revolutionary transformation to a party of consensus politics, virtually led to an implicit refutation of its theoretical orthodoxy. Now I will look into the CPI(M)'s politics in a particular village to study its electoral tactics and forms of negotiation at the micro level.

Amrakuchi[13] was a Congress bastion in the pre–Left Front days, where the CPI(M)'s entry in 1978 coincided with the on-going preparation for the first panchayat poll. In nearby Chaita and Kapashtikri villages the party had had some influence but in Amrakuchi 'nobody dared even to speak for the CPI(M) let alone work for it.[14] However, a group of agricultural labourers and poor peasants in Amrakuchi dared to establish contacts with the CPI(M) in the late 1970s.[15] The poor in the village were not the only people to invite the CPI(M). On the eve of the election, the village jotdars split into two camps. One of the sections saw benefits in receiving the CPI(M)'s blessings to defeat its local adversaries who clustered around Dhiren Roy, a prosperous man in his forties and the leader of the Congress.

The party made good use of the split within the village rich and Dhiren Roy's falling public standing. Mishkin Khan,[16] the CPI(M)'s man in the village, requested Kalipada Mallick and Sudhir Ghosh, two well-to-do peasant proprietors who had recently broken with Dhiren Roy, to run as CPI(M) candidates. They refused but Manmatha Sen, one of Dhiren Roy's principal adversaries in Congress, accepted the offer. The CPI(M), in turn, was only too happy to get Manmatha as he could drive a wedge into the Congress support. Ajit Rana, a schoolteacher, was nominated as the second candidate of the party.[17]

As in the selection of candidates, the CPI(M) played the personality card in its campaign as well. Thus Sudhir Ghosh was discredited for frequenting a mistress from the Lyak jati whom he allegedly forced to undergo a tubectomy. The party received the necessary support from Sudhir's brothers who, being part of his family, had suffered public humiliation and had once beaten him up. Skilfully manipulating this sentiment the CPI(M) inflicted a defeat on Sudhir. After the election, when the dust settled, it was found that Amrakuchi had sent Dhiren Roy of the Congress and Ajit Rana of the CPI(M) to the gram panchayat. Dhiren Roy's personal equation with the local people, the CPI(M) admitted, made his victory inevitable despite his 'corrupt and scandalous past'. Dhiren, however, committed suicide by consuming poison not long after the election was over.[18]

Three things emerge out of this description. The CPI(M) had no qualm in allying with a section of the jotdars in the village to weaken the Congress. Its campaign was also based more on personal equations in the village than larger political or ideological considerations. As a result, the party was closely identified with its local functionaries. What surprised me was the unabashed nomination of a hated jotdar like Manmatha Sen. Manmatha's defeat was perhaps an indication that at the local level, the personal reputation of a candidate was more important than his or her political lineage. However, a qualification must be offered here. Dhiren Roy won despite his bad reputation because of his influence over the local population which had been cultivated through several political and extra-political means. That he stayed with the Congress and did not swap his position as Manmatha chose to do might also have added to his credibility.

The electoral considerations of the CPI(M) thus cut across class and political divides in Amrakuchi and provided a degree of manoeuvrability to the colossal organisational hierarchy marking its characteristic style of operation. This style became so crucial to the party that some of its actions got progressively ritualised in the process.

Take, for instance, the strike actions on the part of the agricultural labourers as an illustration of this ritualisation. Almost every wage negotiation between the agricultural workers and the landowners in Amrakuchi was routinely preceded by a strike. The labourers usually stuck to a rate supplied by the party; the party regarded such strikes as manifestations of class struggle against the landowners; the landowners usually agreed at the end of the negotiation to raise the wage up to a rate lower than the official rate with the full consent of the party; the workers withdrew their strike in response and got back to work with a sense of gratitude to the party. Such a sequence usually led to the mutual satisfaction of the contending sections of the village.

The meaning of such strikes may be explained in several ways (some of them mutually exclusive). It is suggested that such action was a cover to portray an image of militancy which the CPI(M) had long given up. The CPI(M)'s own claim was that negotiations not accompanied by strikes would lose sting and give the propertied classes an advantage. It was also held that such activities were kept alive as a mode of protest for the future when the CPI(M) might no more have its present political supremacy. Also, the Jharkhandis, who attempted to ignite the political passion of the Scheduled Tribes (constituting the bulk of the agricultural labourers), had to be prevented from eroding the base of the Marxists in the neighbouring areas and such strikes provided a ready mode of agitational politics.[19] Whatever be the explanation, the strikes had little relevance to the actual act of negotiation and they were increasingly seen as a ritual by the khetmajurs. The question remains as to whether this implies the systematic emasculation of an action which had its origin in the radical politics of the Left.

5

Having discussed the rise of the middle section of the peasantry in the Krishak Sabha and the CPI(M)'s discourse on peasant unity, here I take up another important aspect of the regime's rural intervention. The CPI(M) excelled in what may be regarded as the 'politics of middleness', a consensus-evoking unifying politics of mediation between several sectional interests. In this section, I briefly describe various modalities of such politics which, appended to the redistributive policies of the Left Front, assisted the party in widening its electoral base in rural Bengal. I also discuss some of its associated problems.

First of all, given that the dominant middle and rich peasant and non-peasant groups in the Krishak Sabha proposed a separate khetmajur organisation to purge the Sabha of the agricultural workers and retain their entrenched grip, the CPI(M) was identified less and less with the underprivileged. It was of little surprise that the agricultural workers felt marginalised within the Sabha. When the leadership of the CPI(M) proposed to uphold peasant unity for electoral reasons, it hardly paid attention to the fact that such unity might have an adverse effect on the agricultural labourers' struggle for land. There were clear tendencies to accommodate a plurality of demands within the structures of inequality and the rural poor welcomed such moves as long as they were perceived as routes away from the domination of the local landlords. No doubt the Left regime had put an end to the unbridled 'tyranny and oppression' perpetrated in the villages by the local landlords and the police in the early 1970s (GOWB, 1973: 37–8). But beyond that the doctrine of peasant unity had little to offer, especially from the point of view of the party's declared strategy of 'class struggle'.

Second, the rise of 'middleness' in CPI(M) politics was accompanied by the consolidation of a powerful literate section in rural society (see also Ruud, this volume). This section, attached mainly to the primary (and also secondary) schools in the villages and small towns (in the Keshpur local committee area of Medinipur, for instance, more than 60 per cent of all teachers were members of the CPI[M]'s primary teachers' organisation; in some areas the rate was as high as 92.5 per cent [see CPI(M), 1984]), occupied a distinctive place in rural politics. During my visits to villages in Medinipur, Bardhaman and Birbhum districts I found that many people regarded village teachers as the only people with a detached attitude to village conflicts and with the ability to reach solutions acceptable to all. As the principal source of income for the teachers was not agricultural, their tangible interests were perceived as external to those of other peasant classes. Their formal education in a poorly literate society, familiarity with the local complexities as well as with the 'higher' and distant institutions and agencies, ability to act as the most articulate link between the peasants and the party and as interpreters of legal niceties so important for running the gram panchayat were determinate sources of their enormous power and prestige. The schoolteachers almost always placed themselves in various committees, established personal contacts and developed interests not always consistent with the declared goals of the party.[20]

Third, the CPI(M)'s mediatory role was also evident in bringing the bureaucracy close to the village in a manner that no previous regime could accomplish. The pre-Left Front official initiative in the form of a revised settlement of land records brought in its wake frantic evictions of bargadars while the land administration stood as its mute witness from a distance.[21] The Estate Acquisition Act of 1953 and Land Reform Acts of 1956 could only have a success rate of 9.5 per cent as the landlords concealed their holdings in the face of an apathetic, and at times collusive, bureaucracy.[22] The way in which officials in the state government were activated during the Left Front was a case in contrast. The panchayat as an institution was attached to bureaucracy at every stage of policy implementation to 'overcome the inbuilt inertia of the administrative machinery' (Bandyopadhyay, 1980: 7). This made bureaucracy more responsive and accountable. Moreover, certain bureaucratic functions such as the recording of sharecroppers were physically moved from the towns to the villages. So-called 'reorientation camps' were set up in public places in the villages where the Kanungos and Junior Land Reform Officers took special initiative to record all bargadars. These camps were a contrast to the traditional methods of recording at the landlords' premises where the sharecroppers never came without fear, if they came at all. The Left Front held about 8,000 such camps to record 675,000 bargadars between October 1978 and June 1982 (Ghosh, 1986: 88).

Fourth, it is possible to view the gram panchayat as a mediating mechanism for implementing the rural policies of the Left Front, an organ strategically placed to encapsulate and adjudicate a multitude of tendencies. It stood between the bureaucracy and the peasant as a body of administration, and as a dispenser of resources it played a very crucial role in the village economy. Politically, it provided the most proximate institution of democratic representation, and as a rostrum for popular participation it created space for groupings of various hues. In addition, it helped to give the village a common identity by formally unifying the entire village into a single legally identifiable locality and diminishing the divisions between smaller locales based on castes.[23] As an arbitrator between parties in dispute it tended to involve the entire population and helped to make adjudication quicker, cheaper and more transparent.

Fifth, panchayats marked a crucial departure from previous forms of governmental practices. They initiated the rural population in knowledge about the basics of governmental functioning through their activities at the local level. Village politics, due to the familiarity of the issues and the actors involved, always carried ample visibility; the panchayat aligned it to the procedures of planning and disbursement along the principles of rational administration (see Williams, this volume). As a result, local governance, demystified and intimately understood, turned the issues of right and entitlement, bureaucracy and law, corruption and malpractice, from obscure and uncontrollable areas of 'high' politics into subjects of candid discourse and popular scrutiny. This not only made the rural population politically more attentive, it also incited them to be politically more significant as genuinely conscious, self-respecting and dignified actors, who refused to be benign objects of cynical manipulation. The population attained a voice, and also an urge to articulate its various complaints and requirements, making the intermediary role of the village panchayat more efficacious.

Sixth, entities such as 'government for relief' and the 'bourgeois state', which were counterposed in the debate on section 112 of the CPI(M) programme, assumed different connotations in a changed context. On the one hand, an increasingly articulate population identified the panchayat as practically the most relevant and decentralised form of governance at the village level. On the other hand, the CPI(M)'s rhetoric of government as an 'instrument' against the (bourgeois) state reinforced the opposition between the two. 'While the state in New Delhi collects substantial revenue from the people of West Bengal, very little is reinvested in the state' was the standard Marxist argument in the period. West Bengal was discriminated against and exploited, yet the government was helpless because the state in New Delhi was constitutionally empowered to destroy it if it was considered as a threat to law and order or to the country's territorial integrity and sovereignty. Therefore, while the state was shown as coercive, the state-government in West Bengal presented itself as more dispersed and decentralised, concerned with the management of the population and, ultimately, its welfare. While the state made and imposed laws, the state-government used laws as tactics. While the state penalised when its laws were violated, the state-government through strategic mobilisations made violation difficult or unpopular.[24] State-government functioned through wisdom, state through force.[25] Such perceptions helped to create an image of the LFG as a genuinely struggling coalition which needed support in the interest of West Bengal as a whole. Indeed, there has been no marked cessation of support for the LFG in the 1980s and 1990s despite the decelerating redistributive reforms and its failure to satisfactorily make public services (e.g., health and education) available in the rural areas.[26]

Irrespective of its claim to the contrary, the CPI(M) has brought about a gradual deradicalisation in the character of redistributive reforms in rural West Bengal over the 1980s. During the same period the state witnessed an unprecedented growth in agricultural output (see Gazdar and Sengupta, this volume). The CPI(M)'s policies of making administration more accountable and decentralised, diluting the concentration of land ownership, providing tenurial security and creating opportunities for peaceful negotiation for the rural poor along with more transparent and participatory modes of distributing scarce resources and benefits (such as credit, production inputs, employment and various subsidies) created the conditions for both agricultural growth and the party's sustained electibility. This, however, was attained at some cost. The party had to shear off some of its more extreme political goals and tune itself to the principal requirements of the 'bourgeois' democracy. A clear manifestation of this was seen in the CPI(M)'s intra-party debates on issues like parliamentary democracy, peasant unity and land reforms and in its changing attitude to landlords, tenants and agricultural labourers. In addition to such shifts in orientation, the organisational penetration of the CPI(M) was informed by its ability to manoeuvre local issues pragmatically. The party was able to accommodate various versions of politics which otherwise fell outside the ambit of its ideological proclivity. It drew strength from the existing balance of political forces by conducting what has been described as the 'politics of middleness'. Despite economic growth and electoral vindication, the rural reform policies of the CPI(M) in West Bengal continue to suffer from a central dilemma, which may no longer bother too many sections within the party: how far can they contribute to the Left's original project of revolutionising the dispossessed?

NOTES

1. Since 1977, while other Indian states have been changing governments every successive election, West Bengal showed no sign of switching loyalty from the CPI(M)-led Left Front government. In fact no comparable instance can perhaps be found anywhere in the world where a Marxist regime played such a long and uninterrupted innings in an otherwise bourgeois terrain marked by a competitive party system and legally sanctified private ownership of the means of production.
2. Panchayats in West Bengal are local institutions democratically elected to coordinate development requirements at the village, block and district levels.
3. For an account of the economic and political compulsions behind the formation of the United Front in Left politics see Bhaduri (1993).

4. Dasgupta (1989: 23). It is interesting how the party almost reconciled itself to joining the union cabinet after the defeat of the Narasimha Rao government in April 1996 and retreated only after prolonged sessions of a divided central committee.
5. West Bengal Pradeshik Krishak Sabha (WBPKS) (1986: 14) says: 'It is impossible to advance the peasant struggle without the support of all sections of the peasantry including the rich peasants and the democratic people in the countryside.'
6. The Cooch Behar session of the Pradeshik Krishak Sabha held in 1986 defined 'dialectical unity' as encapsulating the needs 'to hold tight the middle peasants' and attract the rich peasants 'by transforming their demands into movements' and to oppose any compromise on 'the struggle for wages in order to advance peasant unity.'
7. This is contrary to Kohli's observation that '[i]t indeed makes a funny type of communism which treats pre-commercial, "feudal" landlords as enemies but not those involved in the capitalist mode of production.' See Kohli (1987: 100).
8. CPI(M) 1973. For functional reasons, like that of determining ceiling limits for the landlords, the CPI(M) evolved a contingent definition of the class in terms of landholding. If in an area having similar agro-economic characteristics 90 per cent of the holders of a particular size of holding did not physically cultivate their land, for the purpose of land ceiling legislations they were assumed to be landlords.
9. Struggle for people's democracy or 'people's democratic revolution' was conceived by the CPI(M) as an intermediate revolutionary phase between the existing 'big bourgeois-landlord state' apparatus and the socialist state of the future, to be accomplished by a contextually progressive coalition of the working class, peasantry, petty bourgeois intelligentsia and national bourgeoisie under the predominant leadership of the working class. It differed as a scheme from the CPI(M)'s 'national democratic revolution' which acknowledged the 'composite leadership of the working class and the national bourgeoisie'.
10. Demesne lands which zamindars retained for direct cultivation by labourers.
11. The percentage participation of agricultural labourers and poor peasants was 8.48 and 21.46; 6.58 and 16.12; and 7.46 and 23.38 respectively in the 26th (1982), 27th (1986) and 28th (1989) sessions of the Krishak Sabha. See WBPKS (1982: 6; 1986: 5; 1989a: 4).
12. Here the proportion of agricultural labourer participation was merely 4.85 per cent and that of the poor peasants 24.28 per cent.
13. I did my fieldwork in Amrakuchi, a village off the Chandrakona–Medinipur bus route in the southern part of Keshpur police station in the Medinipur district of West Bengal, between December 1990 and March 1991. The information gathered and presented here has mainly been collected from interviews conducted in the village and surrounding areas.
14. Interview with Panchu Hemaron, a khetmajur (agricultural labourer). 'We went to them and said, "Babu, the Congress is giving us a very tough time, we want to join your party. Please look after us."'
15. Interview with Santosh Fauzdar.
16. Mishkin was in close touch with district CPI(M) leaders like Tarun Roy and Jamshed Ali. The district leaders reportedly frequented the village regularly around this period. Mishkin hailed from the nearby Charka village within the same gram panchayat area and was elected as the president of the panchayat samity after the 1978 election.
17. In Ajit Rana's words: '. . . they pointed at me and said, "this chap would make a good candidate, he has a clean image, he is educated and unemployed, the people would accept him readily." So was I made a candidate for the CPI(M).'

18. The cause of the suicide is unclear and interpretations vary. The most circulated story, again not necessarily true, tells that Dhiren, at 45, drew himself into an adulterous affair with Kajal, the 16-year-old daughter of his friend Gopal Ghosh. Earlier, Gopal had sold about 4 bighas of his land for Rs 8,000 to get Kajal married. While the negotitions for the marriage were on, Dhiren advised Gopal to lend him the money instead of keeping it in a bank and promised to return it once the marriage was settled. Dhiren used the money to set up a poultry farm which failed. Meanwhile, Gopal fell ill and had to be admitted to a Calcutta hospital where Dhiren's assistance was crucial because of his connections in the city. This was the time when Dhiren frequented Gopal's house and developed an intimacy with Kajal. Her parents allegedly turned a blind eye to the affair at its initial stage although it was a source of humiliation and pain for Dhiren's wife who vehemently opposed her husband's adventure. Finally, when Gopal broke with Dhiren and joined Kalipada Mallick, a plan was hatched to catch Dhiren unawares and force him to wed Kajal. Dhiren somehow got the news beforehand and committed suicide to escape the ordeal.
19. The Krishak Sabha explained the expansion of the Jharkhandis (especially in Jhargram *mahakuma*) by its own failure to attract the khetmajurs and commended the 'comrades of Binpur and Belpahari' for successfully bringing the khetmajurs back to 'the mainstream of the peasant movement'. See AIKS (1990: 20).
20. During my recent (May 1997) field visits to Purulia I observed resentment among some members of the CPI(M) about the schoolteachers who were powerful in the party. The pay packets of schoolteachers are incredibly high for the rural setting, it is felt, which undermines the party's claim to be the representative of the poor. If both spouses of a family are teachers, which is not uncommon, the combined income is often higher than that of a local trader. Moreover, with competition to get into teaching running high, only those with better academic qualifications are recruited. This often makes things difficult for the poorer aspirants and helps the relatively well-off candidates with better access to education.
21. For a graphic picture of helpless administration during the Congress days see GoWB (1977).
22. For a summary of the ways of subversion of the ceiling laws see Bhattacharyya (1994: 62–5).
23. In rural West Bengal caste distinctions are present but they seldom generate political divisions or categories for political mobilisation (though see Ruud, this volume). 'There is a substantive basis of a "caste question" in West Bengal . . . . Yet there is no "caste question" in a politically significant way . . . . The panchayat elections have formally given the middle class access to power in the sprawling countryside and the "caste-in-limbo" situation persists today' (Samaddar, 1994: 56–9). 'Caste has become more of a psychological phenomenon [in West Bengal] . . . here caste relations cannot be interpreted directly in terms of power relations similar to the pattern of many states in India' (Halder, 1994: 71). '[T]he absence of caste articulation of political demands does not mean that caste authority and caste linkages have not proved useful to various political identities as instruments of gathering electoral support in the relatively unmobilized areas. But the considerable fragmentation among the middle castes, and the overall dominance of modes of culture and thought of the urban intelligentsia, have prevented any successful aggregation of caste interests in the state election scene' (Chatterjee, 1997: 82).
24. For an empirical account of the Left Front's tactical use of agrarian land laws see Bhattacharyya (1994).
25. Concerns with the diverging principles of state and government and 'govern-

mentalisation of state' can be found in Foucault's inventory of the 'problematic' of government (1991).
26. See Gazdar and Sengupta (this volume) regarding West Bengal's differentiated performance in health and education along lines of caste and gender. We do not intend here to get into the issues related to public provisions where West Bengal's records are rather poor. An unsatisfactory performance on this front may actually not be politically damaging because electoral choices are not necessarily made on rational administrative considerations.

REFERENCES

All India Kisan Sabha (AIKS). 1982a. *Proceedings and Resolutions*, 24th Conference, Medinipur, 8–11 November 1982.
———. 1982b. M. Basavapunnaiah, 'Land to the Tiller,' *Souvenir*, 24th Conference, Medinipur, 8–11 November 1982.
———. 1986. *Towards a Country-Wide Peasant Struggle: Proceedings, Resolutions, General Secretary's Report and Statement of Policy* (approved by the Thane Meeting of the Central Kisan Committee), 26–28 September 1986.
———. 1989. *Report of the General Secretary*, 26th Conference, Khammam, Andhra Pradesh, 27–30 April 1989.
———. 1990. *Statements*, 29th Session of the Medinipur District, 2–4 November 1990.
Asian Age. 1997. Interview with Jyoti Basu, 1 January 1997: 12, New Delhi.
Bandyopadhyay, D. 1980. *Land Reforms in West Bengal*. Calcutta: Government of West Bengal.
Bhaduri, Amit. 1993. 'The Economics and Politics of Social Democracy,' Pranab Bardhan (ed.) *Development and Change: Essays in Honour of K.N. Raj*. Bombay: Oxford University Press.
Bhattacharya, Dwaipayan. 1993. 'Agrarian Reforms and Politics of the Left in West Bengal.' Ph.D thesis, University of Cambridge.
———. 1994. 'Limits to Legal Radicalism: Land Reforms and the Left Front in West Bengal,' *The Calcutta Historical Journal*, 16(1): 57–100.
Chatterjee, Partha. 1997. *The Present History of West Bengal*. New Delhi: Oxford University Press.
Chowdhury, Benoy. 1985. 'Swadhinata Uttor Yuger Krishak Samaj-e Poribortoner Swarup,' *Marxbadi Poth*, 5(1): 27–31.
Communist Party of India (Marxist). 1973. *Central Committee Resolution on Certain Agrarian Issues and an Explanatory Note by P. Sundarayya, Calcutta*.
———. 1981–82. *Bibhinno Front-er Report*, 14th West Bengal Conference, Calcutta, 27 December 1981 to 3 January 1982.
———. 1984. *The Political Organisation Report*, Keshpur Local Committee No. 1, 2nd Session, 17–18 June 1984.
———. 1989. *Programme* (Adopted at the 7th Congress of the CPI, Calcutta, 31 October to 7 November 1964, with Amendments by the 9th Congress in Madurai, 27 June to 2 July 1972). New Delhi: CPI(M) Publication.
Dasgupta, Sudhanshu. 1989. *Proshongo Jonogonotantrik Front, Bam o Gonotantric Front*. Calcutta: Nishan Prakashani.
Eco, Umberto. 1995. 'Political Language: The Use and Abuse of Rhetoric,' Robert Lumley (ed.) *Apocalypse Postponed*. London: Fleming.

Foucault, Michel. 1991. 'Governmentality,' Graham Burchell, Colin Gordon and Peter Miller (eds) *The Foucault Effect*. Hertfordshire: Harvester Wheatsheaf.
Frankel, Francine R. 1971. *India's Green Revolution: Economic Gains and Political Costs*. Princeton, New Jersey: Princeton University Press.
Ghosh, T.K. 1986. *Operation Barga and Land Reforms*. Delhi: B.R. Publishing Corporation.
Government of West Bengal (GOWB). 1973. 'Report of the Mukherjee–Maitra Commission.' Mimeograph.
———. 1977. 'Kanungos of Midnapore District: Case Studies of Bargadars.' (For official use only.) Mimeograph, January 1977.
Halder, Srijnan. 1994. 'Caste-Class Situation in Rural West Bengal,' K.L. Sharma (ed.) *Caste and Class in India*. New Delhi: Rawat Publications.
Herring, Ronald J. 1983. *Land to the Tiller: The Political Economy of Agrarian Reform in South Asia*. New Delhi: Oxford University Press.
Kohli, Atul. 1987. *The State and Poverty in India: The Politics of Reform*. Cambridge: Cambridge University Press.
Konar, Harekrishna. 1979. 'The Problems of Land Reform and its Solution,' *Agrarian Problems of India*. Calcutta: Harekrishna Konar Memorial Agrarian Research Centre.
Samaddar, Ranabir. 1994. 'Caste and Power in West Bengal,' K.L. Sharma (ed.) *Caste and Class in India*. New Delhi: Rawat Publications.
West Bengal Legislative Assembly (WBLA). 1967. *Assembly Proceedings: Official Report*. Vol. 64, 44th Session, March–April 1967.
———. 1977. *Assembly Proceedings: Official Report*. 66(1), 25–31 August 1977.
———. 1989a. *Assembly Proceedings* (manuscripts). File no. 92, 27 March 1989.
———. 1989b. *Assembly Proceedings* (manuscripts). File no. 76, 3 April 1989.
West Bengal Pradeshik Krishak Sabha (WBPKS). 1979. *Shompadakiya Report o Prastab*. 25th Session, Bankura, 24–27 February 1979.
———. 1982. *Shompadakiya Report o Prastab*. Pandua, Hooghly, 1–4 October 1982.
———. 1984. *Shompadakiya Protibedon*. 6th Session, Bolepur Krishak Sabha, Bolepur High School, 2–3 June 1984.
———. 1986. *Report o Prastab*, 27th Session, Cooch Behar Town, 19–22 February 1986.
———. 1989a. *Sonkhipto Report o Grihito Prostababoli*, 28th Session, Udainarainpur, Howrah, 10–12 March 1989.
———. 1989b. 'Krishok Andolon o Gonotantrik Jonomot,' *Deshhitoishi*, Autumn Number: 89–93 (by Benoy Kumar).

PART 3

Changing Agrarian Structures

11

From Farms to Services: Agricultural Reformation in Bangladesh

GEOFFREY D. WOOD

AGRARIAN DISCOURSES IN BANGLADESH: AGENCY AND STRUCTURE

In one sense there is a relatively simple story to tell about agrarian change in Bangladesh. I have told it in various forms before, but the need to restate it remains. However, the purpose of this chapter is rather different, and much is therefore assumed about the basic trends. In Bangladesh, are agriculture and agrarian change being fundamentally understood? Are the concepts through which agriculture is familiarly analysed relevant and appropriate? Am I the only one who experiences a strong sense of unreality when I witness 'experts' discussing agriculture and making plans for it? The countryside seems to be discussed in terms which function to uphold or create a fictional fantasy of rural behaviour: prescription posing as description. The purpose of this chapter is therefore to question the language (concepts, categories, presumptions) through which the agricultural system in Bangladesh is represented, and to suggest an alternative set of concepts which might better reflect behaviour and its rationale.

Such a purpose instantly brings us into the familiar 'agency and structure' problematic. The main contrasting ways in which agriculture in Bangladesh has been represented both derive from 'structure' positions: the colonial legacy of coexisting modes of production producing janus-faced, hybrid farmers in transition from quasi-feudalism; and the neo-liberal view of rural behaviour as increasingly conforming to market stereotypes. Neither gives much credit to agency in the sense of rural actors creating reformed agendas as a result of managing the various constraints and opportunities offered to them in the globalisation process.

Of course, others have noticed the paucity of observers' models of farming systems, and have placed emphasis upon the significance of indigenous knowledge and its relationships with formal science as represented by experts, research centres and bureaucracies. Such approaches have spawned much participatory rhetoric as embodied in the recent religion of Participatory Rural Appraisal. Some of this contemporary discourse appears as old wine in new bottles as anthropology gets recruited to development, or perhaps a belated acknowledgement on the part of new professionals that they have been guilty of plagiarising the knowledge of their informants in developing their conceptions of the rural development problem. Connected to this is the empowerment theme, which attempts to remove the necessity for people's knowledge to be brokered by specialist intermediaries to those with power through the development of independent voice. The jury is still out on this process with mobilising agencies remaining as patron intermediaries.

Many of these themes have been brought together in 'actor-oriented' epistemology, giving more explanatory significance to actors' views of their universe but also recognising that they operate within constraints to their room for manoeuvre established by the history of their political economy, as well as by culture and resource endowments. In such a philosophy, people are more or less constrained depending on such obvious variables as gender and class, but also by the breadth of their livelihoods options portfolio as well as the particular complexity of their culture. (The Maithili and Bhojpuri cultures of north Bihar, for example, represent a contrasting set of actors' codes enabling more fictional behaviour in the former context.) Crucially for the arguments presented here, this epistemology nevertheless tries to proceed from the question 'what do people do?', rather than the question 'what constrains people?'. Furthermore, it replaces the prospect of forms of analysis via comparison with some implicit model of actors' behaviour, with the idea that people are thwarted from pursuing their real interests by external forces. The very notion of 'constraint' implies a priori knowledge of the behaviour which is being altered by these constraints. This approach presents a positive image 'x' of the farmer, and an analysis of the agricultural system which takes the form of asking why the farmer cannot behave like 'x' but is forced to behave like degraded 'y', and this constitutes the agrarian problem. By contrast, the form of analysis favoured here makes fewer assumptions about what the farmer ought to be and sees behaviour in a more open-ended way as a calculative search for opportunities, while recognising the problems that such actors face in managing portfolios successfully and in gaining quality information as a basis for such calculations. In this way, the teleological problem in the analysis of capitalist intrusion is directly avoided, while recognising such intrusion as a context for behaviour alongside other contexts.

While acknowledging the significance of these themes to an understanding of the process of agrarian change, the question remains open of how to select the actors for the purposes of building theory, intervention with policy, or mobilisation. But if we start from the epistemological premise of enquiring what actors do, then we have to revise our assumptions about who the significant actors are within the reformation of agriculture in Bangladesh. And we certainly have to question the assumption that farming families represent the key signifier of agrarian change, or, of course, that there is a single archetypical model of the family farm whose behavioural change can be traced.

LIMITATIONS OF THE FARM AS A METAPHOR FOR AGRICULTURE: AGRICULTURAL REFORMATION

The title of the paper addresses this issue directly. To understand agriculture almost anywhere, but certainly now in Bangladesh, we have to acknowledge many sets of actors besides the farming family as contributors to the reproduction and evolution of an agricultural system. These actors pursue livelihoods and operate within the triangle of market, state and community—not just as separate formal arenas, but as arenas with discrete rules and entailed cultures. The farm as a central organising concept can no longer (if ever) bear the load of this explanation. Wherever there is an intensification of land use deploying selected and adapted items of new technology, it is accompanied by a proliferation of formal and informal actors (e.g., in Bangladesh, as well as in Indonesia, India, parts of Pakistan, Philippines and other parts of South-east Asia). It is familiar to argue that we are observing the disarticulation and re-articulation of agricultural systems via the connection of production decisions at the farm level to wider market forces. If there ever was a Chayanovian primacy of the internal logic of production behaviour (dominated by consumption needs and the household dependency ratio) this has been disarticulated by the necessity to include other actors' values and prices in one's calculations, as exogenous inputs of various kinds are required. For most landholders in Bangladesh, their behaviour only makes sense within a wider framework of transactions—i.e., re-articulated via exchanges of produce for inputs.

So much is familiar for many societies, but a further logic applies to eco-social conditions like those experienced in Bangladesh with constrained livelihood opportunities outside agriculture (though not outside the farm) and strong socio-cultural attachment to the principle of patrilineal, multiple inheritance. Such conditions are accompanied by continuing population growth, increasing overall population density which is not disproportionately absorbed into urban growth.

The outcome of this process is to increase rural population density, entailing the increasing fragmentation of landholdings as cultivable land is divided among inheriting sons, and plots themselves are divided. Thus, scattered, smaller plots are a feature of smaller scale farms overall. Intensification of land use is therefore particularly accompanied by declining farm size and greater fragmentation into ever smaller plots. This process is reinforced by the piecemeal release of land to others as mortgaged collateral for debt. There are offsetting trends to this simplified version of the fragmentation process via forms of tenancy (discussed in the section on Disarticulation Avoidance).

One of the implications of this combination of eco-social conditions is that the concept of the farm, as an integral unit of production merely adjusting to new circumstances, is increasingly vulnerable. The concept of the farm assumes that the decision makers within it have some control and choice over the way they organise the combination of land use, labour inputs and scarce capital. While the notion of autonomy was always heavily circumscribed, in the past the family leaders had some prospect of making such decisions on the basis of endogenous considerations within the family (managing a portfolio of consumption, debt repayment, dowry savings) with a greater reliance upon the unvalorised labour of family members. In this context, the concept of the farm was uncontroversially valid. Such family leaders, of course, continue to pursue livelihood strategies but now they try to manage portfolios which involve more exogenous variables and other actors with greater significance. A crucial qualitative change has occurred in that a family's land is now dispersed into ever smaller units *while simultaneously* requiring an intensification of investment, management and new technology all of which entail the strong presence of other actors in control of capital, variable biological and chemical inputs, aspects of the new mechanised technologies especially in irrigation and ploughing, and even labour power (as gang *sardars*). This involves a shift of power away from the holder of land per se towards these other asset holders and service providers, who may incidentally also hold land, but this is not of structural significance to the argument (except as the basis of their entry into these other asset or service provider roles; see Wood, 1973).

The agricultural system therefore consists of an interlaced network of transactions in which cropping patterns and labour operations on the land increasingly reflect the disposition of power among such agents, rather than a single category of dominant agents labelled as farmer. In this way, the agricultural system is reformed, but less by policy-led intervention and more by the combination of circumstances noted in the preceding discussion. And in this process, the farm may no longer be an adequate metaphor for the agricultural system in terms of policy intervention either to increase land and labour productivity, or to redistribute the gains from agricultural growth more evenly throughout the population.

LIMITATIONS OF THE HOUSEHOLD MODEL

The main objective of this paper is to sustain an argument in terms of the dynamic combination of forces in the wider agrarian society external to the family. However, it is also now recognised that the household itself is increasingly problematic as a unit of social action. The presumption of the farm as an indivisible social entity within which production and consumption decisions are made is no longer tenable. There is a combination of problems relating to gender, population pressure, intra-familial trust, family division, remittances and diverse income portfolios. All these limit the relevance of a patriarchal, patrimonial image of institutionalised, age-related male power, i.e., the farmer, behaving altruistically on behalf, at least, of the agnatic line within the family.

Nowhere are the epistemological problems of understanding actors' behaviour more acute than within the South Asian 'peasant' family. Nowhere is theory stronger in the interpretation of facts. How are real interests to be attributed? What is robust evidence for consciousness (individual, gender, position in the demographic cycle)? Can we assume gender and/or age symmetry in the subordination of individual interests to the collective household good? What is the value of the inter-generational deal between labour and inheritance in the context of dynamic shifts in prices, debt, cost structures, sustainability of fragile portfolios, and expectations about off-farm, possibly urban, employment? Does rationality lie within the individual or the collective kin unit? Can family leaders be trusted to operate on utilitarian principles? What principles operate in family calculations about crop choice, land disposition (operated, leased out, mortgaged out), and allocation of family labour to valorised activity? These issues are compounded in the context of extended families contemplating division of assets before or after the patriarch's death.

Obviously there have been various attempts to model such intra-household relationships (new household economics, bargaining models and household/gender relations). The strength of some of the debates about the validity of such models can be acknowledged, such as how to impute value to unvalorised reproductive labour functions when no market equivalent exists to calculate opportunity costs; or, how to identify bargaining parties' fall-back positions to assess their respective bargaining strengths and potential gains or losses from different deals (Sen, 1990); and the need to distinguish between the different positions and ages of family members in any gender analysis (Folbre, 1986; Shrestha, 1994). However, none of these critiques directly confronts the problem of determining real interests, and the link to consciousness and behavioural outcomes. How far can dependent men and women have interests and express opinions about them contrary to those of the culturally defined family leader? Do different family members make investment and income-generating decisions which optimise individual or collective interests? Under what conditions can the patriarch regulate or arbitrate upon the degree of compatibility between individual and collective family interest (e.g., birth control, treatment of female infants, investment in education, or trading off land for investing in off-farm business)? According to White, the family farm in Bangladesh is the site of much more debate, strategising and reconciliation of conflicting interests than hitherto imagined (White, 1992). Her argument was gender based, but in the context of dynamic agricultural change, it is also important to understand inter-generational discussions within the family about land use, risk taking, crop choice, optimal dispositions of family labour, diversification, and rental choices (especially in the context of a proliferation of irrigation command areas across the spatial fragmentation of a family's plots). I am not aware of specific research on the discussions between adult sons and their landholding patriarchs over such issues, yet these discussions and their outcomes would appear to be crucial to any policy discourse about increasing land productivity and agricultural growth for food security or commercial objectives.

A more precise point, relevant to the overall argument in this paper, with respect to treating the household as a single actor for analytic purposes concerns the problem of how different actors within the family fundamentally conceive of it. Is it regarded in classical peasant terms as the ultimate provider of welfare and long term security, obliging individual members of all ages and gender (including the patriarch) to subsume in the larger collective interest any short term, more individual interests or opportunities? Does such a conception continue to translate (if it ever did) into risk averting behaviour in which lower yielding rainfed aman remains the key insurance crop, even if it competes with cash crop (e.g., jute or sugar) and winter irrigated boro rotation options, with the related assumption of farmer autonomy in pursuing subsistence objectives (or at least being able to retreat back into them)? Or, do the general conditions of agricultural involution and capital intrusion induce a stronger sense on the part of some or all family members to treat the disarticulation of the 'peasant subsistence farm' as a positive process to be embraced rather than resisted? In particular, is there a pattern of inter-generational difference over these issues, with younger adults more prepared to deploy land and their own labour in innovative combinations to maximise the opportunities available, i.e., conceiving land more in commodity, objectified terms rather than affectively as part of a deep-rooted psyche in which all status and self-identity resides? How we understand this matters a great deal, for example, when we consider attitudes towards reverse and seasonal tenancies as a rational land use option, or the willingness to use land to raise capital not just for desperation and/or consumption expenditure but for strategic, forward looking investment.

CLASSIFICATION

The upshot of this discussion is to remind ourselves that the household is a problematic unit of analysis when considering explanations for agricultural change and policy support to the sector. This problem is further compounded by the way in which 'farms' are classified. The agrarian discourse in Bangladesh has moved a long way from an earlier image of relative homogeneity, although the 'smallholding peasant' remains the dominant ideological stance for most competing political forces in the country, which has some implications for policy.

While this chapter is making a general argument about the need to see features of a wider agricultural system as increasingly determinant of farming behaviour for all types of farms, variations between farming families are important to the overall thesis. The most obvious point to make first is the extent of landlessness among families reliant upon rural sources of incomes and employment. Estimates and definitions vary, but we think in terms of approximately 60 per cent landlessness among rural families which themselves represent about 90 per cent of the total population. Clearly, there are definitional problems, with the category 'effectively landless' including such families as may have some land but are mainly reliant upon hiring out labour for income. The availability of rural labour from such families as well as of land under various forms of 'reverse' tenancy and debt-tied relations certainly affects the cultivation options available to other landholding families.

Within those families trying to manage land (their own and/or the plots of others) as the main source of their income, there are obvious classification possibilities in terms of size of landholding and in the portfolio of owned and rented in/out land. But such a basis of classification itself is complicated by other factors, such as size of family, proportion of family members reliant upon the joint family land as principal source of income (a family diversification index), location of land within overall cropping zones, flood and drought prone status of the land, and the irrigated proportion of the holding. Such variations are familiar enough and limit the ability of planners to predict farmers' responses to prices and incentives (i.e., subsidies and credit remissions).

To this complexity, we may add a raft of issues about the importance of network maintenance in the context of imperfect, interlocked markets and uncertain state services, themselves pervaded by similar market features (see Wood, 1994: Chapter 21). Families do need to be classified by their position within such networks, for therein lies their predictable access to key inputs and the cooperative support of others. Their fortunes and strategic options are as much determined by such factors as the more formally accessible data about landholdings.

However, the particular argument of this paper relies strongly upon classifying farms by the observable data of plot distribution among landholders (whether the land is owned or rented in on long or short term leases) since this affects their land management options, and thereby much else about their farming behaviour. This paper is not able to report detailed statistics on the spatial characteristics of plot distribution since these are yet to be collected purposively in this way on a large scale (though there are data on numbers of plots per farm, up-to-date figures were not available to me at the time of preparing this chapter). In part, the argument here represents a plea for such data to be systematically generated. Nevertheless, much information exists from micro-studies (e.g., Glaser, 1989; Lewis, 1991), from data gathered in pursuit of the landless irrigation model (Proshika, in Wood and Palmer-Jones 1991), data arising from UK Department for International Development–supported Government of Bangladesh irrigation projects, and continuing interviews by myself with farmers all over Bangladesh in the context of various consultancies. The particular argument to be drawn from these observations is that the more land a family holds (setting to one side the significance of any variation in the demographic composition of this 'family'), the more complex are its land management options deriving from the spatial spread of its separate plots. At the same time, it is important to remember that larger landholding families in Bangladesh are generally not large landholders who can afford to cultivate high proportions of their holdings extensively. In other words, despite being larger landholders, they are still obliged to seek intensive strategies for land use, which crucially involves a reliance upon irrigation for a high proportion of such plots. It is this interaction between non-consolidated landholdings and irrigation which is especially significant. This entails a certain counter-intuitive irony with larger farmers having, in certain respects, less autonomy over their production decisions and labour commitments than smaller farmers. Perhaps a way to illustrate this dramatically is to consider findings from Glaser's doctoral research (Glaser, 1989) in old Rajshahi district. There, in the late 1980s, a larger landholding family had land distributed over eighteen different irrigation command areas in the village. While this was an extreme example, plots distributed over ten separate command areas were common. In the landless irrigation programme, we have been at the 'receiving end' of similar data, with 20-acre shallow tubewell (STW) command areas each having up to approximately fifty different plot-holders who also have plots in other command areas (operated by landless groups or other ownership sources). Of course, the lower the number of plots held by a family, the fewer the transactions required to make the land work. In considering the reformation of agriculture, therefore, it is important to distinguish between categories of farming family by the correlation between the number of plots owned and the complexity of transactions required to cultivate intensively. How does such complexity function to remove many management decisions from a farming family, and thus restructure our ideas about how agriculture is to progress in Bangladesh, and about its redistributional outcomes in terms of value-added gains from increases in land productivity?

DISAPPEARANCE OF THE BANGLADESH FARM

Let us proceed via the polemical proposition that the Bangladesh farm, in its 'family farm' form, will effectively disappear as a significant actor and that the larger the farm, the more likely this is to occur. Obviously we have to qualify this assertion by defining what can be meant by 'disappear', and to recognise that there are gradations of 'disappearance'.

To 'disappear' means that the family farm will cease to be the primary decision making unit over a range of decisions on the land formally held by the family: crop rotation; price and cost responsive crop choice; scale of investment in inputs; timing and therefore cost of raising capital to fund operations; timing of labour operations (ploughing, plot preparation, initial applications of fertiliser, transplanting, later fertiliser application, weeding, spraying, even guarding and harvesting); comparative advantage utility calculations on use of family labour; wage levels of hired labour; timing of crop sales and allocation of the crop return between consumption, storage and meeting other obligations.

If such families are losing control over this range of decisions, where are such decisions being relocated and why? As asserted in the previous section, the increasing reliance upon irrigated cultivation is a crucial part of this process. Total irrigated hectares increased from 1.5 million in 1978 to over 3 million in 1990. This increase was dramatically concentrated within boro rice, rising from just under a million irrigated boro hectares to over 2 million in the same period. Production of boro more than tripled in the same period; wheat remained surprisingly constant; aus declined; and aman rose by 30 per cent (which partly reflects a doubling of the irrigated contribution to aman output). Although aman still represented the largest contribution to foodgrain output at 9.3 million metric tonnes in 1991-92, boro was reported at 6.8 million metric tonnes in the same year (with aus at 2.2 million). In the decade 1981-91, the number of shallow tubewells rose from 40,000 to 275,000 (MOA/GOB, 1993); for more details see the contributions of Palmer-Jones, Adnan, Shahabuddin to this volume).

Collectively, these figures reveal the dependence of national food security on irrigation and this is reflected at the individual farm level, with overall cropping intensity standing at 171.7 in 1990–91, with the highest figures concentrated in the areas of highest boro production (viz. Bogra at 202, Rangpur at 198, Mymensingh at 197, and Jamalpur at 196). In 1990–91, almost 50 per cent of the net cropped area was double cropped, and over 10 per cent triple cropped (GOB/BBS, 1992).

The task is to consider the organisational implications for managing land which lie behind these figures. While it may still be the case that rainfed aman rice remains the largest, safest crop for most family farms, the viability of these farms at the margin relies crucially upon the irrigated component, including of course supplementary irrigation for aman in late-starting monsoons or early winters. A revealing irrigation figure is that while the number of STWs has increased dramatically over the decade up to 1995, the number of DTWs has remained virtually constant. Various explanations can be offered for this, some of them concerning the technical and operational characteristics of the technology and appropriate hydrological conditions. However, there is a strong consensus in Bangladesh (especially at the grassroots level) that the organisational complexity of operating DTWs under Bangladesh's fragmented farm conditions is the principal explanation for both the lack of increase in and the under-utilisation of existing capacity. DTWs, with their heavier capital costs and larger command areas, involve more complex forms of cooperation among farmers with irrigated plots than alternative forms of irrigation supply (with the exception of large scale surface water systems; see, for example, Wood, 1995; Wade, 1988). Of course, there have been a variety of organisational experiments with DTWs (e.g., landless services, initiated cautiously by Proshika and then abandoned, but adopted by CARE's Lotus programme in conjunction with Bangladesh Rural Advancement Committee [BRAC] and Grameen Bank, inter alia; Grameen Bank's own staff service model in Tangail; and other 'franchise' type options) with no spectacular success. The major lesson from the main DTW cooperative model is that farmers do not easily cooperate with each other when attempting to supply themselves with a crucial service.

The expansion of STWs reflects not just cost and technological appropriateness, but also organisational flexibility. But even STWs are mainly too lumpy to be considered solely for self-provisioning. There is evidence of their gross under-utilisation under circumstances of private purchase, especially with defaulted or written-off credit. But most private owners of STWs are involved in some sale of surplus capacity, if only on a casual basis to close kin. With command areas ranging between 10 to 25 acres, fully functioning STW command areas may have up to seventy farmers with plots to be served, although numbers of twenty to thirty are more likely. There are now many studies of the economics of providing irrigation services under such conditions, including an analysis of the landless service in Wood and Palmer-Jones (1991). Such findings are not the focus of the present discussion. We also have to acknowledge the ownership and operation of STWs as a deliberate service business activity, where self-provisioning is not the primary motive. Such a deliberately entrepreneurial approach to STW ownership is relevant to the later part of the present argument and the profitability of such operations is certainly crucial to an overall understanding of irrigation-led agricultural growth. This analysis is also pursued by other contributions to this volume (see especially the papers by Adnan, Palmer-Jones and Webster).

The particular concern here is to see the prospects for sustaining or expanding irrigation opportunities from the viewpoint of the family farm operator. Even a pumpset owner will face the problem of having many plots potentially ripe for irrigation but which cannot be serviced by his own equipment, which is located in a command area of most convenience to the spatial distribution of his own plots. He requires services from others in other command areas where he has plots. However there is little prospect for a symmetry of cooperative exchange among a pool of pumpset owning farmers in which they all supply each other in the respective command areas where they have plots. Private pumpset owning is too restricted to a handful of richer actors (not necessarily farmers) to allow for such cooperative exchange.

Our main set of actors consists, therefore, of farmers in need of irrigation from STW sources (as well as from other sources as appropriate) who do not own pumpsets and borings but have to enter into agreements with other providers. Furthermore, they will typically have to enter into contracts with different sets of providers and with different, though overlapping, co-consumers with whom they will have to participate in some aspects of cooperative management (if only in the sense of agreeing to standard patterns of water distribution). They may have to enter different types of arrangements with different providers. For example, farmers receiving an irrigation service from a landless group may accept a set of conditions which varies from those set by a private provider on another of the farmer's plots. Payment systems may vary along with value, and the quality and reliability of service may be differentially insured or secured.

Within this scenario, it then seems that water is the more consolidated asset compared to landholding. Each command area requires an integrity of management with respect to the mobilisation of inputs and subsequent activity, establishing a set of operational imperatives external to the individual landholder with one or more plots within it. The initial availability and sequential timing of irrigation requires a management agreement either cooperatively among self-provisioning plot-holders, or by negotiation or even authoritative compliance with the service provider. With some exceptions, the management of individual plots will then conform to the requirements of the collectivity receiving the service. Indeed the price of non-conformity is high if it relies upon a series of one-off negotiations between individual plot-holders and the service provider. If one farmer claims a right to special treatment, then he must expect others to do likewise. Such negotiations would approximate anarchy, involving unacceptable transaction costs for all concerned. The mutual concern to reduce such transaction costs is reflected in the widespread use of 'rough equity' in the distribution of water in STW command areas. In the landless irrigation programme, we have certainly found that a strict cycle of weekly distribution (two days' flow down each of three main channels in sequence, with one day for maintenance of the equipment) may not have been the most rational in reflecting real differences of need (in terms of moisture retention characteristics of soil) but it did function to create a regular, transparent system of distribution which reduced argument and special pleading. Although aware of its 'crudity' as a finessed distribution system, most farmers preferred this to the alternative of conflict, mistrust and the risk of the management structure breaking down.

When a landholder has plots in several different command areas, these problems of non-conformity to the 'house rules' of irrigation supply are compounded further. It is much more difficult for the farmer to act aggressively in defending either the precise interests of his plot (e.g., in terms of quantity and timeliness of supply) or his own profitability over forms, value and timing of payments for the service. It is important to remind ourselves at this point in the argument that farmers have always been constrained by externalities beyond their control (availability of credit, labour, land to lease in, and so on). But now they are driven by a greater need to raise the productivity of their remaining holdings, and in particular have to commit more of their land to irrigation to achieve this.

Furthermore, the issue now becomes not just whether farmers have to conform to a series of 'house rules' as a precondition of seeking increased productivity, but whether conformity in order to reduce transaction costs is only a necessary and not sufficient condition of such productivity increases. Although this point is empirically more speculative than earlier ones, it is nevertheless crucial for understanding whether existing rates of agricultural growth can be sustained and whether a further production threshold can be overcome. Can farmers actually manage to increase land productivity on diverse plots scattered over different command areas with possibly different sets of 'house rules' even if they are passively conforming? There are several variables to consider in tackling this question. One of the ways in which farmers currently cope with this problem is to under-invest, and underperform in relation to the potentiality of the technology. This farmer policy of underachievement (and risk minimisation) can be obtained along several dimensions: quality of seed; labour and organic materials in plot preparation; extent and timing of fertiliser use; weeding; pesticide use. Clearly, how a farmer operates this portfolio of options will partly depend upon the payment system for irrigation, other related agreements about practices and the credibility of sanctions to ensure compliance, as well as other demands upon the time and capital of family members. So, not only will farmers conform to 'house rules' to reduce transaction costs, but they may also deliberately opt for a low productivity trajectory on all or some of their plots simply because they cannot be in several places at once with the complete high-yielding portfolio of activities and commitments.

Another way to cope with the problem is to turn over more of the management of the scattered plots to others. Such a strategy implies a trend whereby farmers increasingly become passive rentiers on their own land. They may commit family labour as required and as available, but the overall potential for disarticulation of the farm as an integral, self-provisioning unit is high.

Again, there is empirical speculation here. But consider ploughing as a corresponding example to mechanised irrigation. With an increasing intensity of production on cultivable land, land available for grazing is declining along with the availability of gathered fodder. With smaller landholdings per household, it is increasingly difficult to sustain adequate family livestock for ploughing. Some cooperative arrangements between nucleated families of erstwhile extended families exist. But despite this, ploughing services are increasingly acquired in the marketplace outside family provision. Although there have been import hiccups through the later 1980s, power tillers are now increasingly available as partial substitutes for cattle ploughing. In some places, tractors have long been used even among small farmers (Lewis, 1991). Irrigated agriculture, with double or even triple cropping on plots (and irrigated plots tend to be the ones which are double or triple cropped), requires faster turn-arounds in ploughing. Plots on command areas are in a virtually identical time sequence for operations. It is therefore relatively easy to consider ploughing contracts for adjacent plots, coterminous with the command area. Such services may either be provided by the irrigation providers (as is beginning to happen with the landless irrigation providers associated with the Proshika programme), or by some other operator. Of course, power tillers are by definition mobile, enabling richer farmers to consider private purchase to service all their plots, but they are still likely to have surplus capacity, and gaining contracts for the consolidated plots of a command area reduces their transaction costs enormously as such groups of farmers have become used to an element of cooperation among themselves as co-consumers of irrigation.

A command area basis to the provision of other services is less obvious, but no less possible especially if there are capital constraints. Thus there are now examples (within the Proshika experience) of irrigation providers also undertaking purchases of appropriate seed, fertiliser, pesticide and even the mobilisation (and uniform timing) of labour for plot preparation, fertiliser applications, transplanting, weeding, harvesting and even threshing. There are endless possible variations of practice here, with parts of the 'package' being retained by the individual plot-holder according to the overall asset strength of the family and their precise demographic (or strategic business) position in terms of stocks and flows. For some classes of plot-holder, especially the weaker ones along capital and network dimensions, the temptation to 'sign up' for the whole package is increasing as a risk and transaction cost minimisation strategy. In the 1991–92 crop year, labour, irrigation and fertiliser represented 61.4 per cent of boro production costs (MOA/GOB, 1993: Table 5.4).

Before getting too carried away with an argument based on case experience (substantial though it is), it is important to recognise the strong seasonal character of the 'disappearance' thesis. In the non-irrigation season, the prospects for self-cultivation within the family remain higher, as reflected in the higher labour cost element. Nevertheless, in the same crop year, for T-Aman (HYV), 59 per cent of costs were attributed to fertiliser, ploughing and labour, and 58.5 per cent for T-Aman (local) (MOA/GOB, 1993). Families which are poorly networked and confront severe capital constraints in meeting non-family labour costs, face increasing problems in managing a cultivation portfolio even for low cost seasons. It is also worth observing that Tk per hectare profitability for boro in 1991–92 was recorded at Tk 6,424 contrasted to T-Aman (local) at Tk 3,534 (ibid.: Tables 5.3 and 5.2), indicating the significance of the boro crop to the overall economic portfolio of farming families.

DISARTICULATION AVOIDANCE: 'DUCKING AND WEAVING'

However, the seasonal issue draws attention to a wider set of 'ducking and weaving' arguments against the disarticulation thesis. In pursuing an actor-oriented analysis of agrarian structure and change in Bangladesh, it is not our intention to straitjacket actors into a single line thesis about disarticulation, nor to have them overpowered by a single structural process in which all farms 'disappear' to be replaced by a system of command area based agricultural companies combining entrepreneurialism within finance, trade and production with the remnants of rural asset owning classes and their urbanised, business-oriented cousins. The agricultural system is better understood if the activities of farmers to avoid this scenario are also understood; hence 'ducking and weaving'.

The question, then, is how can farming families retain the organic indivisibility of their households as units of action in the management of complex consumption, production, accumulation and survival portfolios? There is a broader question here which needs to be focussed more narrowly upon agriculture. Returning, then, to seasonality: how far does the loss of control over management decisions on scattered irrigated plots in different command areas extend across the rest of a household's cultivation activity?

A central issue is the seasonal character of access to land in leasing markets (see also Webster, this volume, with respect to West Bengal). Practice and analysis have moved a long way from the standard landlord–tenant formula with tenants (of various forms, including share and fixed rents) always acting as 'client' supplicants of richer patrons, and entangled in parallel dependencies of debt and labour obligations. Obviously remnants of such relationships remain, but they have been punctured by a stronger commodity attitude towards land. Reverse leasing is now more common, though it still tends to be seasonal unless we include mortgaging. The figures for the land rent element in the input costs for foodgrain production need careful interpretation (MOA/GOB, 1993: Table 5.4). They are lowest for boro at 21.3 per cent, and highest for traditional, low cost crops like broadcast aus (30.3 per cent) and T-Aman (local) (30.0 per cent).

The story behind these figures is that the renting in of land during the boro season is more likely to be by richer operators, leasing in from poorer ones, with the capital and the networks to manage high input cultivation; whereas such land is resumed for cultivation during the low cost seasons by poorer farmers, who also rent in land (from 'landlords') during this season under more traditional forms of hierarchical tenancy with the usual rationales applying. However, even in this seasonal reverse tenancy scenario, we see attitudes to land commoditised as it is differentially placed under forms of management appropriate to the capital and technological conditions for the season and the crop. Of course, there is the further question of whether such richer farmers place rented-in land in the hands of other entrepreneurial service providers. At the same time, such calculations concerning seasonal leasing would seem to reveal a desire among poorer farmers to retain direct control where possible, to absorb family labour and/or to secure some proportion of the family's annual food requirements. For richer families, continued leasing out during low cost seasons may reflect: the need to retain clients; the desire to reduce transaction costs in labour management; the diversified income and employment portfolios of family members (which would be consistent with allocating high cost crop management to service providers); and transferring risk under rainfed conditions.

Another potential disarticulation avoidance strategy may also have ambiguous meaning for the argument. This concerns 'technology brokering', which refutes the proposition that lumpy but necessary technologies (as in mechanised irrigation and ploughing) will always have the impact of bringing land into the control of those who control the technology. It should be clear that this 'lumpy' proposition has referred more to private, individual acquisition of land using technological leverage than to the central proposition of this paper concerning transfers of the *management* of land (and therefore, of course, the distribution of gross returns between dividends and management costs, with attendant risks of unaccountable rent-seeking) to service providers. Technology brokering refers to those socio-economic processes whereby lumpy technological services can be mediated into small scale situations. The strongest test case for this concerns ploughing with tractors. Lewis (1991) found that brokers were able to sign up groups of neighbouring farmers prepared to 'contribute' appropriate plots towards a consolidated ploughing area, thus making it sufficiently attractive for tractor owners to plough the adjacent plots of a number of small farmers over a day or two. Again, this is an example of farmers being prepared to cooperate as co-consumers of a shared service. However this example has ambiguous meaning for the disarticulation argument. On the one hand, it reveals the exercise of initiative by small farmers to increase the intensity of cultivation by taking innovative steps to secure ploughing services; on the other, it reveals their dependency not only upon a service provider but also upon other intermediaries to negotiate the deal. The example might represent an alternative service provider, ploughing lands which are not necessarily coterminous with those of irrigation command areas, but it would be an empirical question whether ploughing and irrigation command areas become increasingly coterminous under the management of the same service provider. If so, and if this is combined with other variable input services which are nevertheless lumpy in capital terms (e.g., fertiliser), then are we witnessing the embryonic formation of agricultural companies, offering either specialist or a combined package of services?

Clearly one of the arguments in support of the disarticulation thesis is that farming families cannot divide their family labour resources (even just supervisory ones) across a number of different plots which require simultaneous attention in response to the time constraints imposed by irrigation requirements. The option to retain a significant number of loyal, tied labourers to undertake actual labour inputs or supervisory ones on behalf of the family is unrealistic given overall landholding sizes. Being effective in local labour markets is difficult under such circumstances, and depends significantly upon the position of the family within village networks (a variable of not only economic class, but also *bangsho* [lineage], imputed caste, communal status and religio-cultural stances towards gender relations). As Rogaly (1994) has argued for West Bengal, labour markets are socially embedded.

While richer farmers may be in the stronger position to bring labour together as required with optimal timing, by definition they have more plots to service, especially across different irrigation command areas in the boro season. Labour management is, after all, one of the reasons why they are still more prepared to lease out marginal land to tenants in low cost, high risk seasons. Under these conditions, for both richer and poorer classes of farmers (even if for different reasons), the increasing reliance upon contract labour gangs is attractive. If such gangs are operating independently of the technology holders, then farmer employers are remaining active in negotiating sets of services from different providers. Thus a farmer may contract with the same gang to service his plots (typically transplanting, weeding and harvesting) across different command areas. It is again an empirical question whether technology providers (e.g., owners of pumpsets and borings) are using the leverage of the key resource (water) to control who ploughs and who has labour contracts within the command area. While they would have to accept, culturally, that farmers are entitled to use their own family or to directly employ casual labour on such plots, they are more likely to seek direct management (unitary companies) or indirect rent-seeking over other 'contractors' (i.e., labour gangs, tractor and power tiller owners) seeking to offer services within the same domain. Service cartels might be the other organisational option.

A further avoidance strategy, but again with ambiguous meaning, may be for farming families to cooperate among themselves to 'franchise out' the services which they collectively need. This solution has been proposed for the operation of DTWs, where the immovability of the asset increases the danger of placing its owner (state or private) in perpetual monopoly over water distribution in the command area. (STWs operate under different physical conditions, so that both consumers and providers, especially the latter, can use the threat of withdrawal to ensure compliance to contracts or to moderate rent-seeking.) However, the drawbacks of such an option are clear. First, it assumes a capacity for cooperation among farmers to establish a contract and monitor it which most experience disproves. But, as a counter-argument, if farmers clearly perceive themselves as consumers rather than directly as part of supply management, then the prospects for cooperation may be greater. Second, and here the ambiguity of meaning arises, does the franchise solution merely amount to another variant of the disarticulation thesis? Third, it is an option which is more likely to be applied only to DTWs which are not expanding in Bangladesh. And fourth, since DTWs irrigate larger command areas, farmers may end up having more of their irrigated land in the command of a single monopoly provider, over which their effective control is slight since the transaction costs of mobilising voice with other consumers would be very high (rather like the numerous but small shareholders in the privatised utilities in the UK). Quangos (or NGOs?) would be required to intermediate!

A further possible argument against the disarticulation scenario concerns a farmer's crop mix. So far the argument has been based upon foodgrain production, reflecting the principal growth areas and major land use. The increases in foodgrain production have been achieved partly as a result of higher yields (with all the attendant analysis of commercialisation of agriculture), but also by declines in jute acreage (competing with rice) and overall declines in oilseeds (only groundnut acreage has risen) and pulses (competing with wheat) though there has been a rise in potato and sugarcane acreage. (All figures are up to 1991-92, from MOA/ GOB, 1993.) With yields virtually constant in oilseeds and pulses, overall production of these commodities declined between 1983-84 and 1991-92. Sugarcane production increased over this period despite declining yields (from 41.7 mt tonnes/ha to 39.7), but increasing potato yields delivered significant increases in production.

The evidence on other crops is therefore mixed. Jute as the major cash crop fluctuated yearly in acreage, mainly in response to the previous year's prices but also as a function of localised rice conditions (e.g., boro failures increase aman acreages). There was a trend decline in acreage, but this was partly offset by yield increases (from 1.374 mt tonnes/ha in 1983-84 to 1.835 in 1991-92), so that production trends are difficult to identify. Potato and sugarcane production seem unambiguously to be increasing either via acreage or yields. However, it is appropriate to speculate on the basis of potato and sugarcane, and vegetables and fruits in terms of other demographic trends in the country.

Bangladesh, like other societies in South Asia, is experiencing rapid rates of growth in urban and peri-urban populations. On average in Bangladesh, the rate of urban population growth has been 8 per cent between 1961-81, and 6 per cent during 1981-91. Even with a projected declining trend, the urban growth rate is unlikely to be less than 4 per cent until 2010 (Task Force, 1991: Vol. 3). Because of the large national population, the absolute size of the urban population will be substantial. At the current 20 per cent rate of urbanisation, 23 million people can be considered as living in urban areas. By 2000, this figure will rise to 37 million, and by 2015, projections indicate that 68 million (37 per cent of the population) will live in urban areas (ibid.). This rapid rate of urban growth has occurred without a corresponding increase in industrialisation or planned urban development (ibid.; UNDP, HABITAT and GOB, 1993; Alam, 1987; Islam and Muqtada, 1986; Shakur, 1987; 1988), which is more consistent with natural increase rather than migration explanations of urban growth.

At the same time, recent policy initiatives such as *upazila* decentralisation have stimulated the 'rurbanisation' form of this urbanisation process, along with other economic activities surrounding expanded agriculture (variable input supply, supply and maintenance of machinery, processing, transportation, marketing). Over the last twenty years, there has been a considerable expansion of rural market towns in most parts of the country. More of them are linked by metalled feeder roads, displacing labour intensive forms of transportation but creating other opportunities for trade. The purpose of this chapter is not to explore the distributional outcomes of this process, but to consider the implications for farming behaviour of these new centres of demand for agricultural products as well as the expansion of the urban population generally.

One way of appreciating the significance of this rurbanisation process is to conceive of much of rural Bangladesh as consisting of hinterlands of such centres. Such a view is of course close to the growth pole propositions of economic geography. Relying only on casual observation (but again, much of it in most parts of Bangladesh over the last twenty years) both within such centres but also among farmers in these hinterlands (cf. J. Harriss' 'rural rides', 1993), it may be said that there are now increasing opportunities for widening the crop mix to include more horticultural products (vegetables, spices and fruits). A local, commercial demand for such products appears to be rising steadily along with better transportation to such centres but, more significantly, from them towards the major urban centres via the feeder roads. Thus many of the earlier perishability constraints can be overcome by the rapidity of movement. This is assisted by increased rural electrification (especially to such centres) thus improving storage possibilities. The increase in potato production can be seen as partial confirmation of this process (perhaps also sugarcane, although this is less clear). A similar story can be told for fish, entailing the greater commercial exploitation of fish-ponds through intensive fish culture (see Lewis et al., 1996).

A buoyant demand for such products enables small farmers to consider other options for land use, and to bolster the viability of small holdings otherwise threatened by the inability to lease in land during high productivity foodgrain seasons, or by capital constraints which oblige them to lease out irrigated land. It also enables such farmers to bring their homesteads more centrally into their strategic thinking. Through adopting various organic practices, there is a greater prospect of integrating homestead growing and wastage cycles with nearby arable plots. Recent experiments (on 500 plots within the command areas of Proshika group–supplied irrigation) suggest that boro rice yields can be sustained (even increased) at a fraction of the input cost of chemical based technologies, requiring inter alia much less water. Such developments may yet have a profound effect upon the capacity of small farmers to avoid the disarticulation scenario. Ironically, this would apply much less to larger farmers who are locked in, on more land, to a high-tech production strategy from which it would be more risky to disengage.

With such a process, the situation of small farmers in Bangladesh might resemble that of E.P. Thompson's 18th century plebs in England:

> [T]his is the century of the advance of 'free' labour. And the distinctive feature of the manufacturing system was that, in many kinds of work, labourers (taking petty masters, journeymen and their families together) still controlled in some degree their own immediate relations and modes of work, while having very little control over the market for their products or over the prices of raw materials or food (Thompson, 1993: 74).

It is this observation which sets the limits to any distributional optimism which might arise from such a ducking and weaving scenario. A degree of autonomy over management practices at the point of production may nevertheless amount to little, especially where returns to the producer from such new (postmodern?) forms of production (new crop mixes, involving horticultural products and organic practices) rely precisely upon emerging classes of deregulated 'rurban' merchants. Since these are products which incur no state intervention in price management, such merchants can operate in imperfect markets with localised purchasing monopolies (often secured via credit tying) and benefit from the advantages of imperfect information about prices among producers (see B. Harriss, 1993, and her earlier chapters).

CONCLUSION

If we are observing forms of agricultural modernisation in densely populated rural regions with a high incidence of fragmented small farms, then we are looking at a proliferation of services to agriculture involving a range of off-farm agricultural activities and different sets of actors. Furthermore, in the era of structural adjustment, we are looking not just at farmer–state relations but at a whole continuum of private/public agencies. Obviously the study of agriculture is not the study of the farm (land and labour management) but of an extended system of transactions referring to inputs and marketing of output. The particular question of whether the farm is 'disappearing' arises in societies like Bangladesh characterised by processes of agricultural involution and intrusion of new capital.

Major issues of theory and policy arise from this possibility. Processes of disarticulation and re-articulation of the agricultural system have to be analysed in terms of the structural implications of this intrusion of capital into minifundist farming conditions. In a longer analysis, this would entail an understanding of the sources of such capital, the options internationally and nationally available for its investment, and the explanations, therefore, of its investment in the minifundist conditions of this part of South Asia. We will also need to understand more thoroughly forms of investment (land, technology, credit), its structure (markets, institutions, from 'kin' to 'company'), its distributional consequences (differential rates of return on different activities and therefore actors), and its impact upon social relations in terms of market moralities (see Wood, 1995 for a parallel discussion for water markets in north Bihar).

Many policy issues stem from such analysis for Bangladesh: the distribution of new opportunities arising from these processes in terms of region (dry/wet, upland/lowland), communal status and cultural stance, classes of farming family, and gender; the content of these opportunities (market activity, delivery services, transportation, agricultural processing, equipment servicing and maintenance, agro-industrial production); the implications for state activity (appropriate targets for investment, information, training, as well as macro-economic management of prices, exchange rates, and fiscal strategy) and donors; the extent of commoditisation in attitudes towards land use; the prospects for absorbing surplus rural labour; and the implications for rural–urban migration and attitudes to family size.

To pursue some of the detailed argument further: the argument has been that an understanding of agriculture has to move away from the family farm as the analytic focus (entailing the form of argument: is it more or less peasant or capitalist?) to a stronger conceptualisation of an agricultural system in which sets of actors are changing, along with functions and relationships. Such a notion of *system* crucially involves actors outside the village (both historically, and in the present and future). This draws our attention to the problem of appropriate concepts and language for analysis. The more 'nationalist' mode-of-production discourse concentrated upon relationships of a class production kind (to labour, to tenants) and upon forms of surplus extraction (absolute or relative surplus value, through extra-economic relations, antediluvian forms of rent, moneylending and other forms of commercial entrepreneurialism). Variants of mode-of-production analysis have disputed the relative significance of exchange and production in producing change and determining the forms of change. Historically, the significance of imperialism, colonialism and contemporary international capitalism has entailed the distinction between formal and real subsumption of labour under capital. However, that discourse is also essentially locked into the teleological problem of assuming a development of capitalism of some kind, so that the criteria of the projected end-state (e.g., wage labour) are used as ex post benchmark analytic tools.

The main theoretical purpose has been to critique these teleological and perhaps ethnocentric forms of analysis of agrarian change by adopting a more actor-oriented perspective. This offers more explanatory significance to the ways in which different sets of actors perceive the options available to them and calculate the odds to the best of their ability and within their complex rationalities where many objectives may be hidden from view. There are several issues here. First, there is a universal analytic point to make about the nature of economic relationships and transactions. Neo-classical economic analysis of advanced capitalist systems has now been under attack for some time for its domination of the intellectual discourse in understanding economic behaviour anywhere. In this sense the neo-liberal discourse shares its characteristics with the mode-of-production one. But even in the so-called advanced market systems, economic relations have an important institutional content through which economic behaviour is structured and mediated. There is a 'structure' beyond that implied simply by acting out the imperatives of price and value. Understanding this institutional content, this social embeddedness, is vital. Second, and this point is only to be made after the universal point has been established, teleological assumptions can be avoided by deliberately being more culturally relativist in understanding such institutional content, which calls for a greater appreciation of culture in the analysis of economic relations.

Cultural relativism does not preclude us from developing a universal conceptual language, but it has to be a language which avoids simply noting transition from one known system to another. This language has to be sensitive to context, and in particular the values held by different sets of actors as well as their perceptions of opportunity and risk. Certainly the language of interlocked markets, networking, managing complex portfolios, and resource profiles all help to explain real processes, rather than models of them, because they contain culture. They enable us, for example, to understand the cultural contingency of landholding and the extent to which perceptions of status (as well as observed class position) can become de-linked from landholding and the direct management of land. They help us to judge whether, therefore, the codes for attributing status and respect are being revised through necessity; or whether we have misread the rural culture in the past, so that the observed propensity to engage in service delivery and trading activity is not a sign of new entrepreneurialism but a continuation of essential roots in Islamic culture.

Instead of lamenting the disappearance of the farm, it may have existed only as a fixed 'land based family management' concept in the eyes of observers, implying that the newly observed management options based on technologically driven service delivery can easily be embraced culturally; that the moralities which govern such transactions are more continuous with, rather than a break from, previous transactional morality; that the reproduction of the local society has always relied upon brokers and fixers; that calculations are always made about avoiding transaction costs and transferring risk where the power exists to do so; and that capital owning has always underpinned a willingness to take risks (e.g., as service providers) when potential profits from similar activity elsewhere have been revealed. Perhaps the Bangladeshi farmer was 'globalised' before his observers.

REFERENCES

Alam, A. 1987. 'Bangladesh Country Chapter,' *Urban Policy Issues*, Asian Development Bank, Manila, February 1987.

Folbre, N. 1986. 'Hearts and Spades: Paradigms of Household Economics,' *World Development*, 14(2): 245–55.

Glaser, M. (1989). 'Water to the Swamp: Patterns of Accumulation from Irrigation in Rajshahi Villages.' Ph.D. Thesis, University of Bath.

GoB/BBS. 1992. *Yearbook of Agricultural Statistics of Bangladesh*, Bangladesh Bureau of Statistics, Government of Bangladesh.

Harriss, Barbara. 1993. 'Markets, Society and the State: Problems of Marketing under Conditions of Smallholder Agriculture in West Bengal.' Working Chapter No. 26, Open University, Milton Keynes Development Policy and Practice Group, Monograph 1.

Harriss, John. 1993. 'What is Happening in Rural West Bengal? Agrarian Reform, Growth and Distributions,' *Economic and Political Weekly*, 28(24): 1237–47.

Islam, R., and M. Muqtada (eds). 1986. *Bangladesh: Selected Issues in Employment and Development*. New Delhi: ILO–ARTEP.

Lewis, D.J. 1991. 'Technologies and Transactions: A Study of the Interaction of "Green Revolution" and Agrarian Structure in Bangladesh.' Centre for Social Studies, Dhaka, and Ph.D., University of Bath.

Lewis, David J., Geoffrey D. Wood and Rick Gregory. 1996. *Trading the Silver Seed: Local Knowledge and Market Moralities in Aquacultural Development*. London: Intermediate Technology Publications.

MoA/GoB. 1993. *Handbook of Agricultural Statistics*. Sector Monitoring Unit, Ministry of Agriculture, Government of Bangladesh, 1 July 1993.

Rogaly, Ben. 1994. 'Rural Labour Arrangements in West Bengal, India.' D.Phil. thesis, Oxford University.

Sen, Amartya. 1990. 'Gender and Cooperative Conflicts,' I. Tinker (ed.) *Persistent Inequalities: Women and World Development*. Oxford and New York: Oxford University Press.

Shakur, M.T. 1987. 'Urbanisation and Housing in Bangladesh,' *Journal of Himachal Pradesh Institute of Public Administration*, 12(2): 27–37.
———. 1988. 'Implications for Policy Formulation towards Sheltering the Homeless: A Case Study of Squatters in Dhaka, Bangladesh,' *Habitat International*, 12(2): 53–66.
Shresta, A. 1994. 'Eating Cucumbers Without Teeth: Women's Participation in Nepalese Agriculture and Rural Development.' Ph.D. thesis, University of Bath.
Task Force. 1991. *Report of the Task Forces on Bangladesh Development Strategies for the 1990s*. Dhaka: Government of Bangladesh.
Thompson, E.P. 1993. *Customs in Common*. London: Penguin.
UNDP, HABITAT and GOB. 1993. *Bangladesh Urban and Shelter Sector Review*.
Wade, R. 1988. 'Why Some Villages Cooperate,' *Economic and Political Weekly*, 23(16): 773–5.
White, S. 1992. *Arguing with the Crocodile*. London: Zed Books.
Wood, Geoffrey D. 1973. 'From Raiyats to Rich Peasants,' *South Asia Review*, 7(1): 1–16.
———. 1994. *Bangladesh: Whose Ideas, Whose Interests?* London: IT Publications, and Dhaka: University Press.
———. 1995. 'Private Provision after Public Neglect: Opting Out with Pumpsets in North Bihar.' Paper to International Conference on Political Economy of Water, Madras, January 1995.
Wood, Geoffrey, and **Richard Palmer-Jones.** 1991. *The Water-Sellers: A Cooperative Venture by the Rural Poor*. London: IT Publications. First published in 1990 by Kumarian Press, West Hartford.

12

Institutions, Actors and Strategies in West Bengal's Rural Development —A Study on Irrigation

NEIL WEBSTER

INTRODUCTION

Any attempt to understand the interplay of institutions and actors engaged in any dimension of rural development at the local level must first locate the political context or setting in which they are operating. Only then can one begin to discuss the space or room for manoeuvre that exists for action at the local level that can bring about a development process with a strong pro-poor orientation. In the case of contemporary West Bengal the present Left Front government (LFG) came to power in 1977 on the basis of a radical movement of the rural poor majority. This was not a

---

This is a revised version of a paper originally presented at the Workshop on Agricultural Growth and Agrarian Structure in Contemporary West Bengal and Bangladesh held in Calcutta, 9–12 January 1995. A considerable debt is owed to Nripen Bandyopadhyay from the Centre for Studies in Social Sciences, Calcutta, with whom I have collaborated on this research.

revolutionary movement despite its clear class dimensions, but it did enable the LFG to set about implementing a range of radical agrarian policies that characterised the period 1977–83.

In this early period it was an agrarian politics that was conflictual and often violent. It involved personal revenge against oppressive local landlords, moneylenders, local bureaucrats and administrators. But it was also a time in which local politics were constituted in broad-based movements such as 'Operation Barga' (registration of sharecroppers) and in the breaking or weakening of many of the hidden forms of dependency and subordinating practices prevalent in local agrarian relations.

Panchayati raj, i.e., the decentralisation of government, was an institutional means introduced in 1978 with which to implement much of the agrarian strategy. Several observers have argued that democratic decentralisation together with the programme of structural reforms has been effective in politically and economically empowering the poor and the more marginal sections of the agrarian social formation.[1]

The period from 1983 to 1985 was a turning-point in the political direction of the LFG's agrarian strategy. In 1983 the local panchayats proved themselves to be extremely effective in drought relief operations, and in 1985 they were formally brought into the process of development planning. Now one could talk of a genuine degree of local participation, popular accountability, and greater effectiveness in the implementation of development programmes at the local level. But the sustainability of the whole strategy has always tended to be perceived by the LFG and its dominant coalition partner, the Communist Party of India (Marxist) (CPI[M]) in terms of the electoral sustainability of the LFG at the state level. Sustainability at the local level through the involvement and mobilisation of the poor in a wide range of local institutions beyond the panchayats has not been a part of the strategy. Here the status of the party has increasingly had priority.

Yet it is not necessarily paradoxical that a strong party and a strong state go hand in hand with a successful decentralisation of government. The CPI(M) argues that panchayati raj is directed towards a people's democracy by which I understand a pro-poor democracy. It has required a strong state to implement the tenancy reforms of Operation Barga, to redistribute land over the land ceilings, to secure access to formal credit for some of those who traditionally could only turn to private moneylenders, and to bring about a very successful development process with a strong pro-poor dimension. Furthermore, the CPI(M) has also successfully pushed the local panchayats into playing a vital role in local conflict resolution and in establishing a rule of law at the local level that has clearly benefited the poor and marginal groups.

This paper is concerned with the unexpected consequences of these successes. Successes resulting from the general programme of reforms have resulted in the localisation of the politics of the poor. In place of the general needs of agricultural labourers, tenants, illiterates, and so on, are the more specific needs that emerge from particular local matrices of problems that delineate local groups of poor in a quite unique way. Local problems give rise to local tactics and local strategies on the part of the poor through which they seek to secure or transform key aspects of their daily lives. It is the growing lack of correspondence between the politics of the LFG and the politics of what I would call these localised constellations of the rural poor that needs to be examined and analysed today in West Bengal.

Because of the success of the LFG at the macro level, the politics that have a growing significance at the local level are those of caste, gender, ethnicity and religion. This process has accelerated as the impact of agricultural growth has intensified local socio-economic divisions. However, the local constellations of rural poor that such conditions give rise to are not passive. There has been a long history of collective action at local levels by groups of the poor and marginalised on the basis of their poverty. These actions are a response to particular effects, failures and biases that find their expression in markets in which the poor are oppressed, exploited, or from which they are excluded. It is the potential for political and economic action that I seek to analyse here through a case study of irrigation in one locality.

THE CASE OF JAYANAPUR VILLAGE, BARDHAMAN DISTRICT

In 1991, research into rural grassroots production cooperatives brought us to a group of villages in central eastern Bardhaman district, West Bengal.[2] Not only was there a long history of collective action in the area, but one of the villages had been the site of a cooperative involving the collective pooling and cultivation of land, the Jayanapur Samobhar Samiti. (The name of the village was changed due to an agreement with some of the villagers.) Before its demise, the cooperative covered a large proportion of the village's lands with more than 120 acres under collective cultivation. The cooperative ran for some twelve years during the 1950s and 1960s and was advanced not only in its organisational nature, but also in its purchase of tractors to aid the production process, and in its construction of storage facilities and the marketing of its production.

The establishment of the cooperative had been very much a local grassroots initiative. It was formed initially with twenty-four households and led by the village members of the local Krishak Samiti, the village-level organisation of the Communist Party of India's peasant front, the Kisan Sabha.[3] But it was not the first such collective action by the villagers. Already by the 1950s there was a long history of radicalism and organisation in the village of Jayanapur including the paving of its roads, the construction of a primary school, and the organisation of a series of grain banks for the poor at times of hardship due to drought. This tradition, and more importantly the experiences it provided, undoubtedly aided the formation of the cooperative.[4]

The success of the cooperative is demonstrated by its expansion during the early years as it attracted more than sixty households to contribute land and to participate. It was also significant that in all the initial descriptions of the cooperative that we were given, its demise was blamed upon the actions of the state rather than upon problems located within the cooperative itself.[5] This lent further support to the belief that here was an institution based upon collective action of some significance and that it was a potentially important case study in the search for alternative approaches to agrarian development involving grassroots cooperative institutions.

However, closer investigation revealed the nature and dynamics of the cooperative's establishment, organisation, and decline to be far more complex. It became clear that the political activists who had first talked to us about the cooperative now lacked the ability to separate the local from the more general in assessing the cooperative's significance and role. The distinction between these two had become blurred in the intervening years with the political and ideological struggles they contained; the history of the cooperative had become entwined with the political history of Bardhaman district and West Bengal as a whole.

We discovered that while the political impetus for the cooperative had its roots in the local villagers' experience of collective political action stemming from their involvement in the anti-colonial struggles of the 1930s and 1940s and the communist-led agrarian struggles of the 1940s and 1950s, its economic impetus was of a quite different nature. There was considerable debate about whether the dynamics lay in the attempt to bring the benefits from economies of scale to individual small owners or in the somewhat less politically progressive attempt to replace local labour with tractor power. Whichever analysis was closer to the truth, there was a clear attempt to increase the productivity of the land through both intensive and extensive means. As for the eventual collapse of the cooperative, it would seem that this was at least as much the consequence of internal conflicts exacerbated by the actions of the state as it was the consequence of the state's actions by themselves.

The experience of the cooperative was therefore more complex and contradictory than had first been suggested. In its day it had combined both idealism and pragmatism in seeking to unite cultivators in the village within a collective enterprise and to challenge both the production problems of small and fragmented landholdings and the disadvantages of being sellers in a grain market dominated by monopsonists. But the cooperative lay in the past. It was a case of collective action to be unearthed and explored for its logic and its dynamics through the memories of its participants and other local people.

In 1991, however, we discovered that land was once again being cultivated collectively in the village, only this time it was the landless scheduled caste households who were involved. A little over 20 acres of land had been 'vested' by the government under the land ceiling rules of the land reform legislation, and while the legalities of the dispossession were being slowly processed prior to the land's redistribution to the landless, those landless were cultivating the vested land in a form of cooperative. Forty-seven households from Das *para* (neighbourhood), thirty-one from Tetuli–Bagdi para, and twenty-four from Dulal para cultivated 8, 7 and 5 acres respectively.

Once more, the initiative was local. The village's Krisak Samiti had organised the households to carry out the cultivation of this land, arranging the division of labour, the timing of inputs, the distribution of the output, etc. But after three years of collective cultivation of the land, its distribution to individual owners, i.e., the granting of 'patta', could finally take place and not one of the landless households appeared to be seriously interested in continuing with the cooperative exercise. Each wished to assume individual control over the little plot of cultivable land, approximately 0.2 of an acre, that they could now call their own.

For us, the situation in the autumn of 1991 presented a paradox. On the one hand the Left Front government's commitment to the implementation of land reform, largely unparalleled elsewhere in the Indian subcontinent and rooted in a class-based appraisal of West Bengal's agrarian social formation, had succeeded in delivering land to the most disadvantaged and most oppressed in the village. Yet on the other hand the response of these households was promptly to seek to pursue a highly individualistic production strategy, despite being encouraged to cooperate. Why the apparent distance in philosophy and in praxis between the state and the parties of the Left Front on the one side, and the oppressed on the other? Or, to ask a complementary question, what was it that the micro was telling the macro?

The answer to this question we find lies not in the frequently claimed failure of marxism to understand human beings' true nature as individual social actors, but in the politics of agrarian development at the levels of both the state and the village. The landless households' response to the cooperative idea was not a rejection of collective action; on the contrary, it was a collective response to the situation in which they found themselves—specifically their political, economic and social positions within the village. The failure to facilitate these households' engagement in other forms of collective action, not least to facilitate the realisation of the potential productive value of their land through the secure provision of water for dry season cultivation, is a failure that extends from local political actors through to the policies of the state and the politics of the CPI(M)'s agrarian strategy.

Fieldwork undertaken in 1991 in five different locations in West Bengal, covering households in sixteen villages and hamlets, reveals that different social groups amongst the poor possess the ability to respond to particular matrices of exploitation and bias by empowering themselves through local collective action time and again (Webster, 1993). In these cases we believe that the logic for such action can be located not in the orthodox logic of the market nor in one articulated by the state's Left Front government through its development praxis, but in the combination of both a collective experience of deprivation, often tied to exploitation, together with a particular common social identity with significant enough meaning to bring the individuals together in collective action.

Elsewhere we have used the concept of social cooperation to describe the basis for such collective behaviour (Webster, 1993). That is, a set of socio-economic structures manifest themselves in the form of a particular matrix of experiences (in some ways similar to Bourdieu's concept of 'habitus'). This in turn gives rise to particular pressures and specific discourses from which the potential for collective action arises.

Tribal women's cooperatives drawing upon the women's experiences of working in migrating labour gangs in areas dominated by landowning cultivators in a brahminical Hindu culture are one good example.[6] Here the overlaying of different sets of social relations rooted in different social forces can be seen in terms of particular identities—e.g., as 'tribals', women, labourers, etc.; social experiences are recorded in particular bodies of social memory and knowledge—e.g., traditional forest skills, modern labour techniques, ethnic and sexual harassment, etc.; and in turn become the basis for action, potentially collective, at times of specific crises brought about by excessive poverty, exploitation, etc.

Such a potential takes on a greater significance when the political leadership of the poor increasingly takes a macro, programmatic, approach to the issue of poverty. The role of local action and local mobilisation then takes on a new dimension. It can be a part of such a programme; it can be seen as a deviation or local obstacle to such a programme; or else it is simply not seen at all.

The rest of this paper seeks to explore weaknesses within West Bengal's pro-poor agrarian strategy through a further exploration of the case of Jayanapur village and the politics of irrigation. It is a discussion that leads to the suggestion that there is, *at the present point in time*,[7] the political space for a third actor in the pursuit of rural development in the state; that the political requirement of the CPI(M)/LF to be elected and to govern, together with the absence of mass-based political movements, has created both the need and the space for a third group of actors who can facilitate local action by local social constellations of the poor, and project their interests at different levels of policy formulation within the state.

IRRIGATION AND MARGINAL CULTIVATORS IN AGRARIAN DEVELOPMENT

At the core of the high agricultural growth rate achieved in Bardhaman district during the past three decades lies the growth in rice production. In 1962 Bardhaman became the first district in West Bengal to be covered by the Intensive Agricultural Development Programme. But it is not just the success achieved in production that is important, but that it has been based upon an apparently unchanging small producer base over the past thirty years.[8]

Until the beginning of the 1990s, it would appear that in rice cultivation the small and marginal farmers had found security through land and labour. But today, access to water would appear to be slipping away from them in this part of Bardhaman district and in parts of the districts of Hugli, Medinipur, Murshidabad and Bankura. Power rooted in the ownership of land has been directly confronted through the imposition of land ceilings and the registration of sharecroppers, and earlier, during the United Front coalitions of the late 1960s, through direct land seizures. Both have provided forms of security to the marginal and landless either through the receipt of a little land or through greater security of tenure. Higher real wages have further enhanced their position within the local village economy, especially with the increase in agricultural employment from double and triple cropping,[9] and the introduction of a revitalised and politicised system of local government in the form of panchayati raj has brought a degree of participatory development for the poor that they had never experienced previously. In this way the landlordism that dominated the lives of the majority of rural Bengalis, the reaction to which shifted votes towards the left parties in 1967 and ultimately brought the Left Front back to power in 1977, has been broken.[10]

These policies were directed towards the poor in terms of their relationship to land and their sale of labour power as a commodity. They are policies that have facilitated a more egalitarian distribution of the benefits of the recent agricultural growth. There is also strong evidence to suggest that they contributed to that growth through breaking many of the dependency relations formerly oppressing and impoverishing the small and marginal agricultural producers.

For the past twenty years or so, irrigation has not been a significant point of intervention in this pro-poor agrarian strategy. Instead, irrigation was viewed as one more, albeit important, element in the technological package of the green revolution.

West Bengal fairly consistently produced between 14 and 16 per cent of India's total rice production between 1981–82 and 1991–92. In 1980–81 Bardhaman district produced 13.4 per cent of West Bengal's rice. By 1991–92 this had risen to 14.7 per cent. For these reasons, Bardhaman district has long been known as the grain basket of West Bengal. Here, *aman* paddy cultivation is based upon the monsoon, supplemented by irrigation from canals and tanks. The seed beds are maintained by tank water. *Boro* paddy, or summer paddy, is based upon canal irrigation supplemented by deep tubewell irrigation in some areas, otherwise by diesel powered shallow tubewells. The agricultural growth in foodgrain production that began in the late 1960s was based upon canal water provided by the Damodar Valley Canal System. While aman yields have increased with improved seeds and fertiliser use, it is the dry season boro paddy cultivation enabled by shallow tubewells which has been the real base for the rapid growth in foodgrain production in the district in the 1980s and early 1990s. The importance of the increase in irrigated boro paddy cultivation to the growth of agriculture in the district is shown in Tables 12.1 and 12.2.

Table 12.1
Rice Production in Bardhaman District for Selected Years Between 1969 and 1992

	1969–70	1977–78	1978–79	1979–80	1980–81	1981–82	1985–86	1987–88	1989–90	1991–92
Area ('000 ha)										
Aus		25.0	27.1	26	27.4	33.8	23.8	26.3	33	31.5
Aman		444.8	321.3	403	428.1	435.3	417.6	396.2	403	402.2
Boro		65.8	99.9	31	93.9	33.9	97.5	141.0	109	106.7
Total	461.0	535.6	448.3	460.0	549.4	493.0	538.9	563.5	545.0	560.4
Production (metric tonnes)										
Aus		40	40.7	48.3	48.6	43.5	45.2	58.8	113	79.3
Aman		760.8	686	699.1	737.7	585.0	735.8	757.6	975	1062.2
Boro		183.8	282.7	80	216.4	80.3	232.0	424.8	325	391.1
Total	651.3	884.6	1009.4	827.4	1002.7	708.8	1013.0	1241.2	1413.0	1532.6
Productivity (kg per ha)										
Aus		1597	1499	1837	1774	1824	1903	2234	3392	2526
Aman		1710	2135	1735	1723	1344	1762	1912	2420	2641
Boro		2793	2830	2611	2305	2371	2593	2378	2917	3086
Average	1401	2033	2155	2061	1934	1846	2086	2175	2910	2751

Source: Bardhaman District Annual Plan 1993–94, District Planning Committee, Bardhaman; West Bengal Statistical Abstract 1978–89. Bureau of Applied Economics and Statistics, Government of West Bengal.

Table 12.2
Irrigation Intensity and Irrigation by Source, West Bengal and Districts 1971–72

		Percentage of Net Irrigation by					
	Irrigation Intensity[1]	Canal	Tank	STW[2]	DTW[2]	RLI[2]	Other
West Bengal	1.28	41.6	34.8	6.0	3.0	2.3	12.4
Bardhaman	1.25	65.2	22.8	2.0	2.0	1.4	6.6
Hugli	1.69	40.4	33.3	11.3	4.3	5.4	5.3
Birbhum	1.19	58.3	32.3	0.6	0.6	0.8	7.4
24–Parganas	1.46	10.0	45.1	25.6	9.1	2.2	8.0
Nadia	1.54	2.4	1.3	37.6	41.9	4.7	12.1
Murshidabad	1.26	38.4	25.6	13.3	5.4	5.1	12.2
Bankura	1.24	2.4	54.3	0.7	0.3	0.6	13.1
Medinipur	1.26	26.5	33.2	6.3	0.9	3.4	29.7
Haora	1.38	24.4	46.6	6.1	3.3	4.3	5.3
Jalpaiguri	1.08	78.5	0.7	1.5	0.9	0.7	17.8
Darjiling	1.29	49.7	1.1	0.0	0.0	0.8	48.4
Malda	1.33	5.5	40.5	17.4	1.3	6.3	28.9
West Dinajpur	1.37	1.6	65.9	11.6	5.2	3.8	11.9
Koch Behar	1.10	1.1	6.5	42.4	0.5	2.7	46.7
Puruliya	1.18	2.6	83.4	0.0	0.0	1.7	12.3

Source: Government of West Bengal Directorate of Agriculture, Socio-Economic Evaluation Branch, 'Irrigation Survey in West Bengal, 1971–72', unpublished data cited in Boyce, 1987: 173.
[1]Irrigation intensity = gross irrigated area/net irrigated area.
[2]STW: Shallow Tubewell, DTW: Deep Tubewell, RLI: River Lift Irrigation.

The relatively low irrigation intensity in Bardhaman compared to Hugli district is deceptive as 35 per cent or so of western Bardhaman is primarily '*danga*' or higher land that for the most part can only support a rainfed *aus* paddy crop. The central and eastern blocks will have had a far higher irrigation intensity based upon canal irrigation. Bardhaman district's receipt of canal water over the years is given in Table 12.3 with a cropwise breakdown for the provision of canal water where the data have been found.

The DVC provided water to large tracts of land for both aman and boro cultivation. The impact of poor monsoons has been reduced and the extensive cultivation of land for boro paddy enabled. The low prevalence of DTWs and, more importantly, of STWs during the 1970s and early 1980s is a reflection of the government's view, and the farmers acceptance of the fact, that DVC is the primary source of irrigation. However, from the early 1980s to the early 1990s there was a rapid *increase* in foodgrains output in Bardhaman district as elsewhere in West Bengal

Table 12.3
Crop-Wise and Gross Area Irrigated in Bardhaman District from the Damodar Valley Canal System (in hectares)

Year	Crop-Wise Area Irrigated by DVC			Gross Irrigated Area	
	Aman	Rabi	Boro	DVC	All Canals
1971–72	na	na	na	241,505	na
1974–75	na	na	na	na	300,700
1975–76	na	na	na	288,302	304,300
1976–77	na	na	na	269,800	269,800
1977–78	na	na	na	279,784	294,900
1978–79	na	na	na	319,702	335,600
1979–80	na	na	na	234,707	250,700
1980–81	na	na	na	303,333	319,384
1981–82	na	na	na	242,573	259,400
1982–83	na	na	na	225,596	238,200
1983–84	237,741	8,560	31,327	277,608	294,100
1984–85	241,624	8,017	31,845	281,486	294,000
1985–86	237,734	7,913	46,501	292,148	293,995
1986–87	237,830	7,414	42,624	287,868	na
1987–88	241,601	7,471	48,148	297,220	na
1988–89	240,049	7,781	27,548	275,378	na
1989–90	242,859	8,306	36,139	287,304	293,271
1990–91	239,476	7,665	47,348	294,489	312,363
1991–92	241,414	8,013	42,697	292,124	307,936
1992–93	117,656	na	na	na	na

Source: Annual Plan 1993–94, District Planning Committee, Bardhaman; GoWB Statistical Abstract 1978–1989. Bureau of Applied Economics and Statistics.

based on STW irrigation as well as adoption of HYVs (see Gazdar and Sengupta, this volume).

Canal irrigation has nevertheless been important. It has a pro-poor characteristic in that it requires no capital investment on the part of the user. All landowners within the command area have the option of cultivating with the canal water. In this way it is both scale and wealth neutral. But the DVC canal irrigation has had other somewhat unexpected consequences. The supply of canal irrigation, in terms of both timing and quantity, lies in the hands of the DVC authorities, who must balance future hydroelectric needs with current irrigation needs. Poor reserves or poor management can result in delayed or reduced delivery as the variations in the figures in Table 12.3 suggest. In addition there is the problem of a deteriorating canal system with serious problems caused by the silt carried down by the Damodar river into the reservoir and canal system.

Variations in the quantity and duration of the water provided and problems with delayed arrival cause serious problems for cultivators who must time the ploughing of fields and the transplanting of rice seedlings with some precision. The risks involved are quite considerable given the more capital intensive nature of boro cultivation. The presence of these risks has acted as an incentive for larger landowners to lease out land on a fixed contract '*thika*' lease (fixed rent seasonal tenancy), rather than to cultivate it themselves.[11] Thus, for much of the 1970s and 1980s, small and marginal farmers, even landless cultivators who sharecropped during the aman season, were directly engaged in the cultivation of boro paddy and enjoying the benefits of the agricultural growth occurring during this time, and not just through increased employment.

In the early 1990s the silting up of the reservoir and canal system reached a point at which it was finally acknowledged that it was no longer possible to attempt to supply the whole command area. Instead, in certain areas it was decided that the supply of canal water would alternate each year with first one side of certain main canals receiving water for boro cultivation, and then the other side. Risk was now replaced with a degree of certainty. There would be no canal water at least every second year in some village *mouzas*. At this same point in time, mini-submersible tubewells (MSTWs) began to take over from the earlier diesel powered tubewells in some rural areas.

A standard diesel powered 5 horsepower pump connected to a tubewell can raise water from approximately 8–10 metres. To aid the raising of water, the diesel pump is often dug some 2 or 3 metres into the ground. The electrically powered MSTWs can easily raise water from more than 15 metres below ground and thereby reach more secure water resources. The considerable risk of the tubewell running dry during the period of cultivation as the groundwater level falls is thereby reduced. In addition the electricity is subsidised and, in many instances, so are the loans provided for purchasing the MSTWs under the district's minor irrigation development programme. Banks, for a variety of reasons, are also more willing to lend money to individual creditworthy cultivators for the purchase of MSTWs.

The response of cultivators to the changing supply of canal water has been dramatic. Whereas previously the high risk factor involved in cultivating boro paddy had led to farmers often leasing their land out on thika contracts, the situation had now reversed itself. Owner cultivators have individually or jointly invested in an MSTW and proceeded to offer to all those in the 12 acre command area the option of a thika contract for the lease of the land. What they do not offer is the option of purchasing water from them.

JAYANAPUR AND MSTWs

There are two important points concerning the establishment of these command areas: first, the cooperation involved in their location to avoid competing overlaps; and second, the refusal of the MSTW owner(s) to sell the water, the sole offer being to take others' plots of land in the command area on a thika contract. The first is important because as long as there is only one source of water in the MSTW command area, the owner of that source is a monopolist. The terms of the contract can be kept low and the threat of another owner possibly supplying water to the many landowners instead of seeking to cultivate their lands is absent. The high returns to boro paddy cultivation are thereby secured by the MSTW owners.

Jayanapur and the neighbouring villages are all in the traditional canal command area with the majority of their lands able to utilise it. As with many other locations in Bardhaman district, from 1991 the canal water began to be allocated only every second year. But the deteriorating supply of canal water prior to the 1991 decision had already led to a large number of MSTWs being sunk in the village. By 1991 there were approximately forty operating within the village mouza. The *pradhan* of the local gram panchayat claimed that this was more than any other village in West Bengal at that time. Neighbouring villages still only possessed one or two MSTWs in 1991.

In Jayanapur, the significance of cooperation between owners and the 'water blackmail' being practised becomes apparent when we look at the village's landownership pattern (Table 12.4).

Table 12.4
Jayanapur Village: Land Ownership (Bighas)

Land Owned (in Bighas)ʳ	Number of Households	Percentage	Cumulative Percentage
Landless	75	29.3	29.3
0.01 to 2.5	30	11.7	41.0
2.51 to 5.0	29	11.3	52.3
5.01 to 7.5	25	9.8	62.1
7.51 to 10.0	34	13.3	75.4
10.01 to 12.5	4	1.6	77.0
12.51 to 15.0	17	6.6	83.6

Table 12.4 contd

Land Owned (in Bighas)	Number of Households	Percentage	Cumulative Percentage
15.01 to 17.5	4	1.6	85.2
17.51 to 20.0	14	5.4	90.6
20.01 to 22.5	1	0.4	91.0
22.51 to 25.0	5	2.0	93.0
25.01 to 27.5	5	2.0	94.9
27.51 to 30.0	5	2.0	96.9
30.01 and above	8	3.0	100.0
Total	256	100.0	100.0

As a bar chart the pattern is as represented in Figure 12.1.

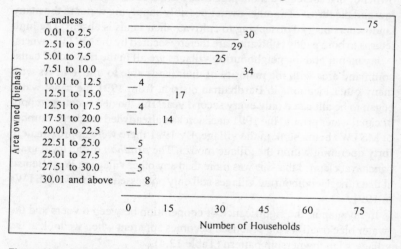

Figure 12.1: Jayanapur village: Land ownership

In this particular locality, 2.5 bighas are equal to an acre rather than the standard 3 bighas per acre found in most parts of West Bengal. Therefore, while 29 per cent of the village's households are landless, a further 13 per cent own less than 2 acres of land while a further 23 per cent own between 2 and 4 acres of land.

We now compare ownership of land with the ownership of MSTWs, operating with four categories of landownership: landless; up to 1 acre; between 1 and 5 acres; 5 to 10 acres; and over 10 acres (see Table 12.5).

The 'private' ownership of irrigation water is in the hands of the larger cultivators. The need for a degree of collective action in organising the

Table 12.5
Jayanapur Village: Land Ownership by Ownership of MSTW

| Land Owned | MSTW Ownership Status | | | |
(Acres)	Non-owner	Co-owner	Sole	Total
Landless	75	0	0	75 (29.3%)
0.01 to 1.0	30	0	0	30 (11.7%)
1.01 to 5.0	72	12	8	92 (36.0%)
5.01 to 10.0	26	11	4	41 (16.0%)
10.01 & above	5	5	8	18 (7.0%)
Total	208	28	20	256

distribution of canal water has ended. Instead a new form of collective action based upon wealth has arisen. In the early 1990s, an MSTW cost approximately Rs 35,000 to install (this can increase to more than Rs 70,000 if the electrical motor is more than 5 horsepower and driven by a separate diesel generator) and a further payment of Rs 1,260 per annum for the electricity consumed. In addition there is the need for a plot of land on which to sink the well, as well as the capital to invest in the seed, fertiliser, pesticides and insecticides necessary for high yielding boro paddy cultivation, and finally the acceptance by others in the village who own land in the command area that 'you' should be the one(s) to have an MSTW in that locality. Here another factor in addition to landownership, another identity, appears to be significant, namely that of caste (see Table 12.6).

Table 12.6
Caste by Ownership of Mini-Submersible Tubewell

| Caste | MSTW Ownership Status | | | |
	Non-owner	Co-owner	Sole	Total
Brahmin	3			3
Uggra Khatriya	98	27	20	145
Other general castes	1			1
Scheduled caste	104	1		105
Scheduled tribe	2			2
Total	208	28	20	256

It cannot be insignificant that only one scheduled caste household is involved in the ownership of an MSTW—in this case as a co-owner—

particularly when looking at the caste composition of the village as a whole, and the ownership of land by caste (see Table 12.7).

Table 12.7
Jayanapur Village: Caste by Landownership (Acres)

No. of Households by Caste (and % of Total)	Landownership in Acres				
	Landless	Under 1 Acre	1.01 to 5.0	5.01 to 10.0	Over 10 Acres
Brahmin 3 (1.2%)		1	2	0	
Uggra Khatriya 145 (56.6%)		6	81	40	18
Other general castes 1 (0.4%)		1			
Scheduled caste 105 (41.0%)	73	22	9	1	
Scheduled tribe 2 (0.8%)	2				
256	75	30	92	41	18

In the short term, it might well be that the most 'rational' course of action for the scheduled caste landless households that received the offer of a small plot of land was to take it and either rent it out to a water-owning cultivator or to use it in order to secure a loan or a capital sum through its (unofficial) sale. Debts could be repaid, ceremonies financed, and life marginally improved. To continue with collective cultivation would yield little material return. But what if they collectively invested in an MSTW and not only cultivated their collective land holdings, but also leased from others?

ARE THERE MSTWs FOR THE POOR?

Interestingly, arrangements exist—with the introduction of a Special Foodgrain Production Programme (SFPP) under the Central Sector Minor Irrigation (CSMI) for the District Rural Development Agency (DRDA)—to provide dug tubewells or filter points:

> ... for poor farming families usually from backward sections e.g. S.C./S.T./Bargadar/Pattaholders/Marginal Farmers etc., with a small extent of holding (even 10/15/20 cents) and who are not usually viewed by Bankers

as credit worthy at all. This scheme we are contriving to make use of as an effective follow-up action for the land reforms beneficiaries by making available to them the source of the critical input, the Irrigation, to make their very small holdings worthwhile and paying as far as practicable . . . The Community of D.T.W. [sic] are identified and initiated by Panchayat Samitis, prima facie verified and recommended by D.R.D.A. and sanctioned and maintained by WBSMIC Ltd. Some Community Schemes of cluster S.T.W.s with submersible pumps are also being implemented, sometimes with linkage of Bank Credit, sometimes with D.P. Fund (Integrated Rural Development Programme & Central Sector Minor Irrigation Under SFPP, Annual Action Plan 1989–90: 27, Bardhaman District Rural Development Agency, West Bengal).

In the 1989–90 plan under this particular programme (SFPP), the target was to provide the following:

Free Bore:	
1) Dug wells	160
2) Filter points	1,831
Community Schemes:	
1) Deep tubewells from minor irrigation	25
2) Mini-submersible tubewells from the zilla parishad	81
3) Mini-submersible tubewells with subsidised bank loans	598
4) Shallow tubewells with pumpsets	1,349
Others:	
1) Pumpsets	1,088
2) Shallow tubewells	255
3) Dug wells	70

The dug wells are in the dry and rocky regions of western Bardhaman where boro cultivation is extremely limited and wells and pumpsets are primarily to provide supplementary irrigation for the cultivation of aman or aus paddy, or for a rabi crop of vegetables. The STWs with either a centrifugal (diesel) pumpset or a mini-submersible (electric) one are primarily for the central and eastern blocks: Memari II, Bardhaman (now divided into Bardhaman I & II), Purbasthali I & II, Mangalkot, Kalna I & II, Jamalpur, Monteswar, and others.

The 598 MSTWs under the community schemes were to be constructed with a 50 per cent subsidised bank loan. The 1,349 STWs/MSTWs with pumpsets were also to be provided through subsidised bank loans, but with a 25 per cent subsidy for 'small farmers' (1 to 2 ha) and a 33.3 per cent subsidy to 'marginal farmers' (under 1 ha). Dug wells are also provided through bank loans with the same subsidies.

What are not covered by these figures are the numbers of tubewells for irrigation that are being privately sunk. The last data covering all MSTWs and STWs were collected in 1988. At that time Bardhaman district was recorded as having the following:

Deep tubewells	390
Mini-deep tubewells	223
Shallow tubewells	
1) Electric	3,130
2) Diesel	33,446
3) Manual (filter point)	471
4) Others	0
Total	37,047
Dug wells:	
1) Mechanical	124
2) Manual	387
Total	511

Since 1988 the only source of data indicating the growth of MSTWs anywhere in West Bengal has been from the requirement since September 1993 that anyone wishing to obtain a subsidised loan and/or a connection to the mains electricity supply must obtain an 'in order' certificate from the State Water Investigation Department. The principal reason for introducing this requirement has been the increasingly rapid depletion of groundwater each year.

The number of 'in order' certificates issued up to the end of June 1994 is as follows:

District	Number Issued
Bardhaman	694[1]
Nadia	11
Murshidabad	360
Hugli	1,100

Contd

District	Number Issued
Birbhum	139
Medinipur	60
North 24–Parganas	3
Total	2,367

[1] This figure is composed of 374 issued plus 320 certificates ready for issue. In addition to these 585 are under consideration and expect to be issued; 234 applications have been referred back due to incomplete data; 326 have been rejected.

In the fourteen months since the implementation of the new order, there have been certificates issued for MSTWs equivalent of three times the total number of MSTWs recorded as existing in the district in 1988.

Diesel pumpsets and, to a lesser extent, electrically powered pumpsets located on the surface have been used for some time in Bardhaman district. They are less expensive, more mobile and can be used for irrigating from tanks, canal and rivers. A mini-submersible structure is a far more permanent source of irrigation. It costs more to sink the bore, to place the pump and, if directly connected to the electricity supply rather than to a generator, to arrange for the link to a local transformer (approximately six MSTWs to one sub-transformer) and the connection to the mains. This means that both the mini-submersible structure (and therefore the command area) is a far more fixed entity. Once the full privatisation of irrigation is achieved, which can only be a few years away in the most important rice growing areas of West Bengal, the small and marginal farmers will have been firmly excluded and a process of polarisation introduced into that section of the agrarian economy based upon production (for comparable analysis with respect to Bangladesh, see the papers by Adnan and Wood in this volume). Are there lessons to be learnt and measures to be taken to prevent or reverse this process?

THE LESSONS: THE ACTORS AND THE STRATEGIES

The most important lesson is not to place too great a faith in a close coherence between policy intervention and outcome. Development is the outcome of interactions between different actors at different levels. Furthermore, the presence and role of actors can change dramatically over time. Therefore, the intended is rarely achieved in development. The unintended also occurs. If we briefly examine the actors involved in the scenario above, the unintended anti-poor dimension of the processes at work in rice cultivation can perhaps be best understood (in this analysis, statements of intent are taken at face value).

The Marginalised Poor

The social constellation of the poor in Jayanapur is defined in terms of their landless or marginal landowning status, their caste, and their political status. (Gender-based poverty is of course present, but is subsumed here within the poverty of the household—there are no female headed households in the village and only the beginnings of positive discrimination towards women in the Left Front's agrarian strategy.) The identities these give rise to are counterposed to those of the landed, high caste, wealthy and politically influential. The processes involved result in dichotomies of poverty and wealth, illiterate and educated, etc. As a group of actors, the options of the poor are limited to the tactics of day-to-day survival rather than enabling a strategy of improvement or transformation. Institutions are found in forms of households, paras (neighbourhoods), occupational groups, etc. These primarily serve the needs of survival rather than of improvement. The collective cultivation of the vested land that we found was not part of any strategic plan emerging from decisions made by these actors or their institutions. It came from outside. Knowledge of how to access other productive assets, specifically water from MSTWs, was not present. In the short term, a small plot of land is an asset that can realise loans, repay debts, finance dowries, etc. It can fulfil an immediate need; its utilisation in this way is a part of a common tactic for survival, no more.

The Local Political Leadership

In Jayanapur, the gram panchayat in 1991 was controlled by landowning Uggra Khatriyas. They also dominate the local Krishak Samiti and the CPI(M). This is not an oppressive local regime practising caste and class oppression and exploitation. Far from it—as land reform has been implemented, sharecroppers have been registered and agricultural wages are close to the legally required minimum. It is merely a local community in which the norm remains that of progressive benevolence implemented by an enlightened local elite who are in turn guided by a leftist political party that sits in government. The investment of these actors in MSTWs and the increase in production that it gives rise to does not run counter to the implementation of land reform, the recognition of sharecroppers' rights, the better payment of agricultural labourers, or the pursuit of pro-poor development programmes through panchayati raj.

Land reform was a response to the concrete demands of the poor in the 1960s and 1970s. The Left Front was bound to take this up once it was elected. At a local level, acceptance of land reform was effectively a condition of coming into the mass fronts and parties of the left, notably the CPI(M). But water? Water is local, it lacks a movement, it requires the granting of rights and opportunities beyond the usual, the norm. Consequently, no attempt was made by the local political elite to take up the issue and to secure water ownership for the poor.

The State and LFG

Just as the local political elite in Jayanapur pursues its collective pro-poor political interests alongside the pro-cultivator economic interests of the individuals involved, so the LFG at the state level pursues a pro-poor agrarian strategy while seeking to sustain its party-specific electoral interest in retaining power (see Bhattacharyya, this volume). The absence of a political policy on irrigation, and MSTWs specifically, has left policy to the administrative side of the state.

For the administrators, irrigation is a technical problem; it is also a basis for corruption. Over-exploitation of groundwater has led to a halt on sanctioning any more MSTWs in one block and a very careful monitoring in thirty-four other blocks in West Bengal. Today, new MSTWs must be 200 metres from any existing MSTW, with some variation according to region. All landowners within a proposed MSTW's command area must be party to the application. Verification that such conditions are met and implemented rests with the panchayats. Today, all applications are processed by the panchayat samitis and passed by the *sabhapati* (panchayat samiti chairperson) to the zila parishad level where the final decision is made in the zila parishad irrigation committee chaired by the *sabhadipati* (zila parishad chairperson).

The technical requirements, while helping to protect them against the increasingly rapid groundwater depletion occurring each year, preserves the water monopoly of the MSTW owner(s). The provision that all owners be party to the agreement is both impossible and blatantly ignored. The high level at which decisions on MSTW allocation are taken might introduce greater political accountability, but it further distances the decision from the poor in the village in terms of both bureaucratic and social space. There are also rumours that both political bias and personal corruption are present, both in getting the support of the panchayat institutions and in obtaining the certificate from the State Water Investigation Department.

At best, we can describe the situation in Bardhaman district as revealing a lacuna in the LFG's pro-poor agrarian strategy. There appears to be little awareness of the nature or extent of the problem. Meanwhile, waterlordism creeps onwards, increasingly affecting the distribution of the benefits occurring from higher agricultural production and eroding the small producer basis of that success of which the LFG is justifiably so proud.

The Facilitator?

The missing actor is the local facilitator. The state, the parties of the Left Front, the various administrative officers, officials from the departments concerned with development, and the individual cultivators—these are the actors most would identify. But then there are also socio-economic actors—the farmers, the landless, the sharecroppers, the landowners—and cultural actors—the scheduled castes and tribes, the general castes, the high castes. These actors exist in the politics of the left as those to be mobilised or opposed, the fermenters of agrarian politics and agrarian class struggles.

But what happens when the state does not act; the parties of the left are the parties of government; and the social groups that provide the broad bases for social movements have been co-opted to the needs of the left parties in government? What other actors can have local agency in such a political context?

The answer can be found in two elements. The first consists of what we call local social constellations of the poor whose experience of poverty, of social and political marginalisation, is experienced in terms of their gender, ethnicity, caste, occupation, religion, etc. One or more of these is central in determining the immediate form of their discrimination in those markets vital to the securing of their material condition. These include markets for labour, for land, for water, for development services, for political patronage, for agricultural inputs, and for social and political rights. The second element is that of the facilitator, the actor who intervenes to bring the means in terms of both ideas and logistical support to help realise the needs and potential to be found within the local social constellations of the poor.

The facilitator might be an individual or a local organisation. It could be a local political or community leader, a grassroots NGO or People's Organisation (PO), a local party, or a workers' or peasants' organisation. There are many possibilities. In the original cooperative in Jayanapur, leaders within the Krishak Samiti appeared to have provided this role. They brought experiences from other movements, ideas from other localities and other countries gained through literature, travel and discussion. But their role was that of facilitating the realisation of something the logic for which was to be found within the local agrarian social formation, and specifically the village of Jayanapur. It was expressed in the need to counter the disadvantages of relatively small and highly fragmented holdings, in the need to counter other players in markets for grain, inputs and credit, albeit at a local level, and perhaps in the need to support households that provided labour during times of hardship.

In our more general research into grassroots production cooperatives we have proposed the following diagram as a visual representation of our ideas.

Diagram 12.1: The Potential for Local Grassroots Collective Action

LOCAL LEVEL

Local social constellations of the poor

Basis: - common identity
- common crisis
- common 'habitus'
↓
Social Cooperation
(e.g., in fishing, chit groups, labour gangs, etc)

↑

Local Facilitator

- NGO local level organisation
- local club
- local party broker
- local social leader
- 'traditional' institutional leader

Strategic institutional intervention
(e.g., credit & savings groups, grassroots cooperatives, irrigation user groups, agricultural production cooperatives, etc.)

Diagram 12.1 contd

MEZO LEVEL

NGOs/POs	*Political Parties*	*Local Government*
Strength: Can mobilise marginal groups based upon specific identities; flexible in adopting new approaches more efficient in resource use; better at participatory rural appraisal (PRA).	**Strength:** Local and mass mobilisation; links to higher levels of party and to the state (government and administration).	**Strength:** State backing with resources (personnel, finance, etc.); popular participation; local input into planning; access and influence over administration.
Weakness: Questionable legitimacy; donor dependency; long term role; ability to leave; geographically limited; internal autocracy; reluctant to network/ cooperate with government.	**Weakness:** Need for electoral support party before local people; popular participation but not necessarily popular initiative.	**Weakness:** Only as good as state permits problems of integrating horizontal councils with vertical departments; lack of political dynamic; 'bureaucratic politics'.
Facilitatory Support: (i) direct: e.g., - financial - training - support staff - institutional mediation - technical expertise (ii) indirect - advocacy as 'macro'/ corporate actors	**Facilitatory Support:** - ideology - political pressure - political mobilisation - access to higher level brokers - etc.	**Facilitatory Support:** - financial - organisational - technical expertise - mediation with other parts of the state - etc.

MACRO LEVEL
(NATIONAL AND INTERNATIONAL)

STATE

Aim:
To generate and structure the political context and space necessary to facilitate the participation of the rural poor in local development. This is to be achieved through:
(i) Structural reform—land reform, rule of law, securing of rights for the poor (land, justice, access to key markets, minimum wages, social security).
(ii) the facilitation, generation, and resourcing (technical, financial, personnel, ...) of institutions and processes that facilitate the poor's own mobilisation for, and participation in, development.

It must be understood that the specific context will be extremely influential in determining the form of facilitation. In West Bengal, during periods when the left parties have been in opposition, it might well be that they themselves played this role in mobilising around local issues. Elsewhere, the total absence of the left and the abandonment of local politics to traditional patrons might well create the political space for local NGOs and POs, especially if the state is increasingly willing to accept their developmental role. The current situation in the state of Karnataka illustrates the latter well. The separation of NGOs/POs, political parties and the institutions of local government at the mezo level is designed to open up the discussion of the different and complementary roles that might exist among these types of organisation.

BY WAY OF CONCLUSION: FROM THE POLITICS OF MASS MOVEMENT TO THE POLITICS OF LOCAL ACTION

In 1977 the political expression of the collective identity of the poor and the marginalised brought the Left Front government to power. The election of this government was the product of a mass social movement. Through its policies, the LFG then proceeded to secure certain key rights and conditions for many of those impoverished through the agrarian structure and marginalised by the power relations found there. Land reform, Operation Barga, minimum agricultural wages, and panchayati raj are the most important examples of an agrarian strategy that aimed at securing certain key rights as well as introducing new processes that could enable a more equitable distribution of the benefits of improved agricultural production and a better allocation of development resources originating from outside the local agrarian economy.

However, the success of this agrarian strategy in terms of both the pattern of growth achieved in the agricultural economy and the extension of empowerment achieved by the decentralisation of the political system has resulted in the end of a mass movement based upon general conditions of impoverishment and exploitation in the agrarian sector. Instead, today, new and more localised social constellations of the poor have taken on a significance for local agrarian development previously not recognised. These social constellations are to be found in the local manifestations of relations rooted in gender, ethnicity, specific occupations, specific migrations, specific forms of marginalisation, specific exclusions from markets and services. While the existence of the local social identities that these give rise to might be acknowledged, this existence is not being reflected in either the agrarian or the political policies of the Left Front government. Most importantly, the development potential that these social constellations of the poor contain, in particular their local grassroots character, is not being realised.

With twenty years of Left Front rule in West Bengal, the ability and the facility for general resistance and revolt within the agrarian social formation has been eroded. It can be argued that the need for broad-based rural revolt has been (temporarily) removed. But due to this, and given the prevailing politics of the agrarian development process, other forms of collective action have assumed a greater potential. Alongside the old actors of class groups, political parties, and the state with all its instruments, new actors have emerged and the Left Front government must openly acknowledge their existence if it wishes to combine the role of government with the role of securing development for the rural poor.

NOTES

1. Assessments of panchayati raj do, however, vary considerably. Some researchers have been far more critical of the effectiveness of the programme. Ross Mallick presents a very negative picture (1992); see also Ruud's discussion of CPI(M) as the new patrons of local politics (this volume). Westergaard (1986) is a more moderated criticism of the performance of the gram panchayats under the LFG. See also Williams (this volume).
2. The author conducted the fieldwork for the research in collaboration with Nripen Bandyopadhyay from CSSS (Calcutta) and with the assistance of Bela Bandyopadhyay, Asis Das and Subendhu Chatterjee.
3. At this time the Communist Party of India (CPI) had not split into the CPI and CPI(Marxist). Bardhaman became a stronghold of the CPI(M) after the split.
4. In fact the Bengal Kisan Sabha had been founded in another neighbouring village in 1933 during the general wave of agrarian resistance to colonial rule at that time. Bardhaman district was the centre for a series of movements against the colonial authorities, in particular the Canal Tax Movement of the late 1920s and early 1930s, which established the basis for its later reputation of being the communist movement's heartland in West Bengal.
5. The government sought to tax the cooperative as a single agricultural holding imposing an impossible burden on the individual members who ultimately withdrew their lands.
6. See Webster (1994); also Rogaly (this volume) for an analysis of the role of such migration in agrarian change in West Bengal.
7. I wish to emphasise that it is at this particular conjuncture of agrarian development in West Bengal—that is, the economic basis of production, the social characteristics of the agrarian social formation, and the nature and balance of political forces—that room for manoeuvre exists for non-party organisations (NGOs etc.) to facilitate the emergence of local grassroots institutions of the poor as part of the process of rural development.

8. See Webster (1989) for a more detailed discussion of the impact of the 'green revolution' on agrarian relations in Bardhaman district in the 1970s and 1980s.
9. A landless household in central and eastern Bardhaman district can now expect at least eight months of agricultural employment in the year, often more. This is then supplemented by employment in wall repairs, house construction, government programmes, etc. Much of the agricultural employment, transplanting and harvesting in particular, will be paid in the form of cash and food/rice at close to, or on occasion over, the minimum required wage of Rs 19.25.
10. I use the term landlordism somewhat loosely to cover the network of dependency relations that most villagers suffered under. It embraces '*begar*' or unpaid obligations, '*mohajans*' or village moneylenders demanding a minimum of 10 per cent per month interest, sharecropping contracts that saw sharecroppers becoming more and more indebted yet unable to break the relationship, paddy hoarding, and much more.
11. In this way they secure a guaranteed income, either in kind from the harvest or through debt owed by the cultivator after a failed season (see Bandyopadhyay, 1978 and Webster, 1989).

REFERENCES AND SELECT BIBLIOGRAPHY

Bandyopadhyay, Nripen. 1978. 'Changing Form of Agricultural Enterprise in West Bengal,' K. Bagchi (ed.) *Problems of Agricultural Growth in India*, Department of Geography, University of Calcutta, Calcutta.
Boyce, James K. 1987. *Agrarian Impasse in Bengal: Agricultural Growth in Bangladesh and West Bengal 1949–1980*. New York: Oxford University Press.
Datta, P. 1992. *Second Generation Panchayats in India*. Calcutta: Calcutta Book House.
Government of West Bengal. 1990. *Statistical Abstract West Bengal 1978–1989 (Combined)*. Bureau of Applied Economics and Statistics, Calcutta.
———. 1993a. *Economic Review 1992–93*, Calcutta.
———. 1993b. *Economic Review 1992–93, Statistical Abstract*, Calcutta.
Harriss, John. 1993. 'What is Happening in Rural West Bengal? Agrarian Reform, Growth and Distribution,' *Economic and Political Weekly*, 28(24): 1237–47.
Korten, D. 1990. *Getting to the 21st Century: Voluntary Action and the Global Agenda*. New Delhi: Oxford and IBH Publishing Co.
Lieten, G.K. 1994. 'For a New Debate on West Bengal', *Economic and Political Weekly*, 29(29): 1835–193.
Mallick, Ross. 1992. 'Agrarian Reform in West Bengal: The End of an Illusion,' *World Development*, 20(5): 735–49.
Saha, Anamitra, and Madhura Swaminathan. 1994. 'Agricultural Growth in West Bengal,' *Economic and Political Weekly*, 29(31): 2039–40.
Webster, Neil. 1989. 'Agrarian Relations in West Bengal: From the Economics of Green Revolution to the Politics of Panchayati Raj,' *Journal of Contemporary Asia*, 20(2): 177–211.
———. 1992a. 'Panchayati Raj in West Bengal: Participation for the People or the Party?' *Development and Change*, 23(4): 129–63.
———. 1992b. *Panchayati Raj and the Decentralistaion of Development Planning in West Bengal*. Calcutta: K.P. Bagchi.
———. 1993. 'Cooperation, Cooperativism and Cooperatives in Rural Production, West Bengal,' *Journal fur Entwicklungspolitik*, 9(3): 309–28.

Webster, Neil. 1994. 'Tribal Women's Cooperatives in Bankura, West Bengal,' *European Journal of Development Research*, 6(2): 95–103.
Westergaard, Kirsten. 1986. 'People's Participation, Local Government and Rural Development: The Case of West Bengal.' Research Report No. 8, Centre for Development Research, Copenhagen.
Wood, Geoffrey D. and **Richard Palmer-Jones.** 1990. *The Water Sellers: A Cooperative Venture by the Rural Poor.* London: IT Publications.

13

Dangerous Liaisons? Seasonal Migration and Agrarian Change in West Bengal

BEN ROGALY

INTRODUCTION

Rapid growth in the output of rice in West Bengal during the 1980s and early 1990s was associated with an increase in the quantity of hired labour demanded by smallholder landowners. In southern central West Bengal, in particular in the districts of Bardhaman and Hugli, double cropping of rice and cultivation of high yielding varieties have become standard, made possible by the widespread use of groundwater irrigation. Yet, transplanting and harvesting continue to be done by hand, although other tasks, including ploughing and threshing, have been mechanised. The increased quantities of rice transplanted and harvested has required more workers and seasonal peaks in demand are now sharper than ever before.[1]

I am grateful to Arjan De Haan and Barbara Harriss-White for comments on an earlier version of this paper, to Barbara Harriss-White and Sunil Sengupta for guidance during the research, and to the Economic and Social Research Council (UK) for funding. Errors and omissions are mine.

The theme of this book is the interrelation between growth in agricultural output and changes in agrarian structures. 'Agrarian structure' was a term commonly used in the past to refer to the degree of concentration in the distribution of landholdings, and to the organisation of labour in production (in particular to distinguish between different types of tenancy arrangements, use of hired labour, and cultivation by the owner of the land). The more inclusive definition used for the purpose of this book expands the concept of agrarian structure to include structures of commerce, of exchange arrangements in land, water and labour, as well as changing ideologies of gender, caste and ethnicity (see the Introduction by Rogaly, Harriss-White and Bose, this volume).

The seasonal migration of agricultural workers in West Bengal is an element of agrarian structure—the organisation of transplanting and harvesting work, with their very sharp seasonal peaks in demand, relies on this form of labour hiring. However, in this paper I will argue that the increase in seasonal labour migration in the presence of growth in rice output has caused further changes in agrarian structure. In particular, I will show that seasonal migration causes changes in other labour arrangements between employers and workers (non-migrant as well as migrant) in the migrants' source and destination areas. Further, increasing seasonal migration involves crossing boundaries associated with ideologies of caste, and is thus a part of the process of changing those ideologies.

Lenin found the seasonal migration of agricultural wage workers in 19th century Russia to be at the same time dangerous and potentially progressive in terms of its impact on feudal employer-worker relations in the source areas. The annual movement south of two million people entailed great deprivation. Boats capsized; employers were able to wait until people's meagre supplies were exhausted and then negotiate lower wages; working days lasted twelve to fifteen hours often in insanitary conditions. Yet Lenin argues that the migration of agricultural wage labour was progressive because workers earned higher wages than in the source areas, bonded labour and labour service were destroyed and people developed 'from an independent acquaintance with the different relations and orders of things in the South and in the North, in agriculture and in industry, in the capital and in the backwoods' (1964: 240–54).

Writing on pre-independent India, on the other hand, Omvedt argues that labour migration was more an expression of relative immobility because migrants remained tied to landlords in their home villages (1980, cited by Teerink, 1995). Teerink's (1995) contemporary study of seasonally migrant men and women from the Khandesh region of Maharashtra and their employment as harvesters by very large sugarcane crushing factories in Gujarat appears to bear this out. Migration for these workers was compelled by the exclusive employment practices of employers in workers' home areas; the fact of returning did not bring about any change in these source area relations.

In a similar vein to Omvedt, Rudra argues that in contemporary India, the seasonal migration of rural workers maintains the rigidity of local labour-hiring practices. Wages are determined within villages according to the relative power of employers and workers, each group combining in its own class interests which are opposed to those of the other. 'Village ideology', a third factor determining who should work for whom, imposed boundaries on workers hiring out to employers in adjacent villages (Rudra, 1984; see also Bardhan and Rudra, 1986). Rudra argues that the seasonal migration of rural labour enabled rigid employment boundaries between neighbouring villages to be maintained. Workers driven to looking for work outside the village to survive were able to find it further afield, while employers suffering a labour shortage could hire in workers from different districts.

This paper investigates the relation between seasonal migration of agricultural workers in West Bengal and changes in agrarian structures. This relation works both ways: different experiences of migration reflect different agrarian structures, but can also influence those structures. First, however, it is necessary to describe the quantitative importance of seasonal migration in the organisation of agricultural production in West Bengal.

AGRICULTURAL EMPLOYMENT IN WEST BENGAL: THE IMPORTANCE OF MIGRANT WORKERS

Of West Bengal's rural workforce of 15 million recorded in the 1991 census, 5.8 million reported their primary occupation as crop cultivation and 4.9 million as agricultural labourers.[2] Of these, 470,000 and 960,000 respectively were women. Compared to the all-India rural average therefore, West Bengal has a high proportion of agricultural labourers and, according to the census figures at least, a low rate of female participation in the labour force.[3]

There has been a continuing decline in the importance of share tenancy (Sengupta, 1981; 1991; Bose, 1993) in contrast to own-account cultivation. Despite the increasing incidence of fixed rate tenancy reported by Webster (this volume), own-account cultivation by a combination of family and hired labour is the dominant form of production organisation in the rice producing areas of the state.[4] Landholdings are fragmented and their sizes are small compared to elsewhere in India. The state has long had a high proportion of very small-scale cultivators, defined as those with less than 5 acres of land (Sengupta and Gazdar, 1997). The intention of smallholders in West Bengal to secure seasonal supplies of workers and to control both local and migrant workforces is a product of their own relatively limited resources.[5]

Agricultural workers come from all over West Bengal and from eastern and southern parts of Bihar seeking employment in the main double-cropped paddy growing districts of Bardhaman and Hugli. Sources of labour include Bankura, Birbhum, Malda, Murshidabad, West Dinajpur and Jalpaiguri districts in West Bengal and Singhbhum and Purnea districts in Bihar (see Chakrabarti, 1986; Banerjee, 1988; Chatterjee, 1991; Choudhuri, 1992; Singha Roy, 1992; Webster, 1993; Rogaly, 1994; Rao and Rana, 1997). Choudhuri observed the activities of all members of forty-four agricultural worker families in two villages in the western part of Bankura district for seven days per week over a twelve month period. 'A substantial part of the landless agricultural labourers, sharecroppers, marginal farmers, etc. go to Hugli and Bardhaman to work as agricultural labourers and as hired labourers for various secondary activities' (1992: 30). Thirty-three of the families took part in seasonal migration during the period of study (ibid.: 40). Peak periods of employment migration in both villages were the months of Baisakh and Jaistha (mid-April to mid-June) and from Agrahayan to Falgun (mid-November to mid-March). These periods included the boro paddy harvest, aman paddy harvest and boro paddy transplanting. Because the aman harvest comes earlier on the red laterite soils of western Bankura district, employment in the migrants' source area harvest is concentrated in Kartik and Agrahayan (mid-October to mid-December). Migration was much less common during the migrants' own villages' peak employment period of aman paddy transplanting in Ashar and Sraban (mid-June to mid-August). Nevertheless, large-scale migration does occur at the time of the aman transplanting from elsewhere in Bankura and Purulia districts, and as we shall see this can be the source of major conflicts between employers in source and destination areas.

This movement of workers is by no means new. Seasonal migration into and across parts of Bengal was common in the late 19th and early 20th centuries (Yang, 1979; van Schendel and Faraizi, 1984: 50–7; Sen, 1992: 26; Bose, 1993: 86). McAlpin's work on the Santals of western

Bengal documents the annual movement of large numbers of Santals at the times of sowing and harvesting paddy (1990, cited by Chattopadhyaya, 1987: 515). Many Santal migrants sought work in Bardhaman, where they earned 5–6 annas per day (ibid.). Yang's research on west–east migration from the north-western part of present day Bihar shows that while at first most of the longer distance migrants were men who left families behind and walked to their destination, with the coming of trains and the improvement of information from returnees, increasing numbers of women and children used to join the cold season movement (1979: 53). In 1891, a year of severe flooding in Saran district, it was estimated that over 80,000 people migrated east for the season.

However, the number of migrants has increased rapidly from the mid-1980s (Banerjee, 1989–90: 200). One survey of agricultural migrant workers from Manbazar II block of Purulia district found that the number of people leaving the block to seek work in the aman harvest had increased from 679 (from 10,573 resident households) in December 1991 to 1,083 in December 1996 (Biplab Basu, personal communication, 11 December 1996). Over a period of fifteen days in June–July 1991, the Bankura correspondent of the *The Statesman* estimated that 70,000 labourers boarded special buses from Govinda Nagar bus stand in Bankura town. They were headed for transplanting work on paddy farms in Bardhaman and Hugli districts (Debajyoti Chattopadhyay, personal communication, 2 March 1992).[6]

Given its importance for both employers and workers, it is surprising how little research has been done to investigate the causes and consequences of seasonal labour migration in West Bengal. The remainder of this section reports on a study carried out as part of a broader piece of research on hired labour arrangements. Field research was conducted in two localities between June 1991 and March 1992 and for a further month in October 1993. The localities were selected to include one source and one destination area for seasonal migrants and will be referred to by the names of their respective districts: Purulia and Bardhaman. Ninety-two households were selected randomly from wealth groups organised via a ranking exercise carried out individually with six key informants. As the localities were deliberately chosen to be relatively far from urban centres, the main business in both was agriculture, also the major source of employment. Almost all the selected households in both localities either hired in or hired out agricultural labour during the year and were thus classified either as workers or employers.[7] The Purulia locality is characterised by unirrigated paddy production, with skewed landownership. The range of landholding sizes extended up to 16 acres, with a median of 1.3 acres. In the Bardhaman locality, river-lift and groundwater irrigation ensured at least two crops per year, usually aman paddy followed by either boro paddy or potatoes. The landholding distribution was more heavily skewed than in Purulia, with holding sizes of up to 24 acres, with a median of 0.4 acres.

Data was collected through informal interaction with sampled workers and employers in both localities, made possible by the engagement of two research assistants, Khushi Dasgupta and Paramita Bhattacharyya, who lived continuously in each locality for most of the main fieldwork period. In addition, sampled workers and employers were asked to record daily the terms and conditions of labour hire. In the aman paddy harvest in December 1991, more days of labour were hired in via seasonal migrant arrangements than any other type in the Bardhaman locality and more days hired out as seasonal migrants than were engaged locally by workers in the Purulia locality. Aman paddy cultivation was common to both localities, although workers from the Purulia locality managed to complete their own harvests before the main migration period in early December.

The main concern of this paper is to analyse agrarian change emerging out of increasing seasonal labour migration in West Bengal. In order to understand how agrarian structures more widely respond to increased seasonal migration, it is necessary to know how migration itself is structured: who migrates, who employs migrants, and how the process is organised.

The fact of large-scale seasonal migration across West Bengal in 1991–92 was not surprising. Migration by some and the hiring of migrants by others reflected spatial differences in demand, supply and earnings prospects. Effective wages varied within each locality, measurement of which was further complicated by the different types of hired labour arrangement (Rogaly, 1996; 1997a). However, while in the Bardhaman locality a single 'going rate' for daily time rate arrangements was a key reference point from which actual payments varied, in the Purulia locality, daily time rate wages were negotiated separately according to season, task and gender. In both localities, however, disagreements over rates of pay were brokered by CPI(M) cadres. The brokered rate for a day's work in the Bardhaman locality aman harvest was Rs 12 with 2 kg of hulled rice (equivalent to approximately Rs 20 to Rs 25 per day), while the rates in the Purulia locality ranged from Rs 10 to Rs 15 per day. Moreover, potential migrants in the Purulia locality knew from previous experience and from the stories of others that they could expect continuous employment from arrival to departure. And the supply of migrant workers in the Bardhaman locality did not displace local workers, who were fully employed during peak harvesting and transplanting periods.

As we shall see, however, the structure of migration was not explainable in terms of market forces alone. Decisions to migrate were embedded in locally specific social and economic structures, and served the interests of employers in destination areas in controlling their workforces (as well as the interests of source areas workers in improving their local labour relations).

Results from regression analysis reported elsewhere strongly suggest that the poorer the labour-selling household, and the fewer the number of dependents (those unable to carry out manual work), the greater the likelihood of migration at the aman paddy harvest (Rogaly, 1997a). Analysis of the uses of remittances suggest clear differences in need among sampled migrants and also provide insights into how the fruits of seasonal migration (and thus to some extent the migration itself) are viewed by workers. Most migrant labourers from the Purulia locality expected to return with a lump sum of cash and some kind 'savings'.[8] They knew that they were likely to receive a kind amount for subsistence each working day and that there was a cash part of the daily wage, which would be paid at the end of the season. Table 13.1 illustrates the uses made of 'remittances' by migrant labourers on their return to the source locality.

Table 13.1
Uses of Remittances by Seasonal Migrants from the Purulia Locality in the Aman Harvest, November–January 1991–92

Migrant	Amount	Use
BS and daughter	Rs 450, 20 kg hulled rice	Daily consumption for 1.5 months (previously in long duration contract); Rs 100 towards repaying shopkeeper NM; medicine, doctor's fees.
GS		Marriage expenses.
KM	Rs 400, 10 kg hulled rice	Cash contributed to the cost of 1 bigha of *baid* land bought for Rs 900 from NB of a neighbouring village. Rice consumed.
JS	Rs 325, 10 kg hulled rice	Repayment of mortgage loan to release 0.75 bigha land from lender of same *jati* and *para*. Rice consumed.
HA	Rs 680	Repaid loans. No kind 'savings' because two children also migrated.
MR	Rs 400, 30 kg hulled rice	Rs 100 spent on clothes for *mela*; remainder on daily consumption.

Table 13.1 contd

Migrant	Amount	Use
AR	Rs 234, 9 kg hulled rice	Clothes: 1 *lunghi*, 1 saree, 3 *jama*.
NR	Not paid	
SM	Rs 345, 12 kg hulled rice	Clothes Rs 100; loan repayment Rs 150; daily consumption Rs 100.
SS	Rs 900, 20 kg hulled rice	Payment to child-minder in camp, Rs 40; Pous mela clothes: saree, Rs 200; petticoat, Rs 15; trousers, Rs 150; gambling Rs 50; goat, Rs 150; in-laws, Rs 30; daughter, Rs 40; other, Rs 50.
MS	Rs 400	Daily consumption; no kind 'savings' because daughter accompanied her.

Aggregate household remittances ranged from Rs 234 and 9 kg of hulled rice to Rs 900 and 20 kg of hulled rice. This included migrations of different lengths ranging from three to five weeks. There was also variation in the number of household workers migrating, and in daily earnings. Remittances were used for daily consumption, loan repayments, marriage expenses, new clothes and the purchase of productive assets.

JS was recently widowed and left with two young children and a debt of Rs 400 against which her late husband's land had been mortgaged. Rs 150 was already owed in interest to the lender, a neighbour and fellow Bhumij (KS). KS had a successful cock-rearing business. JS migrated out of need as part of a mixed-sex gang. Her children accompanied her to the destination and the youngest was breastfed at the workplace. JS used her remittances to repay KS and release her mortgaged land.

BS, another Bhumij widow, previously worked on a long duration arrangement with SM, the owner of the rice husking mill. She had been reliant on regular local employment to provide for her son and daughter. Now that her daughter was approaching puberty, she calculated that together they would be able to survive long periods of scarce local employment possibilities by migrating seasonally. She left her regular employment and migrated with her worker daughter and dependent son for the aman transplanting in 1991.

Agricultural workers in the Purulia locality belonged to a number of jati (social ranks) and lived in para (neighbourhoods) partly defined by either jati, class (worker or employer) or both. Table 13.2 summarises the class and jati composition of each para in the locality. Men and women of caste Hindu households (Goalas, Tamulis and Brahmins) did not hire out their own manual labour, regardless of their wealth. However, both men and women of the other jatis did, including as migrant workers.

Table 13.2
A Summary of the Caste and Class Composition of Para in the Purulia Locality

Para	Jati		Class
Santal	Adivasi	Santal	Labourers
Kodallota	Adivasi	Bhumij	Labourers
Bhuinyadi	Hindu	Bhuinya	Labourers
Mahaltar	Hindu; Adivasi	Brahmin, Goala; Bhumij	Mixed
Mandal	Hindu	Goala	Employers
Namo	Muslim; Adivasi	Muslim; Bhumij	Labourers
Ager	Hindu	Goala, Tamuli	Employers
Pichan	Muslim; Adivasi	Muslim; Bhumij	Labourers

Table 13.3 indicates the number of seasonal migrant workers leaving from each para in the aman harvest of 1991. In total fifty-six women and fifty-four men from the 247 households resident in the locality migrated eastwards that season.

Table 13.3
The Number of Individual Seasonal Migrant Workers Leaving the Purulia Locality for the Aman Paddy Harvest, Late November–Early December 1991

Para	Number of Migrants		
	Male	Female	Total
Santal	13	12	25
Bhuinyadi	9	8	17
Mahaltar	3	3	6
Namo	11	11	22
Pichan	12	15	27
Kodallota	6	7	13
Total	54	56	110

Contrasting ideologies of work, in particular ideas about being tied to a particular employer even for a short time, were evident in the practices of the different labour-selling jati. Santal workers in the Purulia locality valued autonomy from caste Hindu and Muslim employers more highly than other jati. They organised informal grain banks to provide loans to hard up households in the lean season so that there would be less need for recourse to labour-tying advances. Santals' communitarian provisioning was manifest in the different ways in which Santal migration—as opposed to migration by Bhuinya, Bhumij or Muslim workers—was organised. Santal migrant gangs were organised by a *sardari* system, involving labour leaders (or sardars) negotiating on behalf of members of their labour pools. Two men, BM and ST, maintained regular relations with employers MK and PS respectively in the destination areas. When work was offered, the sardar returned to Santal para (or sent word) and recruited individual gang members of both sexes.

Migration by other labour-selling jati was organised along household lines. They did not have such a clear sardari system of recruitment. There were, nevertheless, acknowledged labour leaders, who developed reputations for leading migrant groups, or who had nurtured relations with particular employers. Caste Hindu and Muslim employers from Bardhaman and Hugli districts regularly came to the house of RR in Bhuinyadi, for example, where they based themselves for the search. RR did not have long established relations with any single employer to compare with those of BM and ST in Santal para—this was evinced by the non-payment of some RR-recruited labourers following their premature return during the unseasonal rains in December–January 1991–92.

In contrast to the Santals, who always travelled in Santal-only groups, Bhuinya migrants sought work in mixed gangs with Bhumij jati workers. Bhumij workers from different para also combined together to travel to the bus stand to negotiate with employers, with whom they had no previous connections. For this purpose they were brought together by a leader, who acted as negotiator. These groups also included members of the small number of Muslim households in Namo para.

Although there was greater willingness to form cross-jati and cross-para gangs among Bhumij, Bhuinya and Muslim workers than among Santals, migration was also structured on the basis of para. Bhumij workers of Pichan para were collectively (if informally) obligated to be available when required by the Tamuli and Goala employers of Ager para. In early February 1992, it was these employers alone who were preparing for boro paddy cultivation (on 40 bighas of land—approximately 13 acres). Because of their requirement for transplanting workers, Pichan para was the only labour-selling para in the locality, from which no labourers returned east after the major festival (Pous *purab*) in mid-January.[9]

Purulia locality labourers' decisions not to migrate in large numbers during the aman transplanting season were made partly out of an awareness of the potential loss of employment during the ensuing hungry season in Bhadra and Aswin (late August to early October). They would depend on the employers then for lean season employment and advances of paddy and/or cash to enable them to survive until the demand for labour again rose at the harvest. There is a degree of self-interested mutuality here between employers and labourers from certain para in the locality, particularly Pichan para who, unlike other labourers, hired out only to employers from one para: Ager para.

In contrast to these potential migrants from the Purulia locality, one gang of Muslim workers from Malda district encountered in the Bardhaman locality explained that they did not come at harvest because they had access to harvest share contracts in their own village. They were paid one-sixth of the crop they harvested. However, during transplanting they usually sought work outside. In such ways, locally specific employer-worker relations in source areas influence the flow and pattern of migration.

The labour-hiring practices of employers in the Bardhaman locality also structured the migration process. Of the seventeen sampled employer households in the Bardhaman locality, eleven hired a total of sixty-six migrant workers from nine source districts (see Table 13.4). These included,

Table 13.4
Number of Migrant Labourers Working for Sampled Employers in the Bardhaman Locality by Thana and District of Origin for the Áman Paddy Harvest, Late November–Early December 1991

Employer	Number of Migrants	Thana	District
KM	9	Baharagora	Singhbhum
AA	10	Pothia	Purnea
GB	2	Tapan	W. Dinajpur
DB	5	Ranaghat	Nadia
JM	2	Jiaganj	Murshidabad
NM	3	Ranibandh	Bankura
	1	Puncha	Purulia
MB	8	Tapan	W. Dinajpur
	1	(Dihania?)	S. Parganas
NB	7	Chakulia	Singhbhum
SM	11	Tapan	W. Dinajpur
KG	2	(Gopalpur?)	Malda
	1	(Nijaspur?)	Malda
NB	2	Purulia	Purulia
	2	Baharagora	Singhbhum
Total	66		

moving in a swathe from south to north, Singhbhum, Santal Parganas and Purnea districts in Bihar and Bankura, Purulia, Murshidabad, Malda and West Dinajpur districts in West Bengal. The majority of labour days hired in by sampled employers in the 1991 aman paddy harvest were worked by migrants (1,142 out of 2,013).

The Bardhaman study locality was situated on either side of a river, which also formed the border between two villages and, indeed, two thanas (larger administrative units—literally police stations). Party politics among employers in the two villages, referred to here as Kadapur and Dhanpur, contrasted greatly. While Kadapur employers were predominantly CPI(M)-supporting, Dhanpur employers mainly backed the Congress (I).[10]

Because of the Bardhaman district CPI(M) policy against piece rate contracts and in favour of a uniform time rated wage, the employers of Kadapur were constrained in their capacity to negotiate wages above the locally brokered rate. The Dhanpur Goalas, on the other hand, saw themselves as having much greater liberty to bid up the time rate or to offer piece rate arrangements, should these become necessary in the recruitment of migrant workers.

Twelve sampled employers hired migrant labour at least once between November 1992 and July 1993.[11] Five of these employers used *phuron* (gang-based piece rated) arrangements in either the boro paddy harvest or at aman paddy transplanting. All five were from Dhanpur village. The rate of pay and the work required varied between individual employer households.[12] Phuron rates for harvesting boro paddy ranged from Rs 150 per bigha,[13] with three cooked meals provided by the employer, to Rs 200 per bigha with no food provided. The former arrangement required completion of all tasks from cutting paddy to storing the threshed paddy. The latter included just cutting, binding and threshing. For transplanting aman paddy workers were paid Rs 110 per bigha (with shelter and cooking fuel) by one employer and Rs 125 per bigha by another. In the former case the cash payment was the equivalent of Rs 42 per person per day (fifteen workers transplanted 40 bigha in seven days). All the migrants employed via phuron arrangements were Muslim men from Nadia or Murshidabad districts[14] recruited at Katwa, Krishnanagar or Kusumgram bus stands.

Migrants remunerated at a time rate by sampled employers in the same period received earnings ranging from Rs 14 plus 2 kg of rice (approximately equivalent to a total of Rs 26 at current prices[15] to Rs 25 plus three cooked meals (approximately equivalent to Rs 37 at current prices). Daily earnings above the locally brokered floor were reported by six employers in Dhanpur but by none in Kadapur village.[16]

The allegiance of the Baisnab and Muslim employers of Kadapur to the CPI(M) constrained them from bidding up the wages offered to migrant workers in times of peak demand. In order to ensure a secure supply of workers, several Kadapur employers built and maintained relations with particular migrant source villages over a number of years. In contrast, the anonymous hiring of migrant workers at bus stands was by now common practice among the Congress(I)-supporting Goala employers of Dhanpur. Thus, ironically, not only did CPI(M)-supporting employers pay migrants less on average than Congress-supporting employers, the former also relied much more on personalised arrangements.

In a locality adjacent to the Purulia locality, a wage strike in June 1991 created especially grave problems for employers, mostly high caste Hindu Kayasthas, who were forbidden from manual labour even in their own fields by their caste ideology of work. In contrast, employers in the study locality worked the land alongside hired labourers. It was particularly important to the Kayastha employers that they did not lose 'their' labour force at the technically crucial (and unpredictable) period of transplanting. An attempt by employers from Bardhaman district to employ transplanting labour for aman paddy from those para 'obliged' by their history of settlement to give first option to the Purulia locality Kayasthas, was met in 1991 with the threat of physical violence.

These snapshot case studies provide critical insights into how the process of seasonal migration was structured in the early 1990s. Although both women and men of all labouring jati in the Purulia district study locality migrated, only Santal migrants organised themselves communally. Others travelled as households (especially when women were included in the gang) in combination with other households. Migration was more likely the poorer the labour-selling household. Collective obligations in the source area acted as constraints on when certain groups of workers could migrate. The costs of not fulfilling obligations (implicit or explicit) to local employers, such as possible future refusal of emergency loans or hiring for weeding work in the lean season, would have to be taken into account. The more averse to risk a labour household faced with such a choice, the less likely it was that it would deploy members to migrate. Moreover, the seasonally specific types of labour arrangements available in a source locality are taken into account by potential migrants. The structure of labour arrangements partly depends on employer jatis' own ideologies of work. Different rates were paid in the Bardhaman study locality, depending partly on the state of the market—employers in a hurry to harvest or transplant bid up wages—and partly on party political allegiance—CPI(M) employers stuck to the locally brokered daily rate for all migrant workers.

SEASONAL MIGRATION AND AGRARIAN CHANGE IN WEST BENGAL

Four times each year a regional agricultural labour market of increasing scale emerges in southern West Bengal. Seasonal migration entails change—changes in the structure of rural social and economic relations. Although employers and workers might be assumed to act as opposing classes, workers from the source areas liaise with employers from outside against the interests of source area employers. The possibility of migration strengthens these workers' hand in wage negotiations and threatens a shortage of labour supply. At the same time, migrant workers reside in empty out-houses in the employers' neighbourhoods in destination areas, sheltered from the local workers, who live separately in non-contiguous hamlets. The liaisons of caste Hindu and Muslim smallholder employers from Bardhaman district with the mainly 'tribal' but also 'untouchable' and Muslim worker-cultivators of Purulia district challenge economic theorists' notion of closed village labour markets (e.g., Osmani, 1990) and confirm that caste boundaries take on different meanings over space and time. The liaisons also contribute to the livelihoods of the people concerned. They may, however, be dangerous, threatening the well-being of source area employers as well as that of both migrants and local workers in destination areas.

Because the hiring of migrants involves employers from destination localities staying under the roofs of and sharing food with 'untouchable' and adivasi households, the process entails changing relations between jati and between classes. During direct recruitment of migrant workers in the Purulia locality,[17] Muslim and caste Hindu employers from Bardhaman and Hugli districts would arrive and stay one or more nights with a labourer household. In some cases the employer would be provided with all meals without charge, breaking caste rules that would have operated in his own village. In others, the employer would cook for himself but would sleep in the labourers' dwelling.

At the same time, personalised labour arrangements increased as employers from the Bardhaman locality continually returned to the same source village(s). For example, one sampled employer household no longer recruited directly because a relationship of trust had been established with a Santal *dol* in Singhbhum district. A letter was now sufficient, specifying the dates and numbers of labourers required at the aman harvest.

The knowledge of highly paid continuous employment in Bardhaman and Hugli districts is passed by migrants to non-migrants in the source locality. It changes the discourses around wage setting and the employers' negotiating line over contractual arrangements. Employers are put on the defensive, being forced to explain to local workers the peak season remuneration of less than half that reported by returning migrants (except at transplanting when remuneration is roughly equivalent). They have only one cash crop and even this is highly weather-dependent. In contrast, the cultivators of Bardhaman and Hugli districts can depend on their groundwater irrigation for at least two crops per year. How can 'we' pay the kind of wages 'you' received in Bardhaman and Hugli, when our production is so risky? Local employers' vulnerability in the villages neighbouring the Purulia study locality was evinced by their threat to harm Bardhaman employers during the aman transplanting of 1991. This incident also illustrated the rivalry *between source area and destination area employers*, which was reflected in the contrasting attitudes of the Purulia and Bardhaman district authorities—both CPI(M)-led—to seasonal migration out of and into the respective districts. Because '[t]he overall domination of the privileged classes over the decentralized power structure ... remains' at the district level (Acharya, 1993: 1080; see also Bhattacharyya, 1993 and this volume), we would expect employers to have a major role in determining the strategic priorities of the district authorities. Hence it is perhaps unsurprising that the Purulia District Plan 1991–92, a document indicating spending priorities for the district administration, referred to seasonal migration as a 'menacing problem'. Development efforts were to be dedicated to reducing the 'alarming exodus of labourers' (Purulia District Plan 1991–92: 1, 6).

In the Purulia locality, the possibility of migration for seasonal employment for the aman transplanting and harvest increases the bargaining power of labourers because it overlaps with the peak periods of demand for agricultural labour. This may partly explain the relatively high daily earnings available at aman transplanting and the continued use of credit–labour interlocking with implicit interest (known locally as *berhun* loans) to oblige labourers to work at aman harvest.

In the Bardhaman locality, the presence of migrant labourers and the potential for greater numbers increased the power of employers to control the conditions under which local labourers were employed.[18] Employers argued that because migrant labourers did not object to working longer days, having left domestic obligations behind in the source areas, they were more productive and less trouble to manage than local labourers, who tended to be idle and to find excuses to leave work early.

Yet, local workers were far from powerless. Migrants were not used by employers to undercut the locally brokered daily earnings floor. Moreover, labourers' industrial action for higher daily rates was sanctioned and controlled by the Krishak Sabha (the CPI[M]'s peasant union). On the eve of the one-day strike by local labourers in November 1991, migrant labourers expressed their opposition rueing the inevitable loss of a day's wages as they knew they would not be permitted to work by the strikers (cf. Rudra, 1982: 128). In Bardhaman district, where the CPI(M) has a strong and relatively solid support base, it is in the party's interests to accommodate local labourers, migrants and employers in order to present the 'peasantry' as a single category, untouched by class conflict (Bhattacharyya, this volume).

In the Bardhaman locality, where many 'local' labourers are first generation settlers, potential conflict between local and migrant labourers is further reduced by their common ethnicity.[19] Many of the recently settled households recruit migrant labourers of the same jati from their villages of origin. All adivasi labour-selling households, recently settled as well as migrant, have access to higher earnings than indigenous Bagdi and Muslim households, because the latter do not generally deploy women to paid employment (see Rogaly, 1997b).

Our analysis of remittances, and discussion of how workers use migration to better conditions in the source locality, suggest that migration is not always the menace it is made out to be by the Purulia district authorities. Yet, at the same time, destination employers use migrant workers partly because they are easy to control (cf. Martinez-Alier, 1971; Datta, 1991: 268–9; Rudra, 1982: 127). Data gathered from migrant groups in Bardhaman and returned migrants in Purulia suggests that seasonal migrants in the study area are controlled through the provision of accommodation and payment in arrears. The former renders them beholden for a season to the employer they were recruited by—only if it suits him would migrants find themselves working for another employer in the same village. They cannot shop around for better packages in that village because they depend on their recruiter for shelter.

Table 13.5 shows how arrears payments were used as a means of preventing migrants from leaving before the end of the season. During the aman harvest of 1991, five days of unseasonal winter rain meant that no work could be done. Agricultural workers all over Bardhaman district and beyond sat and waited while employers watched their physical output declining. Many migrant labourers were not even paid subsistence during the waiting period. They were at the same time unable to leave because of the refusal of employers to pay arrears (cases 3, 4, 10, and 11

Table 13.5
How Migrants from the Purulia Locality were Controlled through Late and/or Withheld Payment in the Aman Paddy Harvest, 1991

Migrant Case No.	Days of No Pay	Notes
1	12	It is the end of the aman transplanting season in Bardhaman. They have waited twelve days since the end of their work and are living on the kind part of their earlier payment.
2	3+	As above.
3	5	These migrants in Bardhaman during the aman harvest were unpaid during the unseasonal heavy rains, when they could not work between the 6th and the 10th of Pous. The employer also refused to pay arrears so that they could not leave.
4	5	Left destination after five days of heavy rain, no work and employer refusing to pay; three returned after rain to resume work and collect arrears.
5		Received kind payment during rain in destination village together with all members of his dol. Quantity not clear.
6	3	Received 1 kg hulled rice per day during the rain in the destination village.
7	14	Left destination sick after fourteen days; employer refused payment. GA returned to collect.
8	5	Not paid during rain, but when arrears payment demanded after five days of rain, the employer paid and the whole dol left.
9		Could not leave destination for Pous purab because the employer refused to make arrears payments owed.
10	7	Five days no pay during rains; two days wait for arrears.
11	7	As above.

in Table 13.5). Other workers were forced to consume kind 'savings' waiting for employers to pay their dues after the season was over (cases 1 and 2). However, some migrants were paid a subsistence allowance during the rain (cases 5 and 6).

The experiences of migrant workers from the Purulia locality were thus mixed. For some, including NR, migration was miserable, resulting in a financial loss (having paid his own travel costs, NR was not paid for his work at all). It would seem inevitable that as local political party structures are concerned to resolve *local* conflicts, migrant labourers will be the least empowered, with no recourse to law.[20]

The process of recruitment in source area villages breaks caste boundaries but can involve tensions between recruiters and local employers. The latter raise wages for specific tasks and use labour-tying loans to try to ensure labour supply. However, their bargaining power is weakened by workers' knowledge that higher earnings are available elsewhere. In destination localities the demands of local workers for higher wages and better conditions are contained by the presence and potential presence of migrant workers. Yet in Bardhaman district, local workers are not powerless; among other things they can invoke the wish of the CPI(M) to present an image of peasant unity. The employers of migrant labour gain the most; their power over migrant workers is retained via hiring labourers from diverse sources, the use of arrears payments and even in some cases non-payment.

CONCLUSION

Seasonal migration of agricultural workers has long been part of agrarian structure in West Bengal. Rapidly increasing numbers of migrants for the transplanting and harvesting of aman and boro paddy, the production of which boomed in the 1980s and early 1990s, has brought about wider changes in employer–worker relations. These changes are evident both in the intensively cultivated areas of the state, which attract migrant workers, and in the peripheral areas of rainfed paddy cultivation, where those workers are settled. At the same time there has been a gradual process of more permanent shifts of residence as some regular migrants have established homesteads in the destination areas.

The lack of a comprehensive contemporary study of seasonal migration for agricultural work in West Bengal means that we have to rely on micro studies and anecdotes and are therefore unable to measure the extent of on-going changes. However, by drawing on a simultaneous micro study of a source and a destination area, it has been possible to identify some of the processes of change. The increased possibilities for work in Bardhaman and Hugli districts for periods of at least a month four times per year is changing the nature of employer–worker relations in peripheral areas. Employers there are as reliant on 'their' workers as ever, but the workers' degree of choice about whom to work for and on what conditions has improved. On the other hand, the increasing availability of migrants in intensively cultivated areas enables employers there to contain demands from local workers for better wages and conditions.

The shared interests of employers of migrant workers and the migrants themselves result in the crossing of boundaries associated with caste and class. Workers shelter recruiting employers on their visits to peripheral areas, while accommodation for migrant workers in destination areas tends to be inside employer neighbourhoods in disused buildings, away from the residences of local workers. In West Bengal, where the smallholding peasantry has successfully found political representation in a communist party with the skills also to keep the electoral support of agricultural workers, it might have been expected that the smallholders as employers would act in the interests of their class. The rivalry between recruiting employers and local employers in peripheral areas suggests that at the state level there are conflicting interests within this class. These may be being aggravated by the increases in seasonal migration.

Thus, contrary to Rudra's analysis admittedly made at a time when seasonal migration was quantitatively less significant in West Bengal, the increase in demand for migrant workers and their growing employment in areas of intensive cultivation has been and continues to be an important part of the agrarian change process. The meaning of this migration for the migrant workers is contingent on their structured position in the source localities and the employment arrangements they encounter in Bardhaman and Hugli districts. As we have seen, these factors are not independent. Migrant Santal workers from the Purulia study locality had established intermediaries with long-term relations with specific employers. Although such personalised relations across districts may have been in the interests of employers seeking a relatively compliant and secure source of outside workers, they also meant that migrants could be more certain of timely and proper payment. This was not always so for migrants of other jatis from the Purulia locality.

Moreover, while all migrant workers could expect higher wages than in their home districts and regular employment in the transplanting and harvesting seasons, their uses of remittances suggest quite distinct motivations for migration. Some in the Purulia locality migrated to repay outstanding debts, while others were intending to save their earnings and invest, for example, in a marriage or in livestock. This suggests that for some seasonal migration has been voluntary. Indeed, even those workers who have not been able to choose whether or not to migrate have seen their bargaining power improve in relation to local source area employers. Yet power ultimately rests with destination area employers, and the evidence shows that in cases where employers simply refuse to pay, workers have no practical recourse to law.

Despite this, and despite reports of sexual harassment of women migrant workers by male employers (Banerjee, 1988), seasonal migration in contemporary West Bengal is less dangerous but still as potentially progressive as that described by Lenin in 19th century Russia. It can be contrasted with the migration of Khandeshis from Maharashtra to the sugarcane crushing factories of Gujarat (Teerink, 1995). In Surat and Bulsar districts of south Gujarat sugarcane cultivators were tightly organised, forming just nine large processing cooperatives between them. Migrants were employed by the factories rather than the individual landowners, and it was the factories which organised the harvest. There was also a sugar factory in the migrants' home area in Maharashtra, but this factory only hired outside workers, increasing the compulsion felt by Khandeshi migrants. The insanitary conditions which migrant workers have had to endure in Gujarat led in some cases to severe illness and even death. Thus, even though the individual sugar growers of south Gujarat are not on their own large-scale capitalists, their combination in cooperatives for the harvesting and processing of the sugar crop sets them apart from the relatively unorganised smallholder cultivators of southern West Bengal. Different experiences of migration reflect different agrarian structures, but only in some cases, such as that of West Bengal, does migration influence those structures. Furthermore, any structural obstacles to seasonal migration in West Bengal, such as collective obligations by groups of workers to particular source area employers, have been too insignificant to stand in the way of the increasing numbers of seasonal migrants, who have been essential to the rapid growth in rice output in the state.

NOTES

1. Gazdar and Sengupta (this volume); Lieten (1996). Agricultural growth and labour demand are not always positively correlated (Egger, 1996: 53–4). However, as in other parts of India where green revolution driven agricultural growth has not been based on labour displacing mechanisation, demand for labour in West Bengal has risen (Harriss, 1994; World Bank, 1995).
2. The population of West Bengal in 1991 was 68 million, of which 49 million was rural.
3. In West Bengal the proportion of rural main workers recorded as agricultural labourers was 32 per cent; the proportion of women among rural main workers was 14 per cent (Census of India, 1991).
4. Rural West Bengal can be divided into five main agro-climatic regions based on soil type and topography (Gangetic and Vindhyan alluvium; coastal alluvium; red, gravelly laterite; terai; brown forest). Rice production is predominant in all but the northern hilly brown forest soils zone, where maize and tea together account for a larger proportion of gross cropped area (Reserve Bank of India, 1984). Tea production tends to be large-scale and organised in plantations.
5. 'Most Bengali employers are middle peasants who, in the face of immiseration, have intensified the exploitation of wage labourers' (Basu, 1992: 11).

6. Basu's data was collected by Integrated Child Development Scheme workers who, as part of their responsibilities for the vaccination of infants, keep monthly registers of village residents. Chattopadhyaya derived his estimate of the numbers involved in the aman transplanting migration from discussions with conductors on the specially commissioned buses. Large-scale surveys have failed to highlight the significance of seasonal migrant workers. Very little was made of it in the big surveys carried out by Bardhan and Rudra (e.g., Rudra and Bardhan, 1983) and it is not explicity enumerated nor even self-evident from census or National Sample Survey data (Breman, 1985, cited by Karlekar, 1995: 25), although the decennial census now includes questions on previous place of residence, so that the scale of longer-term migration has become easier to measure.
7. Rural worker households were strictly defined as those containing at least one member who hired out manual labour on at least one day in the year. In many other contexts, this definition would have led to the inclusion of households which hired in as well as hired out labour in the category of workers.
8. Daily time rate and piece rate work in the Purulia locality did not generally pay enough for cash to be accumulated. Furthermore, the arrears payment system used by most employers of migrant workers acted as a savings device in contrast to the sporadic payment timings of local employers in the Purulia locality.
9. Santal migrants did not return at the time of Pous purab as their festival of equivalent importance—Bandhna Purab—fell in mid-November (26th Kartik–1st Agrahayan) (cf. Chatterjee, 1991: 86).
10. Actual and stated political allegiances often differ in rural West Bengal. One study in Birbhum district showed that the same individual may express allegiance to one party in local affairs, including panchayat elections, in order to continue to access resources, but may vote for a rival party in state or national elections (Dikshit Sinha, personal communication, December 1996). However, in Kadapur and Dhanpur, party political differences between employers were an expression of deeper historical rivalry between two peasant castes: the Baisnabs of Kadapur and the Goalas of Dhanpur.
11. This period included the 1992 aman paddy harvest, the transplanting and harvesting of boro paddy in 1993, and the transplanting of aman paddy in the same year.
12. As Egger has shown, in agriculture negotiation over the amount of work expected for a particular time period and the way this should be measured is often carried out on an individual employer-by-employer basis (1996).
13. One bigha, the local measure for land area, is equivalent to one-third of an acre.
14. Except for the migrant workers of one employer, who alleged that his employees were Hindu 'refugees' from Bangladesh.
15. The locally brokered going rate. This had been Rs 12 and 2 kg of hulled rice during the main 1991–92 fieldwork period.
16. Two employers reported paying migrant labourers Rs 20–22 in cash and three cooked meals during the aman harvest of 1992. Three paid Rs 20–25 with three cooked meals at the aman transplanting in 1993. Two paid Rs 16 per day and 2 kg of rice at the 1993 boro paddy harvest.
17. One alternative to anonymous haggling at bus stands.
18. In a study of seasonal labour migration from Bihar to Punjab, Singh found that wages paid to local workers for wheat harvesting had stagnated as a result of the competition with migrants (Singh, 1995: 170).
19. Cf. Burawoy (1980: 159), who reports a similar masking of conflict between seasonal migrants and local labourers in California, where both groups are predominantly Mexican.

20. Also the fate of seasonal migrants for agricultural labour in California (see Burawoy, 1980). Burawoy describes the simultaneous vulnerability of both local and migrant agricultural workers in the face of organised employers with powerful lobbying capability in legislative bodies at both the state and federal levels.

REFERENCES

Acharya, Poromesh. 1993. 'Panchayats and Left Politics in West Bengal,' *Economic and Political Weekly*, 28(22): 1080–2.
Banerjee, Narayan. 1988. *Women's Work and Family Strategies: A Case Study from Bankura, West Bengal*. New Delhi: Centre for Women's Development Studies.
———. 1989–90. 'Family Postures vs Family Reality: Strategies for Survival and Mobility,' *Samya Shakti*, (IV and V).
Bardhan, Pranab, and Ashok Rudra. 1986. 'Labour Mobility and the Boundaries of the Village Moral Economy,' *Journal of Peasant Studies*, April, pp. 90–115.
Basu, Amrita. 1992. *Two Faces of Protest: Contrasting Modes of Women's Activism in India*. Berkeley: University of California Press.
Bhattacharyya, Dwaipayan. 1993. 'Agrarian Reforms and the Politics of the Left in West Bengal.' Ph.D. thesis, University of Cambridge.
Bose, Sugata. 1993. *Peasant Labour and Colonial Capital: Rural Bengal Since 1770*. The New Cambridge History of India, Volume 3(2), Cambridge: Cambridge University Press.
Breman, Jan. 1985. *Of Peasants, Migrants and Paupers: Rural Labour Circulation and Capitalist Production in West India*, Delhi: Oxford University Press.
Burawoy, M. 1980. 'Migrant Labour in South Africa and The United States,' Th. Nichols (ed.) *Capital and Labour: Studies in the Capitalist Labour Process*. London: Athlone Press.
Chakrabarti, S. 1986. *Around the Plough: Socio-Cultural Context of Agricultural Farming in an Indian Village*. Calcutta: Anthropological Survey of India.
Chatterjee, S. 1991. *Poverty, Inequality and Circulation of Agricultural Labour: A Micro Level Study of Birbhum, West Bengal*. New Delhi: Mittal.
Chattopadhyaya, Haraprasad. 1987. *Internal Migration in India*. Calcutta: K. P. Bagchi.
Choudhuri, Anil Kumar. 1992. Report to the Indian Council for Social Science Research on 'Socio-Economic Mechanism for the Survival of Landless Agricultural Labour in Rice Area of West Bengal: Two Case Studies,' Calcutta: Indian Statistical Institute.
Datta, A. K. 1991. 'Control, Conflict and Alliance: An Analysis of Land and Labour Relations in Two Bangladesh Villages.' Ph.D. thesis, Institute of Social Studies, The Hague.
Egger, Philippe. 1996. *Wage Workers in Agriculture: Conditions of Employment and Work*. Geneva: International Labour Office (ILO).
Harriss, John. 1994.'Between Economism and Post-Modernism: Reflections on Research on Agrarian Change in India,' David Booth (ed.) *Rethinking Social Development: Theory, Research and Practice*. Harlow: Longman Scientific and Technical.
Karlekar, Malavika. 1995. 'Gender Dimensions in Labour Migration: An Overview,' Loes Schenk-Sandbergen (ed.) *Women and Seasonal Labour Migration*. New Delhi: Sage.

Lenin, V. 1964. *The Development of Capitalism in Russia, Collected Works*, Volume 3 (2nd edition). Moscow: Progress Publishers.
Lieten, G.K. 1996. *Development, Devolution and Democracy: Village Discourse in West Bengal*. New Delhi: Sage.
Martinez-Alier, J. 1971. *Labourers and Landowners in Southern Spain*. London: George, Allen and Unwin.
McAlpin, M. C. 1909. *Report on the Condition of the Santhals in the Districts of Birbhum, Bankura, Midnapur and North Balasore*.
Omvedt, Gail. 1980. 'Migration in Colonial India: The Articulation of Feudalism and Capitalism by the Colonial State,' *Journal of Peasant Studies*, 7(2): 185–212.
Osmani, Siddiq. 1990. 'Wage Determination in Rural Labour Markets: The Theory of Implicit Cooperation,' *Journal of Development Economics*, 34(1–2): 3–23.
Rao, Nitya, and Kumar Rana. 1997. 'Womens' Labour and Migration: The Case of the Santhals,' *Economic and Political Weekly*, 32 (50): 3187–9.
Reserve Bank of India. 1984. *Agricultural Productivity in Eastern India* (Vol II) (S. R. Sen Report).
Rogaly, Ben. 1994. 'Rural Labour Arrangements in West Bengal, India.' D.Phil. thesis, University of Oxford.
———. 1996. 'Agricultural Growth and the Structure of "Casual" Labour Hiring in Rural West Bengal,' *Journal of Peasant Studies*, 23(4).
———. 1997a. 'Embedded Markets: Hired Labour Arrangements in West Bengal Agriculture,' *Oxford Development Studies*, 25(2): 209–23.
———. 1997b. 'Linking Home and Market: Towards a Gendered Analysis of Changing Labour Relations in Rural West Bengal,' *IDS Bulletin*, 28(3): 63–72.
Rudra, Ashok. 1982. *Extraeconomic Constraints on Agricultural Labour*, Bangkok: ILO Asian Regional Team for Employment Promotion.
———. 1984. 'Local Power and Farm Level Decision Making.' Meghnad Desai, Susanne Hoeber Rudolph and Ashok Rudra (eds) *Agrarian Power and Agricultural Productivity in South Asia*. Delhi: Oxford University Press.
Rudra, Ashok, and Pranab Bardhan. 1983. *Agrarian Relations in West Bengal: Results of Two Surveys*. Bombay: Somaiya Publications.
Sen, Samita. 1992. 'Women Workers in the Bengal Jute Industry, 1890–1940: Migration, Motherhood and Militancy.' Ph.D. dissertation, University of Cambridge.
Sengupta, Sunil. 1981. 'West Bengal Land Reforms and the Agrarian Scene,' *Economic and Political Weekly*, 16(25–26): A–69 to A–75.
———. 1991. 'Rural Economy and Poverty in West Bengal and Public Policy Interventions.' Final Report of the WIDER project on Rural Poverty, Social Change and Public Policy, Santiniketan, September 1991.
Sengupta, Sunil, and Haris Gazdar. 1997. 'Agrarian Politics and Rural Development in West Bengal,' Jean Drèze and Amartya Sen (eds) *Indian Development: Selected Regional Perspectives*. Oxford and Delhi: Oxford University Press.
Singh, Manjit. 1995. *Uneven Development in Agriculture and Labour Migration (A Case of Bihar and Punjab)*. Shimla: Indian Institute of Advanced Study.
Singha Roy, D. 1995. *Women in Peasant Movements: Tebhaga, Naxalite and After*. New Delhi: Manohar.
Teerink, Rensje. 1995. 'Migration and Its Impact on Khandeshi Women in the Sugar Cane Harvest,' Loes Schenk-Sandbergen (ed.) *Women and Seasonal Labour Migration*. New Delhi: Sage.

van Schendel, Willem, and **A. Faraizi.** 1984. *Rural Labourers in Bengal: 1880 to 1980.* Comparative Asian Studies Program (CASP) 12, Erasmus University, Rotterdam.

Webster, Neil. 1993. 'The Role of NGDOs in Indian Rural Development: Some Lessons from West Bengal and Karnataka.' Paper presented at the European Association of Development Institute, 7th General Conference, Berlin, September 1993.

World Bank. 1995. *Jobs, Poverty, and Working Conditions in South Asia.* Regional Perspectives on World Development Report 1995, Washington.

Yang, A. 1979. 'Peasants on the Move: A Study of Internal Migration in India,' *Journal of Interdisciplinary History*, 10(1): 37–58.

14

Agricultural Growth and the Structure and Relations of Agricultural Markets in West Bengal

BARBARA HARRISS-WHITE

INTRODUCTION

The orthodox view that agricultural growth rates in eastern India are hampered by agrarian structure, environmental and infrastructural constraints (GoI, Planning Commission, 1985) has of late been squirming under a pincer movement provoked by the fact that growth rates have accelerated. On the one hand it is argued that neither redistribution nor structural reform solves agrarian poverty and stagnation because agrarian elites find

---

Thanks to Pundarik Mukherjee for his field assistance, Sunil Sengupta and Amartya Sen for their support, to Ben Rogaly for originally cutting me down to size and to Sugata Bose and other participants at the conference for their stimulating comments. The fieldwork was funded by WIDER.

other ways than rent by which to extract surplus. If the agricultural sector grows, it does so because it is in the interests of this surplus extracting elite that it should. They capture a disproportionate share of the gains. On the other hand it is contended that the powerful combination of technical change in crop production and predisposing market prices by themselves not only do offer a solution to stagnation but also may lead to more equitable changes in agrarian structure in the process (J. Harriss, 1992; 1993). *Pace* Bhaduri (1983) who attempted to model an agrarian structure which would perpetuate backward agriculture, and *pace* the Left Front Government (LFG) which has viewed tenurial security as the most important key to unlock productivity, it seems that if appropriate techniques are made available (and this argument stresses diesel and electric powered shallow tubewells, pest and drought resistant HYVs and favourable nitrogen-crop price ratios), agrarian structure does not necessarily impede increases in productivity (Saha and Swaminathan, 1994; Gazdar and Sengupta, this volume; Janakarajan, 1993).

Agrarian structure is conventionally understood as the structure of those land relations and associated contractual arrangements which constitute agricultural production. This understanding of structure neglects relations of household reproduction. (Wood's contribution here explores how household reproduction can affect growth in production.) A narrowly conceived agrarian structural determinism also neglects other factors which affect agricultural growth rates: population dynamics, state intervention and local and national (populist) politics (Boyce, 1987; J. Harriss, 1992: 222-3; Rao, 1995; papers by Bhattacharyya, Ruud and Williams, this volume). These are all also affected *by* agrarian change.

Here it is argued that the most neglected of these other factors which make up an expanded conception of agrarian structure is market structure. Its impact on production and social reproduction has been residualised not only by scholars but also both by the programme of LFG reform and by the central government of India. The historical evidence appears to support the idea that while the structures and relations of agricultural commodity market systems evidently do not *prevent* either technical change or increases in the marketable surplus or transformation of relations of production, they may delay technical change, condition its incentives and systematise changes in the ways those incentives are institutionalised over time in a way which restricts the social diffusion of technology and conditions the distribution of gains from technical change (see, among others, Srivastava, 1989; Majid, 1994).

This essay cannot be a study of the way in which 'the market shapes agricultural growth' but is a snapshot of the character of staple markets at a time of rapid change in a dynamic region, looking rather at the reverse aspect of the dialectical relationship: the impact of agricultural growth upon 'the market' and on the politics surrounding agricultural markets. In order to do this, we look first at the general problems of defining market exchange, of identifying key elements of market structure and establishing a framework with which to study them. Next, we introduce the geographical context in which field research on the impact of agricultural growth upon markets has been pursued. We can then examine actual market structures and the impact of agricultural growth upon them.

What is a Market?

A trawl of classic texts yields no definition. Even the New Palgrave Dictionary of Economics has no entry for 'Market' yet the word appears there 'thousands of times' (Guerrien, 1994: 32). 'The market' is conventionally seen as an operationalised atomistic realm of impersonal economic exchange of homogeneous goods carried out by means of voluntary transactions. These are mediated on an equal basis by large numbers of autonomous, fully informed entities with profit maximising behavioural motivations and able to enter and leave freely. The market is thus the supreme medium for the expression of individual choice (Hodgson, 1988: 178). Much more often than not it is assumed to be perfectly competitive. Models of other stylised market structures (monopoly, oligopoly) alter certain criteria, retain others and predict the consequences for prices and quantities.

These abstractions leave us shorn of means whereby to understand not only how supply is supplied and demand is demanded, but also the structure and behaviour of the real market systems which relate demand and supply. For exactly the reason that markets are highly diverse and complex socio-economic phenomena, real markets have indeed proved awkward to pin down. Fourie's carefully justified definition seems to be one of the more useful: 'an economically qualified, purposeful interchange of commodities on the basis of quid pro quo obligations at a mutually agreed upon exchange rate . . . in a cluster of exchange and rivalry relations' (1991: 43, 48). Here, the social relations unique to market exchange require the combination of 'horizontal' and adversarial competition between populations of buyers (and populations of sellers) on the one hand and a mass of 'vertical', exclusive, bilateral transactions between one buyer and one seller on the other. The implications of this characterisation are that exchange rates mutually agreed on may not be mutually beneficial, that vertical contractual arrangements may prevail over horizontal competition, that purposeful bargaining and the obligations resulting from it may rest on and reinforce a highly unequal base or fall-back position and that relational, repeated contracts, socially determined 'natural prices' and coercive interlocked contracts can and do litter market exchange. The extent of such 'non-market' market exchange will affect the pace at which markets react to deregulation and to other policy levers normally considered to provide incentives.

A Framework for the Analysis of Market Structure

Actually existing markets are thus complex sets of property rights and of their transfer, activities of physical transformation and their associated transaction costs. If they are modelled as commodity chains, transfers of property are assumed to be conflated with stages of transformation; the competitive conditions of property transfer are neglected because they lie outside the analytical framework. If they are modelled in industrial organisation terms, the structures and competitive conditions at multiple points of transfer of property tend to be conflated into one 'market'. In fact agricultural markets are bundles of markets and there is no set relation between the processes of physical transformation, intermediation, transfers of property and types of firm. 'Agricultural marketing systems' are configured in specific ways by means of three types of institution. The first is macro-economic—the regulatory environment, conventionally regarded as the realm of the state but actually also shaped by social norms and collective activity. The second is micro-economic—firms, organisations and contracts. The third is social—those relations of class, caste, ethnicity and religion, locality, gender and age in which economic behaviour is discovered to be embedded.

These 'markets' are not only institutionalised but are also *structured* in the same way as are agricultural production relations. Key elements in the structure of a marketing system are the organisation of assets, and of the complex of physical activities from threshing floor to consumer. Key relationships are those of labour within the system, contractual relations and relations with the state. Let us briefly explain what may seem to some a heretical deconstruction before we turn to the staple food marketing systems of Barddhaman district to illustrate.

Assets represent the means and forces of marketing, concretised in physical plant and buildings (cold stores, mills, parboiling tanks and steamers, drying yards, lorries and bullock carts), and working capital.' The distribution of assets is usually ignored or proxied by the number of firms, from which inferences are made about competitive conditions and stylised processes of price formation (see Maizels, 1984, for a critique). But the distribution of assets says less about competitive conditions than it does about economic power. Unequal access to information, organisational capability and assets is rooted in the ownership of the means of marketing. The latter proxies for power behaviour which is slippery to measure—bargaining advantage, access to force, capacity to bear risk, to press decisions, to hold on to stock, constrain the choices of others and to link and/or segment market exchange. The organisation of control over these assets can be highly varied.[1] It affects the internal incentive structure and the intra-firm returns to trade. By contrast, the consequences of ownership structures for market behaviour are intrinsically indeterminate (Colinet, 1994).

The second structural element is the set of *physical productive activities*,[2] together with the logistical patterns of commodity flows by which physical transformation is patterned and the organisations—firms—which drape themselves over this part of market structure. Trading firms may therefore engage in a range of activities (buying, selling, brokering, storing, transporting, processing, production, finance of trade and finance of production). There are 511 possible combinations of these activities. From the relative simplicity of vernacular classification systems we might assume that activity combinations are distinctively patterned. But this is rarely so. Wherever activities have been vouchsafed by merchants, it is found that diversity, complexity and tendencies to uniqueness emerge, demolishing the notion of a market as consisting of layers or stages and populated by firms which can be compared.

Marketing, like production, is configured by means of *social relations*, as a result of which this highly economically differentiated sphere helps to reproduce not only itself but also other spheres, just as the sphere of production is necessary for the reproduction of the sphere of distribution. Since classically productive activity jostles with necessary but non-productive activity in marketing, the *labour process* is not merely a site of redistribution but also of secondary surplus appropriation. Petty firms and companies with capitalist internal relations coexist in marketing systems just as they do in agricultural production. It follows that *contractual relations* on labour, money and commodity markets take a variety of forms, even in conditions plausibly thought to be transactions-costs constant.

The final key relation of marketing is that with *the state*. States provide a legal framework of regulation. They are implicated in markets as traders and redistributors, via competition with private markets and via invasion by political fiat. Further, state and markets form hybrid institutions when bureaucratic rent-seeking (fuelled by resources garnered from the evasion of regulation) coalesces with pay-offs by the state to fractions of market society which mount challenges to the state's rights over property and its monopoly over regulation.

As a result of these structures and relations, the performance and the 'developmental role' of markets are inherently ambivalent. 'Markets' are at one and the same time self-disciplining mechanisms for the efficient allocation of productive resources, conduits for the intersectoral extraction of resources (whose productive and unproductive use has been a major theme in political economy) and an interface of exploitation.

Institutional Dynamics

It is therefore hardly surprising that there is so little consensus about the dynamics of market systems. Marxian theorists have approached the question of the relations between, and reason for, market structure and behaviour by examining their relation to forms of production: 'a definite form of production logically determines the forms of consumption, distribution and exchange, and also the mutual relations between these elements' (Marx, 1971: 33). Yet Marx also argued that changes in modes of distribution, which he attributed to both exogenous factors (such as the expansion of demand, or the locational readjustment of rural and urban populations) and endogenous ones (such as the concentration of capital) would change production in a process of 'mutual interaction', hedging his bets both over the direction and the nature of determination (ibid.). Contemporary Marxian theorists of South Asia have succeeded in analysing mercantile power as manifested structurally in property relations resulting from specific land relations (see Blaikie et al., 1981, for Nepal; Chattopadhyay 1969; Chattopadhyay and Spitz, 1987, for north-east India; Djurfeldt and Lindberg, 1974; Nagaraj, 1985 for south India). More controversially, the property relations of commodity exchange have been theorised to be manifested in an indirect control over production via a variety of modes of surplus appropriation as well as via control over interlocked markets, such that production relations are determined by exchange relations and the direction of determination is up-ended (Bhaduri, 1986).

Other empirical research into the determination of market institutions has also identified (i) the technical characteristics of crops (Jaffee, 1990); (ii) overarching social institutions, e.g., patriarchy (Harriss-White, 1998 and ethnicity (Pujo, 1996), to be significant conditioners of structure, together with (iii) a certain degree of idiosyncratic institutional autonomy (Harriss-White, 1995a).

We can examine key structures and relationships of marketing systems in a region of West Bengal where marketed surplus has rapidly expanded. By means of business histories we track some of their institutional dynamics and relate them to the conditions of agricultural production and state intervention. There is no space to do more than introduce the social institutions of marketing. Further, the technical characteristics of crops are not a powerful explanation of market dynamics and institutions in West Bengal for the following reasons. First, a variety of technologies of marketing and transformation coexist. Second, the perishability of potatoes has no innate implications for market structure. The polarised control over cold storage capacity is also not technologically inevitable. A variety of types of cooling system and sizes of store are available. Typically concentration of control takes the form of ownership of multiple units of processing and transport. Third, though staples considered here—paddy-rice, potato and mustard—are perishable (the first because an increasing proportion of the annual crop is harvested in rainy conditions), the fact of perishability has not affected market structures and the technological response to perishability (parboiling, which has other nutritional and engineering advantages as well), has taken a variety of forms and scales. About the relationship between agrarian structure and market structure and dynamics and the possibility of institutional autonomy, it is impossible in practice to disentangle the effects of structural reform in tenurial security, technical change and state intervention in labour and credit markets. Structural reform is a state-politically imposed change in agrarian structure with an indeterminate outcome on supply. Whatever its transitional costs and benefits, its impact will be to reduce the obligatory kind rent component of the marketed surplus, increase on-farm consumption (both of which would reduce the marketed surplus) and to lead to changes in the modalities of surplus extraction (which might increase it). Technical change in production, and subsidised directed credit could be expected to increase production (which would increase the marketed surplus) and reduce the incidence of tied contracts (which might well reduce it). In what follows the fact of agricultural growth is taken as given and we examine just one aspect of the hypothesised interaction—the impact of that growth upon markets. That agricultural growth is taken as given does *not* carry the implication that the structure of markets per se causes either agricultural growth or changes in agrarian structure. It is just that we are in no position empirically to investigate such a proposition.

THE AGRARIAN CONTEXT AND THE MARKETING SYSTEMS FOR STAPLE FOODS

Barddhaman district of West Bengal, India, has long been a vanguard district for commercialised food production, especially for rice (Bose, 1987). It was one of the regions of high (but highly weather-elastic) agricultural growth in the 1980s (Saha and Swaminathan, 1994). The region consists of shallow river basins whose water replenishes irrigation canals. Of late, irrigation from private wells has generated water rental markets, transformed the incidence of multiple cropping of rice and encouraged the diversification of the cropping systems. High yielding varieties of the staple crops—rice, potato and mustard—have been widely adopted. Whereas rice production amounted to 449,000 tonnes in 1966 it stood at 1.41 million tonnes in 1989 and the analogous figures for the production of the risky and perishable potato are 137,000 and 1 million tonnes (UCO Bank, 1990; Chowdhury and Sen, 1981). Mustard production in West Bengal has increased from 35,000 tonnes in the early 1960s to 176,000 tonnes in 1986–87. In Barddhaman district during 1979–89 there was a one-third increase (GOWB, 1989: Table 5.6).

The region was the epicentre of sharecropping. This tenurial form covered 30 per cent of cropped land in 1940. As late as 1984, 13 per cent of the entire rural population was registered as sharecropper, though share rents coexist with a wide and stable range of other tenurial forms, characterised by inequality and the persistence of relatively intense agrarian poverty (Webster, 1989; Vaidyanathan, 1992).

Exchange relations for the lowest agrarian classes have tended to be coercive and marked by the interlocked control of a small class of land controllers over input markets, the conditions of pre-harvest debt and post-harvest sales (Bhaduri, 1983; Sarkar, 1981; Webster, 1989).

Barddhaman district was referred to by one rice miller as the 'delivery room of our Marxism'. The increasingly pragmatic socialist programme of the LFG has been marked by a politicisation and decentralisation of development planning, by land and labour reforms aimed at increasing the security of access to livelihoods for the masses in rural areas, by improvements to production credit and credit for the implementation of anti-poverty policy (Kohli, 1987). The Essential Commodities Act guaranteeing food security is more systematically implemented via the public distribution system than in most other Indian states. Markets, however, remain the main mechanism for the allocation and distribution of resources, despite a vigorous and sustained rhetorical antagonism towards the mercantile sector (UCO Bank, 1990, being a recent example).

Rice has been marketed historically through a polarised system with an oligopoly exerting local spatial monopoly powers as its apex and a crowded and rapidly growing petty sector at its base. It is due to this structure, plus low capacity utilisation, plus the distorting impact on the open markets of a relatively larger proportion of marketed surplus which is compulsorily sold to the state, that distributive margins are wider than elsewhere in India (B. Harriss, 1982). Of late, West Bengal's public distribution system, focused on Calcutta and the coal belt cities, has come to depend less upon local procurement and more upon imports from northwest India. At the same time inter-regional movement restrictions have become less draconian (Chattopadhyay and Spitz, 1987).

A rapid rise in the marketed surplus of potato has both responded to and created a need for a reticulation of cold stores, locations of which are highly concentrated. As a result physical wastage rates have declined but off-season price fluctuations are still much in excess of storage costs. The stability of price seasonality (and of the storage tactics based on expectations of it) is now threatened by open market imports of potatoes from other regions of India (Chowdhury and Sen, 1981; UCO Bank, 1990).

Local mustard oil is highly prized, particularly during the festival season from October to February. The 'market' for mustard oil is dominated by open market imports from north and north-west India and controlled by marwaris from the north-west.[3]

In this context field research was carried out in 1990 on sixty mercantile firms in the accessible settlements of Memari and Gulsi and the remote but large market centre of Katwa.[4]

AGRICULTURAL MARKET STRUCTURES IN RURAL WEST BENGAL

The commodity systems are made highly elaborate by the coexistence of a variety of technologies, spatial relations, economic forms and processes and social relations.

Assets Distributions and the Organisation of Ownership

Estimates of the assets distribution are given in Table 14.1. From this it can be seen that assets ownership is both highly concentrated and highly specific. The larger firms are more likely to underdeclare. It can be concluded with confidence that the disparity in assets ownership between the subset of components of these systems studied by us is of a factor of over 200. The origins of the commercial capital which powers the class of local merchant magnates is not to be found in agricultural rents or profit, though both of these were important for a previous generation. Three quarters of the top decile of the assets distribution obtained starting capital from a combination of pre-existing agro-commercial profits and finance from nationalised banks and state development finance corporations. Petty and substantial firms coexist, operating different technologies of transformation.[5] Considerable differences between settlements can also be distinguished in the operational scales of mercantile enterprises. Concentration is associated with a locational specialisation, which is more the product of historical inertia and the power of the commercial elite to segment markets than it is a reflection of the centre of gravity of production of the crops concerned. The characteristics of large scale combined with high degrees of polarisation are by no means structural conditions for effective competition. Capital barriers obstruct entry, even into the small scale processing components of these marketing systems where nearly Rs 1 lakh (£3,300 in 1990) is necessary.

These barriers are compounded by social factors. First, gender (female) prevents entry into all but the pettiest trading activity. Then, though half the traders came from trading castes, caste constrains entry into the largest scale trading firms. One non-Bengali caste alone (marwari) dominates the largest firms in all three commodities in Katwa and is a significant presence elsewhere. The caste and kin basis of their trade finance and informal insurance represents a formidable entry barrier.

To a remarkable extent, the distribution of assets in commerce reflects and exaggerates that in land control. The very wealthiest commercial houses are now investing in urban rather than rural land. In the same way, control over storage capacity and over the duration of storage is highly polarised.

The biggest eight commercial firms declared control over the following empire apart from the case-studied activities.

Table 14.1
Market System Structure

Firm Type	Average Estimated Present Value of Assets Rs '000 (1990)	Land Holdings (Avg. No. of Acres)	Ownership of Firm (% Sample)				Household Population	N
			Single Owner	Joint Family	Partnership	Co-operative		
Cold stores	16,913	25	–	50	25	25	14	9
Potato wholesale	563	15	22	55	22	–	42	9
Oil Mills	437	9	25	66	9	–	25	12
Rice Mills	4,271	10	–	85	15	–	11	7
Paddy Agents	140	–	–	100	–	–	–	–
Husking Mills	111	5	20	80	–	–	26	10
Paddy-rice processors	78	} 1.5	100	–	–	–	'000s	2
wh/ca/ret[1]	84		45	55	–	–	'000s	13

Firm Type	Activity Combinations (% of Firms)								Co-efficient of Combinatorial Uniqueness[2]
	Buy	Sell	Broker	Store	Process	Transport	Finance Trade	Finance Production	
Cold stores	63	63	75	100	–	13	50	50	1.0
Potato wholesale	100	100	100	44	–	44	90	44	0.66
Oil Mills	83	83	–	58	100	25	50	50	0.66
Rice Mills	100	100	100	100	100	57	100	100	0.29
Paddy Agents	100	100	–	66	–	100	100	33	0.66
Husking Mills	30	30	–	60	100	10	60	10	0.70
Paddy-rice processors	100	100	33	100	100	66	33	66	0.66
wh/ca/ret[1]	100	100	82	91	–	18	100	18	0.36

Table 14.1 contd

Organisation of Work

	Average Work Force	% Female Labour	% Unwaged Family Labour
Cold stores	162	30	2
Potato wholesale	7	–	40
Oil Mills	8	–	35
Rice Mills	117	30	2
Paddy Agents	4	–	50
Husking Mills	4	20	30
Paddy-rice processors	3	30	80
wh/ca/ret[1]	2–5	–	70

Source: Field research, 1990.

[1] wh/ca/ret: Small general wholesaling-cum-commission agencies-cum-retail firms.
[2] Cases of unique activity combinations divided by number of cases in sample.

Table 14.2
Approximate/Indicative Net Profits (Rs)

	Potato Cold Store	Potato Wholesale	Oilseed Agents	Oil Mill	Custom Mill	Oil Agents	Paddy Agents	Rice Mill	Husking Mill	Rice Agent	Rice Wholesale	Rice Retail	Kutali
Price net of cash costs/q	3–7	3–8	11	25–50 (seed) (max = 100)	5 (seed)	10 (oil)	1–4 (paddy)	3–6 (paddy excl. bran)	2.5–10 (paddy)	4–6	5–20	4–10	8–25 max = 60 (paddy)
Potential net income from estd average	450,000	825,000		41,000			37,500	250,000 (min)	50,000		93,750	21,000	16,500
	(9,000)	(15,000)		(109)			(1,250)	(5,500)	(800)		(1,250)	(300)	(100)
Output (tonnes)													

15 large rice mills
12 cold stores
2 oil mills
13 wholesale businesses
8 non-agricultural industries (including nails and screws, card board boxes etc.)
12 lorries
   a large amount of urban property and storage space
140 acres of land, some sharecropped (out) but mostly owner-occupied and farmed using wage labour.[6]

Ownership forms are varied. Family businesses predominate. Individually owned firms with unspecialised management and decision making are common only among petty scale rice trading and processing. The joint family in which siblings, or more than one generation, work together is an ownership form dominating all the larger sizes of enterprises. It allows the specialised management of complex activities (and cet. par. the speediest decision making reactions to changing prices). The same is true of partnerships, which are frequently at but one remove from family firms and consist of investors, usually of the same caste. The co-operative may exploit economies of scale, but decision making is decentralised and the salaried management is risk averse. Co-operatives are confined to the operation of cold storage, where (constrained by a different regulative environment from that of private trade) they have not proved successful.

Performance is difficult to measure. Estimated rates of return are comparatively very high. Yet they are much lower in large commercial firms which also process (7–10 per cent net for cold stores, oil and rice mills) than for small scale processing (22 per cent for *kutalis* and 45 per cent for husking mills) than they are for purer trading enterprises (ranging from 27 per cent for paddy agents to 80 per cent for potato wholesaling and brokerage). Estimated net incomes range from under Rs 20,000 for kutalis to Rs 850,000 for potato cold store owners (Table 14.2).

Weekly wholesale price data for two types of paddy-rice, oilseed and oil and potatoes over the period 1988–90 for three settlements have been analysed (Palaskas and Harriss-White, 1993).[7] Though long-term integrated, the markets show poor levels of short-term integration. Spatial price differences far exceed transport costs. The distributive margin is wider in rural West Bengal than in similar regions in south India.

Activities and Activity Combinations

Table 14.1 shows the incidence of each of the activities in marketing. In an earlier study of nine activities associated with trade in south India, 149 mercantile firms reduced only slightly to 108 different activity combinations, most of which were unique (Harriss-White, 1996).

Though it is not so extreme, the same tendency may be seen to operate in West Bengal. One-third of the sixty firms studied have unique activity combinations, including the simplest and the most complex. There are five combinations with two cases of firms, four with three, and two with four cases (buy/sell/store/process/transport/finance production/finance trade/and the same minus transport). The commonest combination is loaded with ten cases: buy/sell/broker/store/finance trade. This is a purely mercantile combination.[8] There is a high level of activity patterning only in the cases of rice milling and general rice wholesale/commission agency/retailing firms. Otherwise the activity structure of firms is both highly diverse and complex, with tendencies towards uniqueness. It is tempting to speculate that such activity differentiation precedes product differentiation but performs a similar role: segmenting markets, establishing niches and constraining competition. Yet it also endows market systems as a whole with a plasticity which defends the system from exogenous shocks (Harriss-White, 1995b).

The Labour Process

The work force ranges from two to four in small scale processing and trading firms to over 100 in rice mills and cold stores. Wage labour takes many forms ranging from a salariat paid on conditions specified by the Factories Acts, through daily, seasonal, migrant contract and piece rate labour paid in a variety of cash and kind, with a range of contractual forms, with varying degrees of incentives, varying degrees of unfreedom through debt bondage and with varying degrees of delinquency. The labour process is more closely related to firm size than it is to the complexity of activity combinations. Family labour firms are restricted to petty scale rice processing (where female family labour is crucial to the parboiling and sun-drying processes) and to rice trading (see Appendix for the gender division of tasks). Accumulating and subsistence firms may be found at any point in the marketing system.

Irrespective of whether the labour process is family or composite, gender relations result in women's being exploited. The domestic reproductive work of women (including the preparation of food as kind payment to wage labour) generally subsidises the work of male family labour which controls the fruits of the process of 'co-operative conflict' in the allocation of family labour within the trading firm. Women form a majority of the casual labour force of large scale agro-processing firms (and are quite deliberately casualised).[9] In other niches within the system, their unvalorised work in small scale processing and trading is uncompensated by control over the results of this work. In accumulating firms, women's bodies are vehicles for the inter-firm transfers of commercial capital at marriage.

Thus the precise modality of their exploitation depends on class position.

Contractual Behaviour

> 'The entire trade runs on verbal contract.'
> 'A known face is the only one I do business with.'
> 'Among equals we behave differently from with weaker parties.'

Rice

We will discuss contractual arrangements for the two main branches of the commodity system for rice separately. Rice mills may receive direct consignments of paddy from large farmers whom they pay directly and instantly. Some mills have hundreds of regular suppliers of this type. These suppliers are not to be supposed to be invariably indebted to the rice mills. Nevertheless 'power premia' exist and (irrespective of debt) weaker sellers may be paid at prices 2-6 per cent below those 'ruling'. At the same time, most mills secure supplies, each via a reticulation of some fifteen to thirty agents. These intermediaries used to be individually licensed and attached to one mill and would operate using money lent by the miller. Of late, accumulation has allowed their illegal delinking. Agents will now operate independently of their accredited mill using mill-money-advances together with their own capital. Advance credit is given to rice producers extensively on a range of terms and conditions and using verbal contracts.

The petty trading system has markedly higher velocities of repayment on verbal contract and most closely approaches spot trading. Normal repayment is between two and seven days. Repeated transactions between regular parties are usual. Credit of up to thirty days may be obtained on a friendly basis by a paddy-rice processor from a husking mill owner and by a consumer from a retailer.

Potatoes

In the market for potatoes, much more complicated contractual forms and ownership transfers are common. Payment velocities are rather slow throughout the system, reaching up to six months at the point of wholesale transfer. Although such repayment may be compensated (at around 2 per cent per month) the asymmetry of periodicity between supply and repayment ensures that the seller of this perishable commodity remains financially vulnerable.

Oil

The uncertain production of mustard seed at long distance has led to the use of brokers. Oil millers and wholesalers develop relations of repeated transactions with small subsets of these brokers on verbal contract and with leisurely velocities of payment and repayment, all evincing high levels of trust.

Advance contracts and long-term, regular patterned and personalised trading relationships pepper these commodity systems. At one and the same time, they reduce transaction costs and express unequal economic power. The enduring nature of verbal contracts made not competitively but on a one-to-one basis testifies to the importance of information about the dramatis personae as much as that of information about prices. Goodwill, for most parties, depends on repayment. Debt behaviour is thus more highly charged in moral terms than is credit behaviour in these commodity markets. Spot prices, open auction and immediate payment are extremely rare and for the most part confined to petty trade.

Contractual Relations of Finance and Credit

'We finance commission agents on a large scale' (potato cold stores owners). The capacity to guarantee supplies is one of the most important results of economic power. It is achieved by locking money and commodity markets. The terms and conditions of such interlocked contracts are also deployed within commodity systems in order to deny choice to the subordinate party in an asymmetrical contract. Table 14.3 indicates the scale of agro-commercial borrowing and lending. Data are assumed to have been underdeclared.

While much large scale rice milling was financed by loans from state banks, husking mill owners tended to borrow relatively small amounts

Table 14.3
Traders' Credit

	Average Borrowed (Rs Lakhs)	Sources (in Order of Importance)	Average Lent Out (Rs Lakhs)
Potato Cold stores	30.5	nationalised banks/co-operatives/private lenders/kin	26
Potato wholesale	5.4	banks/cold store owners/producers/'friends'	4.63
Oil mill	1.1	traders/banks/farmers	2.1
Rice mill	3–75.0	banks/private lenders/kin	10.6
Husking mill	0.24	kin/friends/private lenders	0.18
Paddy/Paddy–rice/rice traders	0.39	private parties/traders farmers/IRDP	0.46

from relatives and friends at zero interest, passing such monies onwards to paddy-rice traders as a means of attaching them to the mill. Paddy and rice trading is carried on at varied scales with associated financial requirements. All traders lent out. About half of them tied production or consumption loans to post-harvest supplies. All traders accepted repayment lags of up to thirty days from agents or other traders. Similar relations characterised credit for mustard seed.

Potato cold stores owners borrow extremely large sums from state financial institutions and *mahajans* (private lenders), most of which they lend onwards along reticulations of agents who disburse pre-harvest credit and post-harvest, pre-sale credit on conditions which ensure repayment in kind and tied storage arrangements. Potato wholesalers borrow from state regulated banks and lend backward to producers (from whom they purchase) and onward to purchasers from them on the long-rolling periodicities described earlier.

It is with great caution that we attempt to place these financial flows into context. But it is likely that the credit distributed by cold store owners, potato wholesalers, rice millers and sundry grain traders in the six blocks comprising our study area amounts to some Rs 73 million (£ 2.43 m in 1990). This can be compared with certain kinds of state finance to production and trade. Co-operative loans on potatoes in cold storage in the entire Barddhaman district amounted to Rs 44 million (£1.47 m) in 1990 according to the Barddhaman district co-operative central bank. Co-operative production credit in the six blocks reached Rs 20 million (£667,000) in 1989–90 (see Table 14.4). So-called informal credit (supplied in part from nationalised banks) can therefore be expected still to dominate production as well as trading finance.

Table 14.4
Co-operative Crop Loans in Study Areas (Rs lakhs)

	1987–88	1988–89	1989–90
Memari blocks I and II	195.26	209.45	162.0
Gulsi blocks I and II	67.8	69.92	17.9
Katwa blocks I and II	29.5	32.4	16.42
Total	292.8	308.77	196.32

Source: Barddhaman Central District Co-operative Bank Ltd, 1990.

Contractual Behaviour: Autoregulation

Recently there has mushroomed a mesolevel set of associational institutions (Table 14.5). These have been brought into existence to establish contractual conventions for non–face-to-face transactions and to respond in a collective way to representations from collective labour institutions and to regulations of the state.

Now, in addition to rate fixing (for labour, transport, sometimes processing), lobbying and responding to the state, these groups perform a variety of other overlapping functions in a rather disorganised way: (i) ownership of market sites as group property; (ii) exploiting scale economies (e.g., in transport); (iii) putting up entry barriers; (iv) colluding over prices; (v) risk spreading and insurance; (vi) reduction in the transactions costs associated with trading (information not just about price but about production, big supply deals, fraud and delinquency [the circulation of which information is confined to the group]); calibration of weights and measures, dispute resolution, enforcement; (vii) expressions of social coherence, philanthropy and piety.

These unsystematically structured, collective institutions are a response to the third relation, that with the state.

Relations with the State

In this section we examine the relations between market institutions and the state, first, as it competes with them, second, as it controls them, and third, as a distinctive mercantile politics evolves. The state is thick with institutional infrastructure. A majority of departments is involved in the mass of law and procedure impinging upon marketing systems.

Table 14.5 Regulation

	Cold Stores	Potato Wholesales	Oil Mills	Rice Mills	Husking Mills	P/P-R/R Trades
Licences	Local government	Local government	Local government	Local government/FCI	Local government but rationed: many unlicenced	Local government but rationed: group membership/ unlicenced
Security	Physical fortification & own private security force	Market architecture organised by landlord/own pvt labourers/local collective action	Own private force/ local collective action	Own private force	Local collective action	Market architecture local collective action
Public hygiene	Privately paid sweepers	Market landlord/ ad hoc and occasional	Privately paid sweepers	Privately paid sweepers	Private sweepers/ farmers & traders	None
Weight & measures	Kg but sackweight varies between 50–60 kg	Calibrated by largest trader in local syndicate	—	Maunds and seers still used by some farmers	—	If at all by Bazaar committee
Contract adherence/ dispute resolution	Political institutions of BDO/syndicate	BDO/Bazaar committee/regulated market officers/ trade association/ mutual agreement	Bazaar Management Committee/Brokers/ mutual agreement/ trade association	Largest trader in association/mutual agreement	Mutual agreement	Bazaar committee/ town business association
Crime Common type	Credit embezzlement	Credit default/ pilferage in transit	Adulteration/ repayment	Credit & repayment default	Theft of stored rice	Default on payment
Detection & punishment	Private agents	Trade association	Collective action/ bazaar committee/ Health dept.	Private agents	—	Trade association

Relations with the State as Trader

There is a strong mutual interest between merchants and the state bureaucracy. Laws on marketing were enacted ostensibly to supply grain at controlled prices to needy consumers, to stabilise prices and to provide price support for producers. They have also been represented as the first step towards the socialisation of trade and the 'eradication of the middleman'. Yet examination of the local implementation of state procurement policy, the establishment of marketing co-operatives and government finance of trade shows the largest private enterprises to have been the main beneficiaries of state action.

With respect to state-traded grain, under the Rice Control and Levy Order of 1960, a prescribed proportion of traded rice (50 per cent) has to be sold by millers to a state trading corporation at administered prices always below those of the local open market. Recent, unlicensed mass entry to the marketing system and the explosion of small husking mills has prevented millers from colluding to hoist open market prices by amounts compensating them for their levy losses. This has led to strictly unlawful renegotiations of procurement arrangements in the form of quotas. This arrangement in turn has led to a scramble for supply at post-harvest low prices at a time when marketed surpluses have been increasing. While procurement agents have been at pains to free themselves illegally from the mills to which they are licensed and to deal independently, the scramble for supplies has involved the rapid multiplication of agencies for rice mills and the rapid consolidation of tied credit hierarchies. Evidently the storage laws specifying ceilings on inventory and on storage periods have been widely flouted. Bribes to the food department and police have ensured the non-compliance of rice mills, and the harassment of small scale traders and paddy-rice processors.

The co-operatives, confined here to the management of cold stores, have to operate in a regulative framework, encouraged—by bribes to inspection officials from private sector competitors—to be strictly enforced. This framework is technologically inappropriate for local factor endowments. Private cold stores threaten the viability of co-operatives not only by competition, but also by having regulations which they avoid enforced instead upon the co-operatives. The experience of collusive sabotage by private cold store owners led one marketing co-operative manager to exclaim: 'We were set up to eliminate the middleman but the middleman eliminates us.'

The stated orientation of local credit policy is towards rural lending for purposes of poverty alleviation (UCO Bank, 1990). There is to be no lending for trade, except for specific purposes. A close reading of these exceptions reveals that the rules are riddled with loopholes and that it is petty rather than large scale trade which is therefore prevented from gaining access to formal sector loans.

Indeed, our data suggest that private merchants are financed on a massive scale by state credit institutions. While only three formal sector loans to traders in the petty subsystem were encountered, averaging Rs 3,000 (got on false pretences), the eight largest firms studied declared investment borrowing from the West Bengal Finance Corporation and from nationalised and co-operative Banks of, on average, Rs 8.9 million per firm. The average borrowing is equal to 2,500 IRDP loans. The eight magnates have borrowed as much as have 20,000 loan-taking households below the poverty line.

Thus the state *qua* trader benefits private capital. Bureaucratic rent-seeking is more than matched by pay-offs to merchants. Non-compliance by the mercantile elite sector is matched by coercive enforcement of procedures for conduct on weaker traders and by debilitating conditions of operation for state trading institutions themselves. Some of the latter contrive to evade these conditions (e.g., banks) while others languish and founder (e.g., co-operatives).

Relations with the State as Regulator

Market structure is controlled by means of licences, access to the rationed supply of which is screened according to the technology deployed (in the cases of all three staples). Conduct is controlled by restrictions on storage quantities and periodicities (rice and potato) and by the imposition of quality standards (potato and oil). Under conditions of mass entry such as has occurred in the rice marketing system, exclusion from licensing oppresses small scale traders in two ways: increasing their vulnerability to a harassment by vigilance forces which is deliberately encouraged by licensed traders; and excluding them from eligibility for credit from nationalised banks. The bribes necessary in the former case and the usurious interest exacted by informal lenders in the latter case prevent small scale intermediaries from expanding. New forms of pre-emptive action develop, e.g., groups of small traders rotating a shared licence. The laws regulating rice processing technology are inappropriate to the factor endowment for local level marketing and are unenforceable. The LFG has responded to this impossible situation by turning a blind eye towards husking mills, refusing for instance to attempt to levy commercial taxes on their operation. Yet this has, as with the other unregulated elements in the markets for rice, resulted in institutionalised bribery and harassment. A parallel system of corrupt private tribute or taxation results.

In the case of potatoes, the exploitation of information asymmetries and legal loopholes enables regulatory law to be reinterpreted and unintended transfers of ownership to occur to the benefit of cold store owners and their associated trading companies and to the disadvantage of small potato producers.

Oil is subject to a cascading tax, levied at each stage of transformation. While taxes can be evaded, private incentives to adulterate are checked by the Food Department's vigilance team and also by a certain amount of self-regulation in immediate local markets (because of concerns for reputation and because of expectations about the costs of future harassment). Technical change is resisted and the commonplace and orderly ranks of rotaries (mechanised pestles and mortars) are technically illegal.

The regulation of the staple marketing systems is tightly specified in complex law, while its implementation is of low priority. At best market regulation is the basis of a layer of petty (and often private) taxation. At worst, it is the basis of a politics of accommodation. The letter of the law is formally pluralist and embodies a contradiction between the state-as-trader and the state-as-regulator. On the one hand unacceptably uncompetitive markets require regulation by means of direct replacement by state-administered distribution. On the other hand markets are competitive and need minimalist control. Laws may have important symbolic functions indicating desirable directions for social change. At present and according to our field material these have no such symbolic function and are merely the basis of a diffused appropriation of bureaucratic rent. Given the record of allocative practice, it is no longer possible to escape questioning political intentions in the making and implementation of marketing policy.

Agro-commercial Politics

The elite magnate class has a distinctive role to play in the making and practice of marketing policy, designed for the most part to curb its power. As elsewhere in the subcontinent, the mercantile elite engages in the risk averting finance of all major political parties. While rarely actively engaged in party politics, the mercantile elite plays an oppositional role when it is roused to action. More importantly it engages in politics by means of organised lobbying and by means of a form of political clientele amounting to patronage. Its defensive, reactive, opportunistic politics of interest is applied at all pressure points in the policy process—in the formation of the policy agenda, in the implementation of law and procedure and, most actively, in the arena of access and allocation of resources. This low-key politics is accompanied by ostentatious social patronage of local institutions of culture and philanthropy. Through federation and networks, the political arena of trade associations is nation-wide in scope. In complete contrast, the elements of these commodity systems which are small scale and partly illegal had a distinctive politics of avoidance and low profile although there was greater open support for the CPI(M) and the LFG among their ranks. Such traders were active in religious institutions and social and philanthropic societies (Harriss-White, 1995c).

The economically powerful in these markets are politically opportunistic while some are openly politically engaged, more often in opposition to the LFG than in support. Such open oppositional support has costs (physical intimidation, being the focus of strikes) and needs protective, pre-emptive investments (private physical security, contacts with politicians and with labour). Local mercantile magnates are also active in business and commodity lobbies often located outside the region in which their firms are sited and in Calcutta itself. Despite the mass politics of the LFG, the locally economically powerful have found ways of being politically powerful too.

We can therefore agree with Kaviraj (1990: 13) that 'since major government policies have their final point of implementation very low down in the bureaucracy, they are reinterpreted beyond recognition'. Merchants benefit from the process of implementation of laws of public provisioning through which the state seeks to control private trade. There is thus a strong mutual interest between merchants and the state bureaucracy in the appropriation of surplus and its distribution via combinations of excess profits and rents, via subsidies, pay-offs and the virement of state financial resources, via compromising ties of kinship and via the micropolitics of interest. State interventions may therefore be implemented as laid down, but this is very rare. More frequently they are implemented by threat, by combinations of corrupt transactions, non-compliance and harassment. Interventions are ignored. Markets may develop pre-emptive structures. Relations between markets and the state are deeply embedded in broader social relations.

THE DYNAMICS OF INSTITUTIONAL CHANGE IN AGRICULTURAL MARKETS

In this region of India, the historically polarised and concentrated assets structures of agricultural markets have been shown to resemble those of land control. Furthermore, the contractual linking between intermediaries within market systems reflects similar practices locking money and commodities between traders and agricultural producers (B. Harriss, 1982; see also Crow and Murshid, 1991, for Bangladesh). Market structure cannot however be read off from agrarian structure. The former is much more polarised and concentrated than the latter, because of the lack of a technical or legal ceiling to the size of non–land-based enterprise. Furthermore the response in market structures to the recent agricultural growth and the more diffuse base to agrarian accumulation is lagged, and the multitude of new small scale entrants have not as yet (by the early 1990s) transformed the economic power relations of marketing.

Yet it cannot be doubted that the marked increase in production and marketed surplus for all crops has been associated with the emergence and consolidation of petty trade. Three kinds of explanation for petty trade link its emergence with changes in relations and forces of production. One emphasises transactions costs: when marketed surplus is generated in extremely small spatially dispersed consignments (e.g., half a 60 kg sack) then a system of bulking is necessary to minimise costs. This latter system has emerged under arrangements where the labour costs of bulking are unvalorised. The other two arguments are from political economy. One stresses that petty trade is the outcome of the poverty-induced search for (seasonal) livelihoods by landless labour and marginal peasants. The other relates the emergence of petty trade to the decentralised, post-reform accumulation process in agricultural production. All these arguments may hold but we have no way of distinguishing them through the type of fieldwork reported here.

It is possible to relate the broader base for agrarian accumulation to two types of institutional change in agricultural markets. These are particularly noteworthy in the case of rice. One comprises *involution*—the increasing internal intricacy—of the formally recognised subsystem resulting in an increasing diversity of roles and contracts. Rice mills are being supplied by a greater diversity of intermediaries than a decade ago. Mahajans, rich farmers and moneylenders, are using agricultural profits to lend in cash and kind (fertiliser) pre-harvest in order to scoop up paddy at harvest at prices 8–10 per cent less than 'prevailing' ones in order to supply rice mills. New itinerant traders are expanding in numbers. They have no fixed costs, no wages to pay and their own funding. These men tie contracts in the way described for mahajans and bulk up to a truckload of paddy (10 tonnes) to supply a mill. Paddy agents, with whom these intermediaries are in competition for supplies, are developing independent trading finance from accumulated commissions from brokerage. Rice mills are increasing the number of paddy agents they try (illegally) to attach, with credit as the medium of attraction.

The second type of institutional development is *evolutionary*—emerging from earlier forms. This applies to the formally constituted marketing subsystem but even more aptly to the informal subsystems formed over the 1980s and 1990s. Peasants with less than 2 acres of land, landless agricultural labourers and economic migrants from Bangladesh (some of the latter of whom have imported capital which is not petty) have entered paddy, paddy-rice and rice trades. Many of the poorest are not independent traders at all but bulk the supplies of mahajans as disguised wage labour on to whom the risks of price fluctuations can be transferred. The English words 'labour' and 'sackman' refer to petty agents who transport sacks of parboiled paddy by bike and on commission to husking mills and who also transport rice to other petty agents in the rice markets. Apart from economic dependence, the important feature of this process of evolution is the meshing of productive activity with trading.

Three forces other than that of the type of agrarian accumulation have also been observed to affect markets. The impact of state regulation in shaping, often in pre-emptive fashion, the development of markets has already been discussed. The state has been instrumental in transforming capacity in the cold stores studied from 3,000 tonnes in 1960 to 75,700 tonnes in 1990, with a doubling between 1985 and 1990. The state provision of infrastructure in the form of highly regulated diesel and rural electrification may have released a limiting constraint on boro production and on small scale post-harvest processing at the historical moment when the marketed surplus from wet-season harvests expanded and investible surpluses could be put into forward linkages. Here the state, in contributing to change in market structure, may have facilitated agricultural growth.

The second force affecting markets consists of other markets themselves. The reproduction of firms (by means of competition, nicheing concentration and the life cycle of the households controlling them) affects both accumulation in market exchange and the formation of contracts, prices and profitability. Profits in the paddy and rice trades are used to hive off new firms. Competition from processing plants newly set up in regions of production threatens installed capacity in oil markets in Barddhaman. The influx especially at cold storage unloading times of imported fresh potato from as far west as Punjab, south as Tamil Nadu and north-east as Assam has affected trends, levels, seasonal minima and irregularities in prices and may be capable of restructuring local exchange relations.

Third, it is plain that markets have a certain institutional autonomy. As national markets develop, an immediate cause and effect relation between the structure of local landholding and the structure of marketing is less and less plausible. The case of mustard oil is instructive. Although local production has rapidly expanded, the bulk of seed is still imported into Barddhaman long distance from north and north-west India. Non-local, non–land-based capital is as involved in marketing mustard in West Bengal as is local commercial capital.

CONCLUSION

In his comparative study of the developmental impact of regime types in India, Kohli (1987: 9, 95 et seq.) argues that West Bengal's Left Front government (LFG) has demonstrated the greatest autonomy. The regulation of sharecropping (of tenurial security and of shares) under Operation Barga, the development of red panchayats (politicised, participative and decentralised local government), credit from co-operatives and nationalised banks for smallholder production and real increases in wages for landless agricultural labour are argued as representing a bundle of 'successful reforms aimed at altering the conditions of the poor... with significant long term impact on the living conditions of the lower agrarian groups in West Bengal'.

Kohli's explanation for this success emphasises the coherent nature of leadership, the appropriate combination of centralised and decentralised organisational arrangements, the exclusion of propertied classes from participation in governance together with a pragmatism in facilitating a stable and non-threatening political atmosphere in which the propertied entrepreneurial classes can invest. For Kohli, as for most other analysts, the distribution of private property is the principal constraint on social restructuring.

Kohli argues that the broadening agenda—from a revolutionary to a reformist one—demonstrates a pragmatism wherein the only enemies are those not productive, identified as absentee landlords and big *jotedars*.

The nature of the accommodation with the propertied classes is unaddressed. Kohli's argument stresses the importance of reforms in production relations to poverty alleviation and political popularity. The nature of the sphere of circulation and of market exchange, where the classic Marxist position is that merchants' capital is unproductive but necessary and where the Lead Bank, as financial agent of the LFG, can state in print its commitment to eradicating middlemen (UCO Bank, 1990) is unaddressed.

Staple food markets in Barddhaman district show a certain adaptive efficiency, consistent with comparatively well-diffused information and comparatively well-developed transport infrastructure and communications. Despite state regulation, the barriers to entry into petty trading are not high enough to prevent a proliferation of intermediaries at the base of the structure. Reports of the delinking of previously attached trading agents, of (albeit limited) competition to mercantile credit from nationalised banks (UCO Bank, 1990), of processes of rapidly expanding long distance flows, of slow decline in tied contracts and of the replacement of certain verbal contracts with written documents, all point unambiguously to a gradual freeing of conditions of exchange. And in the absence of effective state regulation, private institutions of collective action are beginning somewhat haphazardly to regulate these agricultural markets.

Such development notwithstanding, in the short term, all the crop markets are inefficient in terms of their price behaviour. Reasons for inefficient price behaviour are to be found not in transport and informational deficiencies so much as in structural factors—collusive oligopoly in Memari's potato market, spatial segmentation in flows of oil, systemic, opaque segmentation, instability brought about by the explosion of entrants—and in any incentives for inter-seasonal storage with respect to paddy and rice (Palaskas and Harriss-White, 1993).

It has proved impossible to disentangle the effects of tenurial reform, technical change and the expansion of formal credit on market structure and behaviour. They are all associated with increases in the production of each of the three crops considered. Just as these reforms to agricultural production have been concluded to have succeeded in consolidating forms of petty production and to have arrested historical trends towards polarisation (Webster, 1989 and this volume), so our work confirms that the expansion of marketed surplus has enabled the consolidation of a 'livelihood intensive' petty trading subsystem.

But the state not only fails comprehensively to regulate agricultural markets. More seriously for a political programme stressing livelihood creation for a mass base, the state has set up considerable blocks to accumulation in petty trade. The mechanisms of such blocking include:

1. Rationing licences to petty traders. This practice has two important effects. It feeds petty traders to the state's coercive wing, to the vigilance squads which are appeased by bribes. It labels petty traders as ineligible for formal credit and therefore feeds them to private money 'markets', where loans carry interest reflecting both monopoly and the risks of illegality.
2. The continual protection of mercantile magnates, through legal arrangements which ensure the oligopoly of rice mills (though these are increasingly contested); and through subsidised credit for investments (which lead to technological upgrading and to the concentration of ownership) and for working capital. An arrangement which ostensibly minimised the transactions costs of the local procurement necessary to provision Calcutta and the industrial belt has survived into an era of grain imports from north-western India which could have liberated the LFG to address the changed structure of exchange. The power of mercantile magnates has not yet been challenged and the locations of their mercantile politics and financial transactions are in Calcutta and increasingly in New Delhi, quite outside the local political arena and quite invisible from the villages (see the papers by Webster, Ruud and Williams here).

The LFG seems to be relying increasingly on an army of relatively petty traders, emerging from the lower echelons of its voting stronghold, locked into relations of trading and finance and into forms of state and non-state regulation which are costly—certainly to petty trade and arguably to society as a whole. It is this petty sector which controls the distribution of staple food to those sectors of the populations of its provincial towns and their rural environs which cannot rely on statutory forms of rationing. Accumulation in this sector is constrained.

The LFG whose rhetorical stance is to eradicate middlemen is actually presiding over a multiplication of middlemen without precedence in the entire history of Bengal. The 'eradication of the middleman', which is a live issue at the level of political discourse in Barddhaman, would involve the eradication of the local 'monopoly trading houses'—an emerging rural industrial bourgeoisie which is unafraid of battening onto residual semi-feudal forms of exchange, which is diversely endowed and whose material interests cut through the rural and urban economy.

Further, any future reforms to marketing in West Bengal will also have to deal with institutions of civil society through which these markets are coming to be regulated, institutions which have not been observed to play such important economic roles elsewhere in India and which have compensated for defective state regulation. Yet, there is a public interest case for state regulation: the penalties of civil-society institutions on outsiders are often oppressive, arbitrary and extra-legal. They function to protect their own interests in highly differentiated market structures. They are thus capable of preventing the organisation of weaker intermediaries,[10] and of encouraging concentration.

Serious questions of intentionality with respect to policy on marketing cannot escape being asked. Are the LFG's policies clever? Is the regulation of the new 'merchants' capital' to be by suffocation: at the very best a regulation of markets which are widely talked about by politicians, administrators, journalists and bankers as 'unproductive' by the diversion of formal credit to sectors such as agriculture and industry which are 'productive' in classic Marxist terms? Or are the LFG's policies stupid? Are they antisocial because they are anti-livelihoods for a crucial political constituency, and rely on an interpretation of merchants' capital which ignores Marx's own observations on the necessity of 'tendrils of productive activity in the sphere of circulation', productive activity such as quality-maintaining storage, transport and processing which we have seen to be intimately meshed with trading? Is marketing policy deliberate: to let markets develop by neglect, often pre-emptively moulded around defectively implemented state regulation? Or is it by default—and out of ignorance, where research problematising the development of petty trade has not reached the LFG, or has reached it but has been prioritised low on the political agenda. If so, then why?

The economic empowerment of commercial magnates, whose trading is also defectively regulated by the state, is entirely consistent with the incoherent, implemented interventions which comprise policy on agricultural markets, elements of which may threaten this commercial class. The political programme of the LFG has been preoccupied with property in historical conditions of great inequality in the property distribution. It has been preoccupied with poverty in production. This has led to a high position on the reformist agenda for tenurial reforms and land rights. Yet Kohli's illustration of the mechanism whereby the LFG 'acts autonomously' and against the interests of the propertied classes is instructively inadequate. It suffers for two reasons. First, Kohli sectoralises 'poverty', wrenches poverty from the relations which cause and perpetuate it, and classifies a prioristically the subset of interventions used to explain changes in poverty. He is in good academic company in doing this. Second, the subset of economic interventions examined by him to provide explanations for the success of anti-poverty policy is mainly concerned with landed property, with unexamined assumptions about the nature of the property relations being challenged.

In practice the rural classes identified by Kohli as 'losers' (rentier landlords) were probably losing interest in rural landed property anyway by virtue of long having diversified out of agriculture into the professions and commerce. Their non-land wealth is massive compared with the land component of their mercantile–financial–industrial–professional portfolios.

Its focus on production has led the LFG to neglect property relations in the rural non-farm economy and reform of the sphere of distribution—in practice if not in rhetoric. This has permitted the perpetuation and strengthening of an accommodation between the state and the 'rurban', commercial power elite developed under previous regimes.[11] Further, while merchants' capital is a useful analytical concept, Marx cannot be blamed for failing to anticipate or theorise the development of composite forms of merchants' capital and productive capital or for failing to see that markets may dissolve old production relations slower than they add new ones (so that both can coexist for rather long historical periods). The actually existing counterpart to merchants' capital—commercial capital—is not an autonomous independent force floating above society, but is deeply embedded in production relations, and in accommodation with power points of the state. This kind of concrete market exchange is entirely capable both of being changed by agrarian structure and also of changing it.

The impact of the accommodation described here is mainly experienced in the setting of constraints on challenges to these relations. This has repercussions on accumulation by a mass of small peasant/petty traders. These may yet provide a theoretically and politically uncomfortable challenge.

APPENDIX

The Labour Process

Labour Process for Potatoes

The process of cold storage is gendered as follows:

unloading from lorries (m)
grading (f)
weighing (m)
bagging and labelling (f/m)
controlled refrigeration at 36 degrees Fahrenheit, during which period potatoes cannot be lifted from the stores, but ownership of potatoes is often transferred,
unloading (m)
controlled reclimatisation under fans for eight hours (f)
grading (f)
bagging, weighing and loading (m).

There is no by-product but cutpieces from bad potatoes are used for wage payments or given or sold to the poor at 50–70 per cent of the prices of whole potatoes.

Labour Process for Mustard

The processing of mustard oil is hardly gendered and tends to be male. Its operations comprise combinations of several possible technologies:

unloading
optional drying down to 12 per cent moisture content for optional storage (f)
decortication
milling in one of two technologies: rotary (a metal pestle and mortar) or expeller, oil from which has further to be filtered through cloth.
The by-product, oil cake with a 6 per cent oil content, is used by local farmers as cattlefeed and manure.

The casual labour force is comparatively small—on the average eight labourers—and is supplemented with managerial and technical maintenance personnel.

Labour Process for Rice

Larger scale rice mill technology involves:

soaking paddy in tanks for twenty-four hours (m)
steaming for three minutes (m) and repeating this sequence
carting to the drying yard (m)
drying (f) for eight hours to four days depending on weather
bagging paddy at 12 per cent moisture for storage (m)
paddy separation, double milling (m/f) and bran separation (f)
polishing (m)
bagging (m)

The process takes a minimum of forty hours and can take up to ten days.

NOTES

1. It ranges through petty firms (the commercial analogue to petty commodity production), private firms with or without wage labour and with huge variations in the internal division of labour, scale and capitalisation, co-operatives with or without paid labour, state trading institutions varying from those completely state-directed to independent part private joint stock companies.
2. Assembly, processing, transport and storage (which prevents deterioration—productively).
3. 'In Calcutta, almost any North Indian Bania may find himself called a Marwari' (Timberg, 1969: 158). Technically Marwar is the former state of Jodhpur in present day Rajasthan; however, as large groups of industrialists and traders also came from the Shekhavati region of neighbouring Jaipur and Bikaner (loc. cit.), a fuller definition would include all those mainly from a number of different trading castes, presently involved in trade or industry with origins in any of these areas of Rajasthan.
4. See B. Harriss (1993) and Palaskas and Harriss-White 1993 for details of the locations and the fieldwork.
5. The segmentation of these marketing systems by technologies is discussed at length in B. Harriss (1993).
6. We did not ask for data on finance, or for expenditures on education, dowries or (foreign) travel but would expect these also to be considerable.
7. Conventional price series data is a record of spot prices which may be a rare contractual form. A variety of exchange relations are masked by these data. Further, even weekly data masks the daily variations in spot and other prices, engagement with which provides merchants with profit.
8. The co-efficient of combinatorial uniqueness is the ratio between the number of combinations per segment of the marketing system and the number of cases in the segment. Low co-efficients represent highly patterned and similar cases.
9. It is female labour which has been massively displaced by new post-harvest processing technologies (Harriss and Kelley, 1982), although whether the effect is of labour displacement or of drudgery reduction depends on configurations of local classes and on labour markets.
10. We have case study evidence of this for rice.
11. On this accommodation I have been reported as arguing that the LFG has 'sold out' to mercantile capital (J. Harriss, 1993). And indeed the 'buckling' of the LFG to capitalist interests has recently been described at length by Mallick (1993). On the basis of the account presented here, however, there is inadequate evidence for an argument about 'selling out', and it is proper only to conclude that if current policy practice on agricultural markets still serves magnate interests, this is indicative (a) of the constraints posed by this elite to a more radical programme, and (b) of a not very responsible neglect and inconsistency in policy practice on the part of the LFG.

REFERENCES

Bhaduri, Amit. 1983. *The Economic Structure of Backward Agriculture*. London: Academic Press.

———. 1986. 'Forced Commerce and Agrarian Growth,' *World Development* 14(2): 267–72.

Blaikie, P.M., J. Cameron and D. Seddon. 1981. *Nepal in Crisis.* London: Oxford University Press.
Bose, Sugata. 1987. *Agrarian Bengal: Economy, Social Structure and Politics, 1919–1947.* New Delhi: Orient Longman.
Boyce, James K. 1987. *Agrarian Impasse in Bengal: Agricultural Growth in Bangladesh and West Bengal 1949–1980.* New York: Oxford University Press.
Chattopadhyay, B. 1969. 'Marx and India's Crisis,' P.C. Joshi (ed.) *Homage to Karl Marx.* Delhi: People's Publishing House.
Chattopadhyay, B., and P. Spitz. 1987. *Food Systems and Society in Eastern India.* Geneva: UNRISD.
Chowdhury, S.K., and A. Sen. 1981. *The Economics of Potato Production and Marketing in West Bengal.* Agro Economics Research Centre, Visva-Bharathi, Santiniketan.
Colinet, L. 1994. 'The Organisation of Firms and its Implications for Market Behaviour in Developing Countries.' M.Sc. thesis in Agricultural Economics, Oxford University.
Crow, Ben, and K.A.S. Murshid. 1991. *Foodgrains Markets in Bangladesh: Traders, Producers and Policy.* Open University, Milton Keynes, Report to the Overseas Development Administration.
Djurfeldt, G., and S. Lindberg. 1974. *Behind Poverty.* Scandinavian Institute for Asian Studies Series no. 22, Curzon, London.
Fourie, F.C. von N. 1991. 'A Structural Analysis of Markets,' G. Hodgson (ed.) *Rethinking Economics.* Cambridge: Cambridge University Press.
Government of India, Planning Commission. 1985. *Report of the Study Group on Agricultural Strategies for the Eastern Region of India.* New Delhi.
Government of West Bengal. 1989. *Statistical Appendix: West Bengal.* Calcutta.
Guerrien, B. 1994. 'L'Introuvable Theorie du Marche,' A. Caille, B. Guerrien and A. Insel (eds) *Pour Une Autre Economie.* Revue du MAUSS, Eds de la Decouverte, Paris.
Harriss, Barbara. 1982. 'Food Systems and Society: The System of Circulation of Rice in West Bengal,' *Cressida Transactions* 2(1–2): 158–250.
———. 1993. Markets, Society and the State: Problems of Marketing under Conditions of Smallholder Agriculture in West Bengal.' Open University, Milton Keynes, Development Policy and Practice Group, Monograph 1.
Harriss, Barbara, and C. Kelly. 1982. 'Foodgrains Processing: Policy for Rice and Oil Technologies in South Asia,' *Bulletin,* Institute of Development Studies, 13(2): 32–44.
Harriss, John. 1992. 'Does the "Depressor" Still Work? Agrarian Structure and Development in India: A Review of Evidence and Argument,' *Journal of Peasant Studies* 19(2): 189–227.
———. 1993. 'What is Happening in Rural West Bengal? Agrarian Reform, Growth and Distribution,' *Economic and Political Weekly,* 28(24): 1237–47.
Harriss-White Barbara. 1995a. 'Maps and Landscapes of Grain Markets in South Asia,' John Harriss, J. Hunter and C. Lewis (eds) *The New Institutional Economics and Third World Development.* London: Routledge.
———. 1995b. 'Efficiency and Complexity: Distributive Margins and the Profits of Market Enterprises,' G.J. Scott (ed.) *Prices, Products and People: Analyzing Agricultural Markets in Developing Countries.* Lynne Reiner, Boulder and London.
———. 1995c. 'Order, Order . . . Agrocommercial Micro-Structures and the State: The Experience of Regulation,' B. Stein and S. Subrahmanyam (eds) *Institutions and Economic Change in South Asia.* New Delhi: Oxford University Press.

Harriss-White Barbara. 1996. *A Political Economy of Agricultural Markets in South India: Masters of the Countryside.* New Delhi: Sage.
———. 1998. 'The Gendering of Rural Market Systems: Analytical and Policy Issues,' C. Jackson and R. Pearson (eds) *Feminist Visions of Development: Gender Analysis and Policy.* London: Routledge.
Hodgson, G. 1988. *Economics and Institutions.* Cambridge: Polity Press.
———. (ed.). 1991. *Rethinking Economics.* Cambridge: Cambridge University Press.
Jaffee, S. 1990. 'Alternative Marketing Institutions for Agricultural Exports in SubSaharan Africa with Special Reference to Kenyan Horticulture.' D.Phil. thesis, Oxford University.
Janakarajan, S. 1993. 'Triadic Exchange Relations: An Illustration from South India,' *Bulletin, Institute of Development Studies,* 24(3): 75–82.
Kaviraj, S. 1990. 'On State, Society and Discourse in India,' *Bulletin, Institute of Development Studies,* 21(4): 10–13.
Kohli, Atul. 1987. *The State and Poverty in India: The Politics of Reform.* Cambridge: Cambridge University Press.
Maizels, Alfred. 1984. 'A Conceptual Framework for Analysis of Primary Commodity Markets, *World Development,* 12: 25–41.
Majid, N. 1994. 'Contractual Arrangements in Pakistan's Agriculture: A Study of Share Tenure in Sindh.' D.Phil. thesis, Oxford University.
Mallick, Ross. 1993. *Development Policy of a Communist Government: West Bengal since 1977.* Cambridge: Cambridge University Press.
Marx, Karl, 1971. D. McLellan (ed.) *Grundrisse.* London: Harper.
Nagaraj, K. 1985. 'Marketing Structures for Paddy and Arecanut in South Kanara: A Comparison of Markets in a Backward Agrarian District,' K.N. Raj, N. Bhattacharya, S. Guha and S. Padhi (eds) *Essays on the Commercialisation of Indian Agriculture.* Bombay: Oxford University Press.
Palaskas, T.B., and **Barbara Harriss-White.** 1993. 'Testing Market Integration: A New Method with Case Material from the West Bengal Food Economy,' *Journal of Development Studies,* 30(1): 1–57.
Pujo, L. 1996. 'Towards a Methodology for the Analysis of the Embeddedness of Markets in Social Institutions: Application to Gender and the Market for Local Rice in Eastern Guinee.' D.Phil. thesis, Oxford University.
Rao, J. Mohan. 1995. 'Agrarian Forces and Relations in West Bengal,' *Economic and Political Weekly,* 30(30): 1939–40.
Saha, Anamitra, and **Madhura Swaminathan.** 1994. 'Agricultural Growth in West Bengal in the 1980s: A Disaggregation by Districts and Crops,' *Economic and Political Weekly,* 29(13): A–2 to A–11.
Sarkar, S. 1981. 'Marketing of Foodgrains: An Analysis of Village Survey Data from West Bengal and Bihar,' *Economic and Political Weekly,* 16(39): A–103 to A–108.
Srivastava, R. 1989. 'Interlinked Modes of Exploitation in Indian Agriculture During Transition: A Case Study,' *Journal of Peasant Studies,* 16(4).
Tinberg, T. 1969. *The Marwaris.* New Delhi: Oxford University Press.
UCO Bank. 1990. *Annual Credit Plan 1990–91, Barddhaman District (West Bengal).* Calcutta.
Vaidyanathan, A. 1992. 'Poverty and Economy: The Regional Dimension,' Barbara Harriss, S. Guhan and R. Cassen (eds) *Poverty in India: Research and Policy.* New Delhi: Oxford University Press.
Webster, Neil. 1989. 'Agrarian Relations in Barddhaman District, West Bengal,' *Working Paper* 89(2), Centre for Development Research, Copenhagen.

NOTES ON CONTRIBUTORS

SHAPAN ADNAN is presently based at the Shomabesh Institute in Dhaka. He has taught at the Universities of Dhaka and Chittagong and was a visiting scholar at Queen Elizabeth House, University of Oxford, during 1995–96. His theoretical and empirical research publications include topics in political economy, sociology, anthropology, demography and development. He has undertaken fieldwork-based research on the peasantry and the agrarian structure of Bangladesh since the early 1970s.

DWAIPAYAN BHATTACHARYYA is Assistant Professor at the Centre for Political Studies, Jawaharlal Nehru University and Visiting Fellow, Centre for Studies in Social Sciences, Calcutta. His doctoral thesis, completed at the University of Cambridge in 1993, was on 'Agrarian Reforms and Politics of the Left in West Bengal'. He has published a number of articles, including 'Limits of Legal Radicalism and the Left Front in West Bengal' which appeared in the *Calcutta History Journal* in 1994.

SUGATA BOSE is Director of, and Professor of History and Diplomacy at, the Center of South Asian and Indian Ocean Studies, Tufts University, Medford. His publications include *Agrarian Bengal: Economy, Social Structure and Politics, 1919–1947, Peasant Labour and Colonial Capital* (The New Cambridge History of India, Volume 3, Number 2) and (with Ayesha Jalal) *Modern South Asia: History, Culture, Political Economy*.

BEN CROW is Assistant Professor of Sociology at the University of California, Santa Cruz, USA. His publications include *Sharing the Ganges: Survival and Change in the Third World* and the *Third World Atlas*. He is currently working on a book about the interaction of market and class in rural Bangladesh.

HARIS GAZDAR is Research Officer at the Asia Research Centre of the London

School of Economics. His research interests include the role of public action in economic development, and he is currently researching the political economy of mass basic education in Pakistan. His recent and forthcoming writings include work on inter-regional contrasts between the development experience of Indian states, rural poverty in Pakistan, and the analysis of economic institutions and landed power.

BARBARA HARRISS-WHITE is Professor of Development Studies, Queen Elizabeth House, and Fellow of Wolfson College, Oxford University. Since 1969 she has spent a total of six years in the field studying the political economy of agricultural markets in West Bengal, Bangladesh, Tamil Nadu and Sri Lanka. Her other research interests include regional economic transformation and social welfare. Her most recent books are *A Political Economy of Agrarian Markets in South India and Liberalisation and the New Corruption* (co-edited with Gordon White). She has also co-edited (with S. Subramanian) *Illfare in India: Essays on India's Social Sector in Honour of S. Guhan*.

RICHARD PALMER-JONES is Lecturer in the School of Development Studies at the University of East Anglia. He graduated in agriculture and economics from Cambridge and Reading universities, and came to South Asian studies after more than a decade working in Sub-Saharan Africa. He has been mainly involved in studies of irrigation, agricultural growth and poverty. His published works include *The Water Sellers: A Cooperative Venture by the Rural Poor* (co-authored with Geoffrey D. Wood).

BEN ROGALY is Lecturer in the School of Development Studies, University of East Anglia. Having completed his doctoral thesis on 'Rural Labour Arrangements in West Bengal, India' in 1994, he spent two years as policy adviser to Oxfam (GB). His research interests include rural change in India, and poverty and social exclusion in Britain. He recently co-authored (with Susan Johnson) *Microfinance and Poverty Reduction*. He is presently based in eastern India researching seasonal migration for rural manual work.

ARILD ENGELSEN RUUD holds a Ph.D. from the London School of Economics. His doctoral thesis was on 'Socio-Cultural Changes in Rural West Bengal'. He has been engaged on a University of Oslo project on 'Policy and Practice in District Administration in India' focusing on the educational bureaucracy and literacy campaigns of West Bengal. He is currently working on the history of Norwegian development aid.

SUNIL SENGUPTA has been Professor at, and Head of, the Centre for Rural Studies, Visva-Bharati, between 1984 and 1987. Prior to this he was associated for twenty-eight years with the Agro-Economic Research Centre at Visva-Bharati. He has worked on the sex bias and on malnutrition among rural children with Amartya

Sen. He has also been director of a WIDER project on rural poverty, public policy and social change in West Bengal (1987–1992). Apart from publishing numerous books and articles, he has acted as a consultant in recent West Bengal-focused research projects, including an IDRC-funded study of common property resources and rural poverty, a Planning Commission of India-funded study of the working of panchayats, and an Oxfam project on vulnerable women.

QUAZI SHAHABUDDIN is Research Director at the Bangladesh Institute of Development Studies. He obtained his Ph.D. from McMaster University, Canada, in 1982, and served as Consultant-Economist in the Master Plan Organisation under the Ministry of Water Resources, and as Deputy Chief in the Planning Commission, Government of Bangladesh. He has undertaken extensive research and consultancy work in the fields of growth performance and risk management in agriculture, management of water resources, and food policy analysis in Bangladesh.

NEIL WEBSTER is Senior Research Fellow at the Centre for Development Research, Copenhagen. His published works include topics in Panchayati Raj and decentralised development planning, and the role of southern NGOs in rural development. He is currently engaged in research on local institutional interventions in rural development in West Bengal and Karnataka and is also finalising a co-authored book based on a study of European Union aid for poverty reduction in India.

GLYN WILLIAMS is Lecturer in geography at the Department of Environmental Social Sciences, Keele University, UK. He began research in India while studying for his doctoral thesis at the Department of Geography, Cambridge University. His current work investigates the relationships between rural development and the political geography of the local state, and this is the subject of an ongoing programme of field research in West Bengal.

GEOFFREY D. WOOD is Director of, and Reader in Development Studies at, the Institute of International Policy Analysis, University of Bath. He has published numerous books and articles on agrarian change and rural development in the Indian subcontinent, where he has worked for the last twenty-seven years. He has also worked in Zambia, Thailand, Venezuela and Peru. He is currently working on urban livelihoods in Dhaka slums, and is co-editing a volume on resource profiles and poverty in Bangladesh.

Index

absentee landlords, 29, 153, 169, 170, 197, 265, 276n17
absolute poverty, 18, 130n39
agrarian capital 28, 150
agrarian structures, ambit of, 15, 358, 382; and growth, 28, 143–4, 178, 179, 382; and market structure, 404
agribusiness, 180, 193, 219
agricultural credit, 190, 198–201, 210
agricultural growth, and agricultural policies, 140–3, 178; causes of, 20, 62–3, 64, 65–8; constraints on, 178, 183–7, 193, 198, 215, 218; and intensification, 72, 97, 99, 140; and local level enfranchisement, 23; and population, 11–2, 16–7, 18, 20, 42–3; and poverty, 15, 18, 31, 70, 92–3, 114, 131n43; rate of, 19, 20, 55, 61, 63; and rise of real wages, 31; and the state, 48–9
agricultural labour, 289, 290, 292, 293, 331, 359, 376n3; class of, 54, 243–4, 251n23; need for, 23; wages of, 30, 78, 85, 108, 110, 112, 115, 123, 129n35, 265, 349
agricultural operations, labour-intensive, 88n26
agricultural production, slowdown in, 13, 178, 179, 183–7, 210, 215–7, 218; stagnation in, 179 *See* Bangladesh, West Bengal
'agricultural reformation', 28, 301–27
agricultural technology, capital-intensive 18, 54, 151, 152, 218
Agricultural Workers' Acts, 30
agriculture, commercialisation of, 114, 123; innovation in, 149, 150, 187
Aguris, 263, 264, 266, 267, 275
All India Krishak Sabha (AIKS), 284, 289–90
aman rice, 17, 20, 21, 47, 48, 63, 67, 99, 101, 102, 106, 108, 113, 122, 126n11, 128ns25, 27, 29, 130n37, 137, 182, 221ns5, 6, 224n56, 265, 309, 312, 322, 336, 337, 338, 340, 345, 360–74 *passim*
anti-casteism, 23, 255, 274
aratdars, 152, 166, 167
arsenic poisoning, 29
assets, ownership of, 32, 390–3
attached labourers, 240, 242, 243 *Also see mahindars*
aus rice, 47, 63, 67, 95, 106, 107, 122, 126n11, 182, 221n6, 312, 337, 338, 345
Awami League, 129n30

Bagdis, 259–74
Bamun *jati*, 275
Banerjee, Mamta, 249n9
Bangladesh, consumption of chemical fertiliser, 20; disappearance of the family farm in, 312–8, 327; foreign aid in, 25–6, 27, 180, 188, 191, 218, 220, 224n61, 225n68; increase in agricultural growth in, 13, 20, 22, 27, 93, 94, 99, 128n27, 137, 181; increase of irrigated area in, 20, 312; policies in, 14, 25–7, 181, 183; and poverty, 115–21; rich peasantry in, 53; rurbanisation process in, 322–3; stagnation in agricultural growth in, 12–3, 20, 21, 26, 27, 29, 44, 48, 94, 95, 106, 108, 112, 122, 123, 125n7, 137, 182; urban population in, 322–3
Bangladesh Agricultural Development Corporation, 99, 105, 129n30, 142, 188, 189, 210, 211, 212, 224n60
Bangladesh Krishi Bank (BKB), 190
Bangladesh National Party, 123
Bangladesh Rural Advancement Committee (BRAC), 170, 313
bargadars, 55, 294
'barriers to entry', 201
'barriers to exit', 200
Basu, Jyoti, 48, 284
Bengal Kisan Sabha, 354n4
Bengal, agricultural growth in, 55, 61, 84–5, 296, 357, 388; agricultural stagnation in, 12–3, 17, 20, 21, 26, 27, 29, 48, 61–2; change in class makeup of, 24; gender bias in reforms, 24–5; lack of regulation regarding agricultural employment, 30; land reforms in, 14, 22, 23, 29; organised peasant struggle in, 22; population density in, 42–3; rich peasantry in, 53–4; role of the colonial state in agricultural stagnation, 17, 19
beparis, 167
berhun loans, 371
bhadralok, 256, 258, 270, 273, 277n23; decline of, 24, 34n9
Bharatiya Janata Party, 237, 238, 239, 250n13
'big men', 246–7
birth rates, 43

bonded labour 358
boro rice, 13, 20, 21, 27, 28, 29, 48, 63, 64, 66, 67, 87n16, 97, 99, 102, 103, 105, 106, 108, 110, 112, 115, 122, 126ns11, 14, 128ns25, 27, 28, 29, 129ns34, 35, 130n38, 140, 145n2, 182–7, 196–217 *passim*, 221ns12, 13, 14, 15, 224n56, 265, 309, 312, 322, 323, 336, 337, 338, 340, 341, 343, 360, 362, 366, 368, 374
Bose, S., 256
'bourgeois state', 295
Boyce, J., 12, 16, 19, 46, 47, 48, 57, 62–3, 65, 66, 184
Bureau of Applied Economics and Statistics (BAES), 19

canal irrigation, 197, 199, 336, 338, 339, 340, 341
Canal Tax Movement, 354n4
cash cropping, shift to, 43, 45
caste, and barriers into trading firms, 390; and land ownership, 81; and literacy rates, 79–81, 86; and nutrition, 81; and ownership of irrigation equipment, 343–4; shifts in boundaries of, 370
casual employment, 31, 242, 244, 245, 250–1n19, 321, 395
Central Sector Minor Irrigation (CSMI), 344, 345
Chambers, Robert, 229
chashi model, 270–1
chashis, rise of, 24, 34n9, 256, 257, 258, 259, 269, 270, 273
Chowdhury, Benoy Krishna, 280, 288
class, and agricultural output, 155; and distribution of income, 203, 217, 221n16; income distribution shifting towards dominating, 27; and loans, 166–7, 190; and sales of output, 155–6, 201, 202; and surplus product, 162–5, 192, 198, 201–2, 203
class conflict, under-emphasis on, 23–4
class society, emergence of, 149; identification of, 153; and prevalence of 152
class struggle, 280, 281, 285, 292, 293
co-operatives, and cold storage, 393, 400; response of landless to, 333–4

cold stores, 387, 389, 390, 391, 392, 393, 396–7, 399, 400, 402, 405, 406, 410–1
collective wage bargaining, 65
collective water rights, need for, 57
colonial capitalism, 50, 56
colonialism, 325
commoditisation, 197, 325
Communist Party of India (Marxist), 13, 23–4, 25, 27, 30, 31, 34n8, 56, 65, 86n7, 230–3, 237, 239–40, 243, 245–6, 248, 250ns15, 16, 253–77, 279–96, 330, 362, 372, 403
Congress Agrarian Reforms Committee, 287
Congress Party, 14, 237, 238–40, 249n5, 291
contract labour gangs, reliance on, 307, 320–1, 351, 365–6
corruption, 236, 239, 250n11, 268, 349, 350
cost-sharing practices, in agriculture, 54
credit, commercialisation of, 27; loan repayment delinquency in, 27; as a means of exploitation, 17, 33, 52
crop diversification, 143–4, 145n2
cropping intensity, higher, 72, 99, 122, 140, 313
cultural reform processes, 258, 261, 269, 273, 274
cyclone, 137, 151, 152, 171

daily time rate wages, 362–3, 369, 371, 377n8
Damodar Valley Canal System, 336, 338, 339
day labourers, 54 See *munishes*
death rates, 43, 45
debt bondage, 243, 250n18
debt, 306, 308, 310, 355n11, 364, 375, 388, 394, 395, 396; as a means of exploitation, 17, 51, 202, 318
Deep Tubewells (DTW), 99, 105, 142, 183, 188–9, 192, 195, 196, 204, 206, 207, 221n9, 313, 321, 336, 338, 345, 346
democratic decentralisation, 231, 330, 388
deregulation, 22, 27; policy in Bangladesh, 180–220

District Rural Development Agency (DRDA), 344, 345
dol, 236, 237, 238, 240, 246, 370, 373
double cropping, 21, 45, 313, 317, 336, 357
dowry, 25, 35n11
drought, 108, 137, 151, 152, 171, 310, 332

East Pakistan Agricultural Development Corporation (EPADC), 99
education, neglect of, 56
employment, seasonality of, 121 See seasonal labour
empowerment, 229–30, 248, 248ns1, 2, 304, 330, 353
Engels, F., 95
environmental threats, 29
erosion, 152, 153, 171
Essential Commodities Act, 389
Estate Acquisition Act of 1953, 287, 294

fallow lands, reclamation of, 42
family businesses, in trading, 393
family farm, 308, 312, 325
family labour firms, 394–5
family labour, 17, 18, 51, 56, 306, 308, 316, 319, 320
famine, of 1770, 42; of 1943, 46, 51
farm size–productivity argument, 66
farm, classification of, 309–12; concept of the, 306–7
fertiliser distribution, privatisation of, 141–3, 189, 210–1, 216, 218
fertiliser–rice price ratio, 35n13, 66, 110, 112, 142, 144, 211, 216, 382
fertilisers, adoption of new, 54; class aspect of, 215–7; decline in use of, 138–9, 190, 205; fall in prices of, 66; increased use of, 72, 138, 151; rise in prices of, 205, 211, 215–6; subsidy-withdrawal from, 26, 129n30, 141–2, 189, 203, 205, 211, 216, 217, 225n63
fish cultures, 323
floods, 137, 151, 152, 171, 310; control of, 97, 99
Food for Works, 121
food security, 106, 309, 313, 389
food wage, 117
foodgrains production, growth of, 12, 19;

stagnation in, 13, 17, 19 *See* Bangladesh, Bengal
force-mode tubewells, 139, 145n5, 183 *See* deep tubewells
'forced commercialisation' process, 200
formal practices (*sorkari kaj*), 23
Forward Block (FB), 86n7
Friedmann, John, 230, 248n2

gender, and class, 154; and literacy rates, 79–81, 86; and market interactions, 154; and nutrition, 81–4, 86; and trading activity, 390, 394–5, 410–1; and unequal property rights, 24–5, 56, 87–8n20
gender disparity, and allocation of resources, 81; and backwardness in education and health, 71
globalisation, 304
grain banks, 332, 365
Grameen Bank, 170, 313; programmes, 121, 125n6
gramer kaj, 23, 232, 238, 240, 246, 247
Great Depression, 17, 52
green revolution, 28, 54, 72, 130n40, 151, 247, 336, 355n8, 376n1
groundwater, depletion of, 126n13, 349
groundwater irrigation, expansion of, 12, 20, 21, 63, 66, 72, 93, 123, 151, 357; importance of, 28
Groundwater Ordinance, 142
Gulf War, 26, 105

haoladars, 43
health, neglect of, 56
higher crop shares, legislation for, 65, 71
higher yielding varieties, 97, 108, 180, 182, 184, 185, 186, 187, 190, 197, 242, 265, 382; adoption of, 20, 54, 63, 66, 68, 72, 99, 102, 122, 128n25, 140, 144, 145n2, 151, 177, 199, 339, 357
horticultural products, 323, 324
household, as a unit of social action, 307–9

imperialism, 325
indigenous knowledge, 304
infant mortality rates (IMR), rural, 75–7
informal politics (*gramer kaj*), 23
institutional credit, 198–201, 210, 330

Integrated Child Development Scheme, 30, 377n6
Integrated Rural Development Programme (IRDP), 234, 235, 249ns7, 8, 261–3, 264, 273, 345, 401
Intensive Agricultural Development Programme, 335
International Fertiliser Development Corporation (IFDC), 142
irrigated area, stagnation in, 213–5, 217–8
irrigation, class action as dominant in, 194–6, 207, 209, 212, 214; disputes among peasantry over, 194, 195; as a 'leading input', 62–3, 66, 182, 183; liberalisation policies for, 139, 143, 193, 210; and non-consolidated landholdings, 311; private sector involvement in, 26, 27, 48, 54, 62, 66, 67, 122, 139, 141, 142–3, 178, 183, 189, 190, 192, 206, 211, 214, 222n24, 347; public sector distribution systems for, 187, 192–3; reduction in subsidies for, 141, 143, 203, 211–2, 217; subsidy in, 225ns64, 65
irrigation cooperatives, 27, 188, 192, 193, 224n50
irrigation intensity, 338
irrigation payments, 195, 204, 206, 207, 208, 209, 213, 316
itinerant trade, 33

jajmani exchanges, 149
jatis, as social markers, 275n1
Jawahar Rozgar Yojana (JRY), 234, 235, 249ns7,8, 250n11
joint family, 310, 393
jotedars, 43, 153, 251n26, 280, 290–1, 406
jute, 34n4, 45, 47, 321, 322

Kahars, 257
Kayastha *jati*, 275, 369
khetmajur, 292, 293, 298n19
Kisan Sabha, 242, 268, 332
Konar, Benoy, 280
Konar, Harekrishna, 288
Krishak Sabha, 23–4, 284, 285, 287, 289, 290, 292, 293, 298n19, 372
Krishak Samiti, 332, 333, 348, 351

labour markets, social embeddedness of, 320
labour, non-commodification of, 17, 51, *See* family labour
land ceiling, 14, 29, 285, 286, 330, 333, 335
land fragmentation, and intensification of land use, 306
land management options, and spatial spread, 311
land redistribution, 14, 24–5, 29, 30, 32, 56, 57, 65–6, 67, 70, 72, 241, 254, 261–3, 265, 273, 276n11, 330, 333
Land Reforms Acts of 1956, 294
land revenue collection, large scale property rights in, 50
land-lease, exploitation through, 17, 33, 52
landlessness, 28, 52, 310, 348, 355n9
landlordism, abolition of, 285, 286
lease market, class contentions in, 196–7, 219
Left Front government (LFG), 29, 30, 32, 34n8, 55, 56, 65, 69–70, 127–8n25, 233, 234, 235, 241, 243, 248n3, 249n6, 265, 280, 281, 286–7, 329–31, 333, 353–4, 382, 388–9, 403, 406, 408–10
Lenin, V., 130ns38, 40, 153, 358, 375
liberalisation, 12, 21, 22, 26, 27, 33
literacy rates, 70; gender and caste breakdown in, 79–81; rural, 77–8, 79–81
literacy, 23, 255, 258–9, 271–3, 274
loan repayments, 158, 159, 162, 163, 164, 165, 189, 363, 364
loans, lenders of, 165–7
Low Lift Pumps (LLP), 99, 102, 105, 139, 183, 188–9, 192, 195, 221n9, 224n55

mahajans, 243, 355n10, 397, 404, 405
Mahalanobis, P.C., 19
Mahila Samiti, 268
mahindars, 54, 242, 250n18
majurs, 256, 257, 259, 273
Mallick, Kalipada, 291, 298n18
malnourishment, 79, 81, 88n34
malnutrition, 81
Mao Tse Tung, 153
market structure, key elements of, 383–6; and perishability, 387; and the role of the mercantile elite, 32, 402–3; significance of, 382; and the state, 32, 386, 398–403, 405, 407
marketing cooperatives, 400
Marx, Karl, 95, 130n40, 153, 386, 409, 410
Marxism–Leninism, 281
mechanisation, in agriculture, 72, 88n26, 170, 247, 307, 357
mechanised irrigation, 144, 184, 196, 204, 307, 316, 319
migrant workers, labour-hiring practices for, 367–9; payment of, 362, 363–4, 367, 368, 369, 375, 377ns15, 16
migration, to Assam, 34n4, 45
mini-submersible tubewells (MSTWs), 20, 28, 34n7, 340, 341–50
minimum wages, 348, 352, 353; lack of awareness of, 243; lack of concern over agricultural, 30, 188
modern varieties (MVs), of crops, 48
money wage–rice price ratio, 35n13
money wages, 131n44, 205
moneylending, 153, 169, 170, 198–201, 243, 325, 330
mortality rates, 44, 70, 81 *See* death rates
Muchis, 259, 260–1, 262, 263, 264, 269–70, 271–3
multiple cropping, 242, 388
munishes, 54 *See* day labourers
mustard, 32, 338, 396, 406, 411

Napit, 263, 264
non-agricultural employment, 28, 308, 355n9
non-agricultural petty commodity production, 162
non-agricultural wage rate, 243
non-family labour, 317 *See* agricultural labourers
non-governmental organisations, and development, 26, 351–3; and groundwater irrigation, 93, 125n5
non-institutional credit, 199, 200–1, 210, 219
'non-market' market exchange, 384
non-market transfers, 151
non-tied exchanges, 167–8
nutritional status, and female literacy, 83–4, 88–9n35; and household income, 82–4

oilseeds, 322
Operation Barga, 14, 54, 65, 67, 69, 72, 87n13, 128n25, 241, 242, 330, 353, 406
organic practices, in agriculture, 323, 324
own-account cultivation, 359–60

paikasht raiyats, short distance migration by 42
panchayat funds, use of, 236, 247
panchayat members, increasing party politicisation of, 236
panchayat system, 48
panchayati raj, 13–4, 23, 31, 230, 233, 235, 246, 247–8, 330, 336, 349, 353, 354n1, 406; institutions (PRIs), 65, 71
panchayats, 296; and drought relief operations, 330; and local conflict resolution, 330–1; role of, 294–5; SC/ST membership in, 266
Participatory Rural Appraisal, 304, 352
patnidars, 53
payments in kind, 149, 151, 156–8, 160, 162, 362, 363, 364, 373, 394
peasant mobility, 52, 53
peasant movement, 285, 289
peasant movements of 1960s, 68–9
peasant struggle, 284
peasant unity, 281, 284, 293, 296
peasantry, rise of middle and rich, 29–30, 53–4, 192, 212, 215, 255
People's Organisations (POs), 351–3
periodic market places, reduction in use of, 33
Permanent Settlement, 43, 50, 153
personalised finance, 33
petty trade, 32, 404, 407, 408, 409
piece-rated work, 31, 368, 377n8, 394
ploughing services, as acquired in the marketplace, 316–7
potato, 32, 322, 387, 388, 389, 396, 397, 401, 402, 406, 410–1
poverty, 31, 229–30, 243, 348, 350, 388; measures of, 70, 84
power tilling, 316–7, 319–20, 321
Pradeshik Krishak Sabha, 285, 286, 289, 290
procurement centres, 190, 201, 202
productive investment, 149, 150
Proshika Programme, 317

public distribution system, 27, 389
real wage rates, 71–2, 94, 115, 117, 120, 131n44
real wages, higher, 31, 336; maintaining rates of, 188
relative poverty, 18, 35n14, 130n39
rent, 325, 382, 387; as a means of exploitation, 17, 52
rent, change from crop share to, 195, 196, 197
reverse tenancy, 309, 310, 318–9
Revolutionary Socialist Party (RSP), 86n7
Rice Control and Levy Order of 1960, 400
rice imports, subsidised, 114
rice monoculture, 140, 145n2, 184
rice price to wage rate, 110, 112
rice prices, fall in, 31, 110, 113, 122, 128n29, 130n37, 142, 211; stabilisation of, 102, 114
rice wage, 117, 118–9
rice, price support programme for, 201–3, 210, 216
river lift irrigation (RLI), 338, 362
rural credit markets, 128n29
rural population density, increase in, 306
rural poverty, 73–5, 77, 78, 85, 116, 117, 120, 122
rural production cooperatives, 331–3
rural–urban migration, 325

samiti, 166, 167
Sanskritisation, 257–8
sanskritising, 259, 260, 271
Santal migration, 365–6, 369, 375
sardari system, 365 *See* labour gangs
school participation rates, rural, 78
seasonal labour, 394
seasonal migration of labour, 30, 31, 358–9, 360–78, 394
seasonal tenancy, 309, 318–9, 340
secondary surplus appropriation, 385
security of tenure over land, 14, 30, 32, 65, 70, 71, 72, 335–6, 348, 382, 387
Sen, Amartya, 123
sexual mores, and status, 257
shalish, 194, 222n27
shallow tubewells (STWs), class ownership of, 192, 215; command area of,

207, 314; credit for purchase of, 22, 346; decline in number of, 95, 105, 110, 189; decline in rate of return, 206, 224ns52, 53; fall in prices of, 212; increase in, 68, 102, 108, 110, 113, 114, 126n14, 145n5, 183, 185–6, 213, 221n8, 312, 313; lack of monopoly conditions regarding, 321; number of plot holders in command area, 311, 314; privatisation of, 20, 62, 105, 143, 188–9; as a service business activity, 314; siting of, 26, 142; underutilisation of, 313
share crop payments, 151, 156–8, 165
share tenancy, 143, 359 *Also* sharecropping
sharecropping, 150, 152, 153, 169, 170, 171, 178, 196, 197, 198, 256, 287, 288, 289, 294, 330, 350, 355n10, 388
social formation, sexual mores as instruments of, 257
social mobility, upward, 253, 259, 274
sorkari kaj, 23, 232, 233
Special Foodgrain Production Programme (SFPP), 344, 345
specialisation, 149, 150
Srinivas, M.N., 257–8
'state-as-trader', 190
Stay of Eviction Proceedings Ordinance, 287–8
strikes, 292
structural adjustment, 102, 209, 324
subinfeudation, 43
sugarcane, 322
Sunderbans, reclamation of, 44, 46
Surjeet, H.K., 284–5
surplus, appropriation of, 17–8, 21, 33, 43, 49, 51–2, 147, 149–51, 163–4, 165–71, 218, 219, 220, 325, 382, 386, 403; concept of, 148, 158; impact of, 149; reinvestment of, 53
surplus labour time, 149, 150
surplus product, 149, 150
surplus value, 149, 150, 167–9

Tagore, Rabindranath, 248
tank irrigation, 338
'technology brokering', 319
teetotalism, 23, 255, 270, 274
Thorner, D., 49
traders, as lenders of credit, 201

trading activity, contractual arrangements in, 395–9
tribal women's cooperatives, 334–5
triple cropping, 313, 317, 336
tubewells, siting of, 22, 26, 28, 139, 142–3, 211, 213

underachievement, farmer policy of, 316
underconsumption model of agricultural stagnation, 113–4, 130ns41, 46
undernourishment, and female literacy, 83–4
United Front, 14, 29, 68–9, 87n10, 232, 284, 286–7, 335
United Nations Development Fund (UNDP), 105
United States Agency for International Development (USAID), 26, 127n20
untouchability, 253, 254, 255, 259
Upazila decentralisation programme, 121, 322
urban poverty, 116, 117
usury, 199 *See* non-institutional credit, moneylending

Vulnerable Groups Feeding Programmes, 121

wage labour, 162, 196, 326, 394–5, 405, 412n1
wage labourers, 71
'water blackmail', 341–2
water control, as a constraint in expansion of production, 12
water market, 48, 54, 57, 127n24, 178, 192–6, 213, 222n24, 350; monopoly in, 208, 209, 213, 321, 349
water table, drop in, 29
water-holding, 28, 34
'waterlord', 127n24, 193, 222n24, 350
Wellesley, Lord, 43
West Bengal Finance Corporation, 401
wheat, 322
women's activism, 25
women, improvements in position of, 255; position of, 71
World Bank, 26, 102, 105, 110, 125ns7, 21, 129n30, 181, 191

zamindari system, 50, 153
zamindars, 153

ABOUT THE EDITORS

BEN ROGALY is Lecturer in the School of Development Studies at the University of East Anglia, UK. He has been engaged in research on rural change in India since the mid-1980s and in West Bengal since 1990. After completing his doctoral thesis on 'Rural Labour Arrangements in West Bengal, India', he spent two years as policy adviser to Oxfam (GB) and recently co-authored a book on *Microfinance and Poverty Reduction* (with Susan Johnson). He is currently based in eastern India researching seasonal migration for rural manual work.

BARBARA HARRISS-WHITE is Professor of Development Studies, Queen Elizabeth House, and Fellow of Wolfson College, Oxford University. Her research interests include the political economy of development, particularly agrarian markets and social welfare in India. Her most recent books are *A Political Economy of Agrarian Markets in South India* and *Liberalisation and the New Corruption* (co-edited with Gordon White). She has also co-edited (with S. Subramanian) *Illfare in India: Essays on India's Social Sector in Honour of S. Guhan*.

SUGATA BOSE is Director of, and Professor of History and Diplomacy at, the Center of South Asian and Indian Ocean Studies, Tufts University. He has previously been Fellow of the Program in Agrarian Studies, Yale University (1991–92), and Fellow of the John Simon Guggenheim Memorial Foundation (1997–98). Professor Bose's published works include *Agrarian Bengal: Economy, Social Structure and Politics 1919–1947* and *Peasant Labour and Colonial Capital* (The New Cambridge History of India, 3[2]). He has also co-authored (with Ayesha Jalal) *Modern South Asia: History, Culture, Political Economy*.

Also from Sage

DEVELOPMENT, DEVOLUTION AND DEMOCRACY

Village Discourse In West Bengal

G.K. LIETEN

Indo-Dutch Studies on Development Alternatives—18

As the longest surviving democratically elected communist regime in the world, the performance of the Left Front government in West Bengal has excited considerable attention. This book critically examines the achievements of the LFG in the areas of land reforms, rural development, devolution of administration to the grassroots level, and above all in ensuring the democratic participation of the poorest classes. The on-going process of socio-economic change is understood on the basis of the village-level discourse which illuminates the worldviews and ethical assumptions of West Bengal villagers. Overall, the study provides enough evidence, based on extensive primary field research, to support the view that the West Bengal government has not only assisted agrarian development but has done so with a strong pro-poor orientation.

...the work provides a vivid account of how at the micro level the processes of development and democracy interface. The implications of his findings are clearly of great significance....
—The Indian Historical Review

CONTENTS: *List of Tables/List of Graphs/List of Abbreviations/Preface/* Introduction/Contextualising/Changing Environment in Local History/ Panchayats: Caste, Class and Gender/The Interventionist State/Power and Enfranchisement/A Class Account of Economic Benefits/Social Renewal/ Summing Up/*Glossary/Bibliography/Index*

220mm x 140mm/252pp/Hb/1996

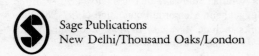

Sage Publications
New Delhi/Thousand Oaks/London